DATA IN MEDICINE:
COLLECTION, PROCESSING AND PRESENTATION

INSTRUMENTATION AND TECHNIQUES IN CLINICAL MEDICINE

Volume 1

Data in medicine: Collection, processing and presentation.
A physical-technical introduction for physicians and biologists.

edited by ROBERT S. RENEMAN and JAN STRACKEE

future volumes

Angiology
edited by M. VERSTRAETE

Otorhinolaryngology
edited by B. H. PICKARD

Nuclear Medicine
edited by K. H. EPHRAIM

Diagnostic Radiology
edited by P. P. G. KRAMER

and others

DATA IN MEDICINE: COLLECTION, PROCESSING AND PRESENTATION

A PHYSICAL-TECHNICAL INTRODUCTION FOR PHYSICIANS AND BIOLOGISTS

Edited by

ROBERT S. RENEMAN, M.D.
Professor of Physiology,
University of Limburg
Maastricht

JAN STRACKEE, PH.D.
Professor of Biomedical Informatics
University of Amsterdam
Amsterdam

1979
MARTINUS NIJHOFF PUBLISHERS
THE HAGUE / BOSTON / LONDON

The distribution of this book is handled by the following team of publishers:

for the United States and Canada

Kluwer Boston, Inc.
160 Old Derby Street
Hingham, MA 02043
U.S.A.

for all other countries

Kluwer Academic Publishers Group
Distribution Center
P.O. Box 322
3300 A H Dordrecht
The Netherlands

ISBN-13:978-94-009-9311-2   e-ISBN-13:978-94-009-9309-9
DOI: 10.1007/978-94-009-9309-9

PREFACE

Nowadays clinical medicine is to a great extent dependent on techniques and instrumentation. Not infrequently, instrumentation is so complicated that technical specialists are required to perform the measurements and to process the data. Interpretation of the results, however, generally has to be done by physicians. For proper interpretation of data and good communication with technical specialists, knowledge of, among other things, principle, advantages, limitations and applicability of the used techniques is necessary. Besides, this knowledge is required for critical comparison of systems to measure a certain variable. Critical evaluation as well as comparison of techniques and instruments ought to be an essential component of medical practice.

In general, basic techniques and instrumentation are not taught in medical schools nor during residencies. Therefore, physicians themselves have to collect practical information about principle, advantages and limitations of techniques and instruments when using them in clinical medicine. This practical information, focussed on the specific techniques used in the various disciplines, is usually difficult to obtain from handbooks and manufacturers' manuals. Hence a new series of books is started on instrumentation and techniques in clinical medicine.

The aim of these series is to provide a clear and critical survey of what technology has to offer to clinical medicine in the way of possibilities and how the latter should be used. For example, limitations as well as advantages will be presented when feasible. The authors have been asked to write their chapters with the above mentioned consideration in mind and to use a more or less instructive style. Volumes on instrumentation and techniques in diagnostic radiology, angiology, otorhinolaryngology, nuclear medicine and cardiology are on their way.

The specific volumes are preceded by an introductory volume which covers the three main aspects of instrumentation, being collection, processing and presentation of data. Although most chapters are devoted to these aspects, some chapters, e.g. chapters 2, 3 and 4, stress even more fundamental points. In these chapters it is shown that properly measuring a phenomenon is heavily tied to prior knowledge of the way the phenomenon comes to us and of the disturbance factors, the latter going under the universal name of noise.

The allocation of the chapters in relation to three different aspects is roughly as follows. Chapter 1 is a more or less philosophical approach to data collection. The chapters 9 and 10 will deal with processing. Chapter 11 is purely about presentation. The chapters 5 and 6 cover all three aspects and the chapters 2, 3, 7 and 8 are a mixture of collection and processing. Chapter 4 about radioactivity, which deals with collection and processing, was included because we considered the basic information presented in this chapter to be of interest to the users of most of the specific volumes of the series.

Techniques using, for instance, ultrasonic waves or monochromatic light (laser) and techniques based upon electromagnetic induction are not included in this introductory volume. It seemed more appropriate to discuss these techniques, which are in fact applications of relatively simple physical laws, in the specific volumes.

Formulas were used by some of the authors. This has been a point of discussion because in general physicians are not acquainted with the use of these mathematical relations. However, for the application of certain principles, formulas have to be used. *A compromise was found by modelling the content of the chapters in such a way that one simply can skip the formulas and still retain the scope of a chapter.*

It is the aim of the introductory volume to supply the reader of the specific volumes of the series with some basic information which might facilitate their use. Besides, this book may be of interest to physiologists, pharmacologists, physicists, electronic engineers, technicians and to all those who enter the domain of biomedical research for the first time.

We wish to express our thanks to the different authors and to the publishers for the care they have taken in producing this book. But most of all we would like to thank Mrs. Els Geurts and Mrs. Mariet de Groot for carrying the heavy secretarial load with fortitude and for their help in preparing the manuscripts.

March 1979   ROBERT S. RENEMAN
                                                                JAN STRACKEE

TABLE OF CONTENTS

CONTRIBUTING AUTHORS

JOHN H. ANNEGERS, M.D., Ph.D., Professor of Physiology, Northwestern University Medical School, 303 East Chicago Avenue, Chicago, Illinois 60611, U.S.A.

JAMES B. BASSINGTHWAIGHTE, M.D., Ph.D., Professor and Director, Center for Bioengineering, University of Washington, Seattle, Washington 98195, U.S.A.

HENK G. GOOVAERTS, E.E., Laboratory for Medical Physics, Free University, Van der Boechorststraat 7, Amsterdam, The Netherlands

RICHARD B. KING, Research Technician, Center for Bioengineering, University of Washington, Seattle, Washington 98195, U.S.A.

JOHN D. LAIRD, D. Phil., Professor of Physiological Physics, Medical Faculty, University of Leiden, Leiden, The Netherlands,

ADRIAAN VAN OOSTEROM, Ph.D., Laboratory of Medical Physics, Medical Faculty, University of Amsterdam, Herengracht 196, Amsterdam, The Netherlands

ROBERT S. RENEMAN, M.D., Ph.D., Professor of Physiology, Medical Faculty, Biomedical Centre, University of Limburg, Maastricht, The Netherlands

LINDA REYNOLDS, BSc. MSc. M I Senior Research Fellow, Graphic Information Research Unit, Royal College of Art, 6A Cromwell Place, London SW7, England

HENK H. ROS, Ph.D., Laboratory for Medical Physics, Free University, Van der Boechorststraat 7, Amsterdam, The Netherlands

HANS SCHNEIDER, Ph.D., Professor of Medical Physics, Laboratory for Medical Physics, Free University, Van der Boechorststraat 7, Amsterdam, The Netherlands

HERBERT SPENCER, RDI DrRCA Professor of Graphic Arts, School of Graphic Arts, Royal College of Art, Exhibition Road, London SW7, England

JAN STRACKEE, Ph.D., Professor of Biomedical Informatics, Medical Faculty, University of Amsterdam, Herengracht 196, Amsterdam, The Netherlands

L. H. VAN DER TWEEL, Ph.D., Professor of Medical Physics. Medical Faculty, University of Amsterdam, Herengracht 196, Amsterdam, The Netherlands

B. VELTMAN, MSc., Professor of Theoretical and Applied Physics, University of Technology, Delft, The Netherlands

J. VERBURG, Ph.D., Laboratory of Medical Physics, Medical Faculty, University of Amsterdam, Herengracht 196, Amsterdam, The Netherlands

LEO VROMAN, Ph.D., Associate Professor, Department of Biophysics, Graduate School, State University of New York at Downstate, U.S.A., *and* Chief, Interface Laboratory, Veterans Administration Hospital, 800 Poly Place, Brooklyn, New York 11209, U.S.A.

HAROLD WAYLAND, Ph.D., D.Sc., Professor of Engineering Science, Division of Engineering and Applied Science, California Institute of Technology, 1201 East California Boulevard, Pasadena, California 91125, U.S.A.

1
OBSERVATION: HOW, WHAT AND WHY

Leo Vroman

1.1 Introduction

We can now measure picograms of a pure molecular species, and lightyears of nearly empty space. Will we eventually be able to measure everything, and will we then understand it all? Or nothing? Could it be that numbers will prove themselves symbols, not of precision, but of our blind senses and mind? I believe there are properties in both the living observer and the life observed, that we must face and focus on, before we try to focus on the perhaps less real world of numbers beyond. I believe some properties within ourselves may be unavoidable obstacles: our tendency of thinking along lines, of homogenization, circularity and perspectivation are products of the life processes we observe with. The specific functionalities of shaped or patterned interfaces, and of amplification or image intensification in real life, on the other hand, are products of the same processes rendering measurement doubly difficult.

1.2 Problems in the observer

1.2.1 *Linear thinking*

We cannot walk a branched path. Even in thought, we can fully enter only one world at a time, and where more than a few –such as total reality –could await us, uncertainty blends them to a merciful fog, softening this confusing spectacle of our future and leaving only the moving spot clear where something aims our line of steps. We cannot imagine a path as if not taken. You probably know the kind of trick drawing that appears to show a flight of steps seen from above, but that at other moments looks to you like a flight of steps seen from below. We cannot see the steps both ways at once. We stop and rest our minds as soon as we arrive on solidly familiar ground.

1.2.2 *Homogenization*

We seem addicted to returning and repetition as the only form of harmony assuring us that there is no need to observe more than a fraction of reality, and assuring us that eternity, though unfathomable, does exist. We pay for this addiction: to absorb anything new, we must digest it into familiar parts, believing that this completes our process of understanding; and we our-selves, to be understood, must speak only in words defined by earlier words. Repetition – in time as well as in space – easily leads to counting. Con-veniently, our senses distinguish many entities as bulk: a cloud, a noise, a mood, a day, a season, each a mass of grey, sadness, hours, defined by some uniform property of their subunits, and these subunits can be counted to give us a number representing an important property of the whole. But in real life, no two subunits are ever alike; even in a crystal, each atom occupies a different, unique place in relation to the crystal's borders. If we knew everything, each subatomic particle in the universe would show us its very individual history and destiny, and we would have neither the need nor the ability to speak of masses or of numbers; we would no longer add 1 and 1 of anything, to arrive at the number 2. But many of us would not even be happy with that prospect, because our very senses prefer repetition, and so, from grouping and counting, we easily go to ranking. We even rank each other, expressing work in money and sports in points. To be good, we must be better than others, and from rating how good we are, we drift into rating how well we are. Minicomputers with a memory harder than ours can now accept your blood, perform about 20 tests on it, print these values as bars on a background of other bars that indicate the accepted normal ranges ... normal for a pool of blood samples from strangers who may share with you no more than their sex. This unstable hematological profile of yours, or even your entire medical one, is a display as variable as a reflection of the moon on a windswept ocean, its statistical "normal" shape a blurred disc that could just as well be the sun. Your facial profile, rather than your medical one, is the shape your friends will recognize out of thousands, no matter how you turn. Of course your face, like the reflection of that moon, is also a statistical mean; your grimaces around this mean, as well as the microscopic turbulence in it (see below under "perspective") cannot hide it for long. And yet, only a small mask hiding part of your face can make you unrecognized. Without your nose, your eyes do not seem the same as usual.

The numbers representing measurements of your life processes taken out of context often also lack the desired meaning. The minicomputer that looked at your blood is a little bit aware of the complex web of physiological interactions from which its data are plucked. For example, nothing may seem more absolutely clear and diagnostically numerical than a red blood cell count, but it has little meaning without information on the size of the cells and the concentration of hemoglobin in them. The computer adds these to the list it prints, and makes a few calculations to indicate how well the red cells function, but it probably will miss the most essential information: with a sample of blood taken after your death, the computer may still tell you that you are well, because it has been built with our own tendency to count and therefore to homogenize. Not only silent about the red cells still in your body, it also failed to ask how much you weigh, what altitude you are living at, what work you do, how anxious you are – all questions about conditions that strongly affect the number of red cells circulating, and that should affect the number of red cells per cubic millimeter regarded as normal for your own personal situation.

The more forces we know at work, the less satisfied we are with a single value for their summed effect. There is no end: to become more and more meaningful, measurements need other ones of things farther and farther removed. For example, we found that in certain breeds of rats under a certain chronic stress, the adrenal weight increased, while body weight decreased in one breed, increased in another. We measured the ascorbic acid contents of the adrenals in these chronically stressed rats under a mild, acute stress – the latter causing a depletion of ascorbic acid in relation to adrenal hormone production. Usually, the concentration (mg ascorbic acid per 100 mg adrenal mass) is reported. But since the adrenal weights had increased, how could they serve as proper reference? And if the ascorbic acid had performed a function in the adrenals which in turn must have served the entire body, would not the drop in ascorbic acid weight have to be related to the total animal's body weight? But then, should we not first try to find out which parts of the body were served most and with what normal requirements these parts draw on the adrenals?

Let me give you another example of one measurement requiring more and more others to gain meaning. The clotting of human blood plasma involves a series of reactions among 10, or perhaps more, proteins. Five of these have been named after the first patients discovered without: Hageman, Fitzgerald, Fletcher, Christmas, and Stuart factor. Even normally, each of these factors is present in such minute amounts, has such bland properties other than its individual function, and hides its function so well as merely one small link in the smoothly functioning clotting chain, that we can only find it if broken, and can only identify it by matching its broken chain with other ones. Factors VIII and IX, for instance, were distinguished from each other only when someone found that the blood from certain hemophiliacs corrected the clotting of blood from certain other hemophiliacs:

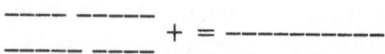

This, finally, is an example of simple measurements (three clotting times) leading immediately to a great discovery (an additional clotting factor). But this also means that at any time, as indeed has continued to happen till quite recently, another patient may be discovered whose blood takes a long time to clot but will clot in a normal time when blood from any human deficient in any known clotting factor is added to his . . . and another factor will have been discovered. In other words, the clotting chain reaction may well contain five times as many links each one fifth as long in reaction time as we had hoped and thought to have measured.

The ultrafine experiment of nature in which it produces malfunction in a single protein, often does not even produce symptoms in its victim. Total factor XII deficiency can only be discovered by accident. Is this victim then really sick, is he an abnormal human, or a newly discovered subspecies? For a while it has even been thought that lack of factor XII, preventing the activation of clotting, would protect the "deficient" person against thrombosis. Certainly the word "deficient", though founded on firmly established measurements, has a very far from homogeneous significance. The plasma of Peking ducks has been found to prolong the clotting of human factor XII deficient plasma. I found it also prolongs the clotting of human factor IX deficient plasma. People with no factor IX, or worse, with an inhibitor against factor IX, often bleed severely. Why then do Peking ducks not have to be hospitalized after a tooth extraction? Because they are not deficient at all within their own blood clotting system, or else they would not have survived on earth. In our attempts to correct the system of one species with that of another, even if less remotely related one we commit a typically homogenizing act. Perhaps it is true that we are getting more cautious the more we fail. When blood was blood, transfusions of sheep blood into people seemed reasonable. Now, so many blood groups and substances have been discovered that perfect cross-matching requires a careful search among those we thought our equals. Blood groups, like fingerprints, come close to proving that we have a constant individuality. Why, if the sum of these quiet features is so important as well as constant, why do we have so much trouble expressing it in numbers . . . and we do want numbers from our measurements to prove they were real measurements, don't we?

1.2.3 *Perspective*

Perhaps we should make sure that things are really sitting still to be counted. A distant house may look more dormant than our own, simply because we want it to be. If I ask you to imagine (or even to look at the picture of) an amino acid molecule, say glycine ... close your eyes and you will think it just far enough from your face to fill your imaginary field of vision. However, if I first ask you to imagine starting to build a protein molecule out of 200 amino acid residues, and *then* I ask you to imagine starting off with one residue of glycine ... you face a problem. You must leave room in your mind's vision for the other 199 residues. So you step back, until the glycine takes about one two hundredth of your imagined space. The single residue has become so small that I dare you to distinguish its structure. Imagine a protein you may know the overall shape of, for example, a gamma globulin molecule. You will see it as a somewhat fuzzy Y – fuzzy with wires or with tiny bumps representing the forces and forms of atoms in whatever model impressed you most. If you now imagine a cell covered with gamma globulin, you must step back again to see the whole cell, and again the detail you know so well becomes temporarily erased. Our minds are as limited as our senses. We will never imagine the complexities of life, even within us, at their true size. Each time we step back (or "zoom out") a bit, we lose track of the interactions among substructures, not only because of their small size, but also because of their high speeds of vibration and interaction that blend them into a statistically much more stable whole. By the time we have stepped back far enough to see our whole body, it has become a distant and tranquil scene: we are half-blind giants to our own molecules. Finding tranquility by totalling turmoil allows mountain-climbers to look down on their village of origin as a place in peace, and allows us to believe in statistically stable normal values as permanent points of reference. The less we know, the easier it is to discard detail, and to believe our head is the one stable thing in a world of meaningless litter.

1.2.4 *Circularity*

To think that even our own brain is no reliable mass of reference points, is as staggering as an earthquake. There is no escape from uncertainty. Does it matter that A will measure B differently from the way B measures A, and does it matter that even the most accurate, abstract numbers become less absolute in value, and less international as a language, when they enter the real world to which they must be applied?

I accept that I am a product of life, describing itself, as if looking at roses through rose-colored glasses, and seeing nothing but the frame. Therefore, I warn you that what follows is merely my own description of, perhaps, one mirage meeting another, but just as possibly it may be a correct description of what a living human sees of life. Only the perfectly lifeless reader can tell how wrong I am.

1.3 Problems in the observed

To our own mental problems of linear thinking, unrealistic sensitivity to repetition, the blur of perspective and the circularity of our conclusions all taking the glory out of our ability to count and to measure life processes, we must add certain properties of life itself that make the life scientist's life even harder: the problems of interfaces, and of amplification.

1.3.1 *Interfaces*

Every living thing, no matter how bulky, is packed with interfaces. Concentrating and focussing on a 2-dimensional rather than a 3-dimensional world is not a 33.3% simplification for those chemists bent on measuring. The simplicity of a bulk reaction between enzyme and substrate* molecules can be rendered less predictable by an interfering interface, in 2 × 3 ways. The reaction rate will depend on whether either the enzyme, the substrate, or both will be adsorbed, how fast, and in what position.

Most simply interfering is a surface coating itself immediately and entirely with active enzyme molecules. Even then, the coated subphase will offer the liquid bulk of substrate a sheet of enzyme accessible at a rate depending on the shape of the sheet and the flow of substrate solution. The relationship between substrate and adsorbed enzyme is easily interfered with in the laboratory, and such "solid phase" techniques are now quite popular. Nature, however, is always similarly active on such a minutely complicated scale that it usually helps to confuse rather than to clarify.

A finite rate of adsorption, affecting the measurement of adsorbate/bulk interaction, is a further disturbance of conventional techniques especially if other substances variably interfere with this adsorption.

But the predictability of a reaction rate is most interfered with if the enzyme's activity (or, e.g., its antigenicity) depends not only on whether it is

---

*"Substrate" here means substance that can be acted upon by an enzyme; matter on which the molecules discussed are adsorbed, will be called "subphase".

adsorbed and hence concentrated, but on *how*, in what position or con-
formation, the subphase forces it to be adsorbed. A well studied case is that
of blood clotting factor XII again. It becomes much more active when
adsorbed, and normally exists as a few dormant molecules per liter of blood
plasma. Say that factor XII becomes $m \times$ more concentrated at a glass
surface, and then becomes $n \times$ more active. It activates a complex of 2 other
factors after adsorbing it, so we can say it complexes with the complex – as a
result of its own activation – at an $n \times m \times$ higher rate than in bulk, increasing the complex's activated activity by another factor. The activated complex causes further activation of adsorbed factor XII, so that the final stage,
if it exists, will show an overwhelming amplification of the reaction rate.
Now, what if the adsorbing subphase was not uniform, and only able on
certain interfacial spots to adsorb and activate factor XII?

1.3.2 *Image intensification*

The answer to that question is clear: the pattern of activating spots on the
subphase will be "painted" by the self-amplifying blood clotting system.
However, this paint soon becomes partly re-soluble: factor XIa, its product
of activation formed so explosively on the surface, is released and carries
the next amplifying step to the surface of phospholipid particles via activation of factor IX. On these free-floating, tiny particles, charged surface
areas will adsorb activated factor IX molecules, and neighboring hydro-
phobic inner areas of the particles will adsorb factor VIII molecules, so that
the complex can attract factor X and activate it. In the next amplifying step,
factors Xa and V are brought together by an identical phospholipid pattern,
to create the enzyme that converts prothrombin to thrombin which will
render fibrinogen clottable. We could therefore summarize the clotting
sequence as follows. An image of the subphase is intensified, then scrambled
in solution and transferred to other, suspended images that are then also
intensified and again dissolved.

Meanwhile, however, the blood may superimpose on its intensification of
one subphase image, an entirely different one of other areas on the subphase that affect processes other than clotting. We found that onto many
surfaces blood deposits fibrinogen, and then changes or replaces it at a rate
dependent on subphase properties. Platelets adhere only where fibrinogen
was not "converted". Thus, the pattern of adhering platelets will be, at least
in part, an intensification of the subphase's pattern of non-fibrinogen-
converting sites. In their effect, the two images – that of clot-promoting and
that of platelet adhesion promoting sites – are interconnected: activated
clotting factors promote platelet aggregation so that the platelets change

their membrane·structure and liberate the kind of phospholipid particles that promotes clotting as I described.

Yet another image of the adsorbing subphase may be created through certain plasma globulins being adsorbed at certain sites in such a way that white blood cells adhere, in a pattern that will not be identical to either the clot-promoting or platelet adhesion promoting one.

The sensitive behavior of blood at interfaces can be used as a tool or probe to amplify or intensify surface imperfections on man-made materials. For example, if I want to find microscopic areas of hydrophobicity on a surface, I can coat it first with gamma globulins, then expose it to blood and see where the granulocytes adhere. To enhance the surface image of living matter – matter with a submicroscopic image of great mobility – a probe is required that is fine enough and scans rapidly enough to amplify the surface's living faults. That, perhaps, is usually beyond the blood's power, and that is why thrombosis is relatively rare. For the same reason, we can assume that our usual probes (including the conventional optical ones) are so coarse either in space or time analysis, that only the meaninglessly tranquil averages of the blood vessel wall properties can be measured.

While blood is described here as a large system to probe or amplify minute images, a hormone can be regarded as a molecular probe to enhance the image of certain organs' functions in the entire organism, unless we regard the responding system as known and use it as a probe of the hormone (in bioassay methods). In all of these examples, at least one of the systems used is too complex for computation, but it is often this failure of computation that leads to discoveries. Each discovery of more structure in living things, brings with it the discovery of other structures responsible for its intensification to an observable level. On earth, perhaps the extreme example of structure and its amplifying system is human society, where migration of entire population groups can be provoked by a few electrons of sick thought in the brain of one dictator; where communication among its subunits is almost instantaneous but socially or mechanically channeled, the multiple images superimposed within any single society will create just as falsely dormant a picture as our averaged one of the blood vessel wall.

The closer we come, the more structure we see. A single amoeba will no longer be a number to us once we observe it through a microscope and watch it crawl. One problem of zooming in like this, is that we take our prejudices with us. Even through the microscope, we think to see things as hard and soft, tough and fragile – sensations that on a molecular scale have no probe. Zooming in close enough on the blood vessel wall, we would land in a world that has no walls at all, a hazy mesh that only statistically exists, the way a spinning fan is only statistically a disc of 50% metal in air. Zooming out, and far away again from our subject, all that remains of our memories at closest view of it is a vague discomfort while measuring, in the knowledge of missing the detail that would have explained the whole, and doubting the significance of the number that our measurement yielded. We may well doubt the significance of the word significance itself.

1.4 Significance

In statistics, a significant difference between 2 sets of values is one that is unlikely to have been caused by chance. With a technique that is barely able to measure a wanted effect, statistics can retrieve something important from a large number of often repeated tests. However, if the effect we are measuring is one of importance to human survival, "significance" has a different significance. A drug that extends survival of patients from 2 years to a statistically significantly longer survival of 2 years and 5 minutes, will be rejected by clinicians. I have worked for 8 years with spleen extracts that were intended to lower the number of circulating platelets. Trying to interpret the platelet counts of rabbits after – and even before – injection, we found the animals to have individually different "normal" levels, often with slow but wide variations. I know a scientist who spent so much time finding people with constant platelet counts to serve as subjects for a study of dietary effects, that his grant was discontinued and he dropped all of his plans and went back to full-time practice. We must not become slaves of statistics. Often, what in a curve looks like "background noise" obliterating the desired "signal", may be no noise at all but the music that we should listen to, even if it can be silenced by statistical means.

The data we like most are usually those that fit on a simple line, or show a trend at least. Nothing in life seems easier to imagine than a trend. Weigh yourself twice and you'll think you are getting heavier or lighter. Two points do not make a trend. The trouble with numbers is that the only 3 things they can do: go up, or down, or stay the same, all seem important. It is easier to publish an article describing the increase or decrease of A, than a change in its appearance. True, numbers look more accurate than pictures, but I suspect there is another cause of our love for numbers. If we define beauty as an awareness of unknown structure, then the sequence of discovery: first, sensing the presence of a natural force or law, then proving its presence and then predicting or reproducing its action, has its basis in beauty. To many of us, real life shows no superstructure and is therefore not beautiful, while pure mathematics, puzzles, music, mysteries and mini-computers create a sense of beauty unhampered by the less obvious structures of our real world. Most beautiful, therefore, is the work of art that, even by means of simple repetition, helps us imagine that we can sense the order orchestrating our real, personal life. That order may well exist only in our head. I believe our brain continuously searches among its hallucinations for those that fit its momentary reality best. It keeps trying to match and measure internal against external structure. These fragile bridges of ours are forming, extending and collapsing forever. To reach and feel "reality" – our world without our presence – we try to cleanse all human flavor off our data, so that they can be built up to bone-dry, universal and universally appreciated truths. Fortunately, we cannot help but build them beautifully, because they will always reflect our own cerebral structure of thought. And the more structure is revealed by our measurements, the more love and respect should be generated for all: the measured, the measurer, and the intangible structure between them.

2
PHYSICAL CONCEPTS

L.H. van der Tweel and J. Verburg

2.1 Introduction

2.1.1 *General*

The aim of this chapter is to show how physical principles and concepts may contribute to a better understanding about measurements of acoustic cardiac phenomena and of blood pressures in the systemic circulation (Fig. 2.1). The choice of this subject is not without ratio because certain aspects of the usual method for measuring blood pressure involve physical principles which at the same time are needed to understand what the dynamics of the blood transport means for the interpretation of the measured values. The circulation itself is further a fertile subject for illustrating how physical reasoning and theoretical treatment can contribute to the understanding of, first, the functioning of a biological system, but, also, of the limitations implied in its complexity.

Within the scope of this chapter simplification will be unavoidable with respect to physiological parameters, but also the physical treatment will be based more on a conceptual framework than on intricate mathematical

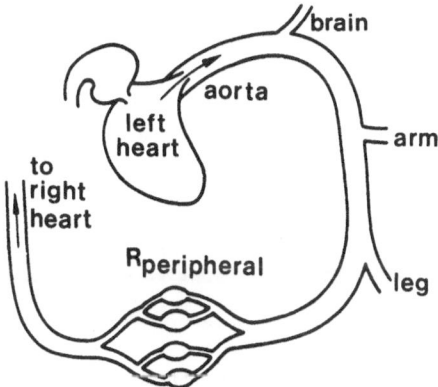

Fig. 2.1 Schematic representation of the circulation. Only the left ventricle with aorta, some branches and the branching of the abdominal aorta into smaller arteries, arterioles, capillaries and small veins are given.

descriptions of the many factors involved. Where contributing to the understanding, analogues will be presented.

2.1.2 *Circulatory aspects*

Let us first look at the blood flow. The left ventricle as a pump (Fig. 2.1) delivers in rest approximately 70 times per minute 60–70 ml blood into the circulation in a time span of 200–250 ms (i.e. in rest approximately 4.5 litres blood per minute is pumped through the body. In heavy exercise this can increase to more than 20 litres per minute).

The pulse-like action causes in the aorta and arteries a pressure wave, the pulse wave, superimposed on a continuous pressure level. The lowest pressure reached just before the beginning of a new cycle is called the diastolic pressure. The crest of the pulse wave is the systolic pressure.

Whereas speaking of *the* diastolic pressure seems justifiable considering the low resistance to the continuous transport of the blood in the big arteries, for the systolic pressure this is not so much the case as can be demonstratively seen in Fig. 2.2 in which the blood pressure course is represented at two different parts in the arterial tree. Whereas the peak value of the pressure is approximately 14 kPa ($\cong$110 mm Hg) in the aorta near to the exit of the left ventricle, this value gradually grows to 20 kPa ($\cong$160 mm Hg) in the lower arm; in many small arteries higher pulse pres-

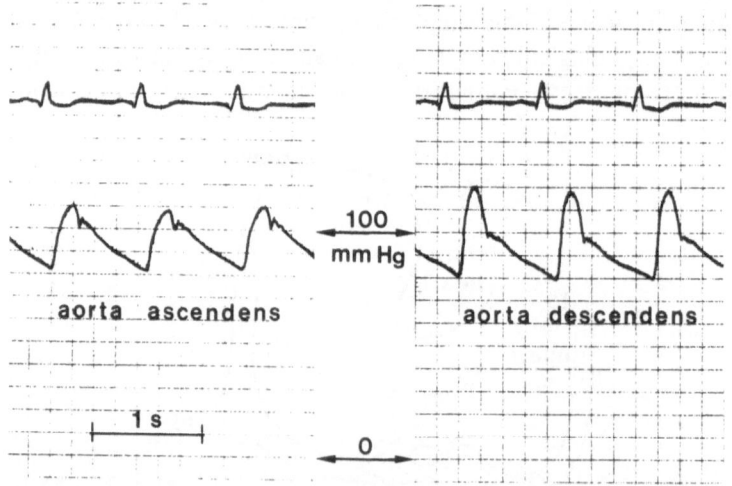

Fig. 2.2 The pressure wave in the aorta ascendens and aorta descendens, illustrating the propagation and increase of the pulse wave towards the periphery. The upper traces represent reference EGG's. (Courtesy of Dr. Schuilenburg, Cardiology Department, University Hospital, Amsterdam.)

sures are found. It can safely be assumed that the pressure conventionally measured with a cuff will not always faithfully represent the value at the root of the aorta, probably the most representative value of the systolic blood pressure, although in practice the differences will not be of too much importance. The pressure pulse spreads with a high velocity through the arterial system which can also be deduced from Fig. 2.2. It may reach a value of 10 m/sec which is considerably more than the average velocity of the blood. With a flow of approximately 75 ml/sec through an orifice with a mean diameter of 3 cm, an average velocity of blood particles of 10 cm/sec will result in the aorta. Thus, whereas the pressure pattern is moving rapidly the movements of the particles transmitting the wave themselves are restricted to smaller regions.

During one cycle the pulse wave is subjected to reflections at branches and transitions into other anatomical structures such as arterioles and capillaries. Reflections will be treated in more detail to show their implications for measuring blood pressure. The results will also be used in dealing with the problems of measuring blood pressure with catheters. For the build-up of the blood pressure and for the generation of reflections it is important that the arterioles and the subsequent capillaries exhibit a high resistance to flow, adjustable by nervous and hormonal control. This is especially true for the arterioles. Although the total area of the 3×10^{10} capillaries (3000 cm^2) is much larger than that of the arterial tree (< 10 cm^2) the resistance of all these very small vessels together is much higher than that of the larger arteries as can be deduced from Poiseuille's law ($R_p = 8\eta \, l/\pi r^4$). The result is the peripheral resistance. It should be realised, however, that Poiseuille's law is but an approximation for blood flow; several factors can disturb the applicability. For instance under normal circumstances blood viscosity has a value of approximately 4×10^{-3}N sec m^{-2} (4 centipoise), but it should be realised that blood is a non-newtonian fluid. Which fact, however, only becomes important when the radii are less than 100 μm. In the capillaries the blood particles move more or less individually and in these cases the use of Poiseuille's law is questionable. A next complication is that at higher pressures and caused by the distensibility of the smaller vessels the Poiseuille resistance will drop, owing to the $1/r^4$ term.

In the great arteries and even in smaller arteries it can be found that the velocity profile of the flow is not parabolic as should be expected from Poiseuille's law. This effect is due to the dynamics of pulsatile flow, further complicated by the inlet length, which may have dimensions that are not small compared to the artery length. However, all these factors do not negate the use of a total peripheral resistance defined by R = pressure/flow which in man has a value in the order of 17.10^7N.s.m.$^{-5}$($\cong 1.3$mm Hg ml^{-1}s).

2.1.3 *Acoustical aspects*

To obtain a better insight into the requirements for objective measuring devices and methods in cardio-acoustics the following more general considerations are presented.

The pump function of the heart is accompanied by vibrations precisely as with other mechanical pumps. For instance vibrations are produced at the moment that valves close and open. Stenoses of the valve orifices can lead to turbulence and vortexes in the adjoining vessels which will again set mechanical structures in vibration. These vibrations propagate through the tissues which surround the heart and can be observed at the chest wall.

Initially auscultation was performed by application of the ear to the skin. The French physician Laennec invented the stethoscope, an instrument without with no practising physician would care to be seen. After the invention of carbon microphones Huerthle (9) and Einthoven (3) used these transducers to register heart vibrations (Fig. 2.3) and in this way phonocardiography was born. A phonocardiograph will consist in principle of a

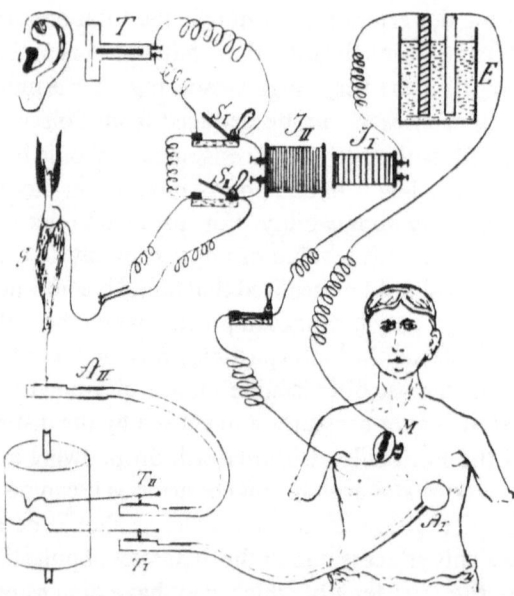

Fig. 2.3 A phonocardiograph 85 years ago. This equipment was used by Huerthle. One recognises the same basic elements as in a modern phonocardiograph: a transducer (carbon microphone), and amplifier (a very non-linear one, consisting of an "Inductorium" and a frog muscle), and a registration unit (a sooted drum).

microphone (a mechano-electric transducer), electronic filters and amplifiers, and some registration unit.

The physical problems connected with phonocardiography are found in the genesis, transmission, and transducing of the vibrations. The last problem is a measuring one and an understanding of it is most important for the study of the first two. With regard to the origin of the vibrations there is a vast literature (14) which however, is, not very conclusive. Most authors agree on a description which says that the sudden de- and acceleration of the blood in the ventricles at the beginning and end of the systole generate vibrations of the whole cardiohaemic system (18). These vibrations which have a transient character, can be heard through a stethoscope as more or less sharp sounds, the first and second heart sounds (Fig. 2.4, second trace). Third and fourth heart sounds are also described. They are mostly only observable or are more accentuated under pathological conditions. Other vibrations which generally originate in a pathological situation have a more noiselike character, the heart murmurs. They can be observed at any time during the complete heart cycle, the timing dependent on the kind of cardiac disfunction. A third source of vibration is the movement of the complete heart throughout the heart cycle. These vibrations contain only low frequencies, roughly up to 10 Hz. They therefore cannot be perceived by the ear but instead by palpation of the chest wall. If a suitable transducer is used on a suitable place they can be easily registered (apex cardiogram or ictus curve, Fig. 2.4, third trace).

In the section on cardioacoustics special attention will be devoted to the way in which the transmitted vibrations can be recorded by different devices.

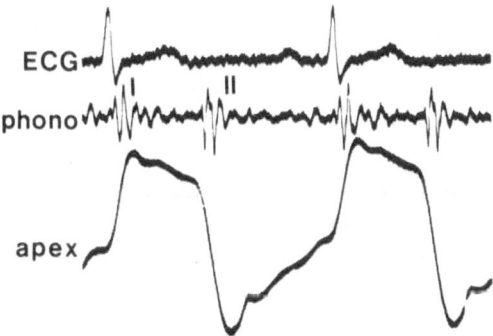

Fig. 2.4 A registration of an electrocardiogram, a phonocardiogram, and an apex cardiogram. I: first heart sound; II: second heart sound. Heart rate approximately 1 per second.

2.2 Mechanical-electrical analogues

The equations of motion in mechanics were well developed long before electric circuit theory was fully established. With the growing knowledge in the latter field it was recognized that in many hydrodynamic and mechanical systems the governing differential equations are much the same as in electrical circuits. Whereas in the beginning there was a tendency to describe electrical phenomena in mechanical or hydrodynamic terms, this situation drastically changed after the rapid development of electric circuit theory, supported by adequate schematic representation of electric circuits and the use of analogue computers. This development is now more or less arrested due to the increasing use of digital computers. Of course, in view of the similarity of the mathematics a schematic representation can also be applied in hydrodynamic and mechanical engineering. A consequence of this was that for certain problems mechanical and hydrodynamic schemes were transferred into electrical models which are more accessible to direct manipulation. These analogues serve two goals:
1) they may give a conceptual insight into the mechanical problem;
2) through hardware models of the analogues which can be built relatively easily, instructive demonstrations can be given and experimentally the influence of changes of mechanical factors can be assessed.
The systems under view are characterized for a part by propagation phenomena. They can in a number of cases fruitfully be approached by a finite not-too-large number of lumped elements.

It should be realized, however, that for complicated phenomena such as inlet length of bloodflow, the profile of non-stationary flow and non-linearities in the hemodynamic system no simple electrical representation may be present. It will always have to be judged, whether deviations have enough physiological significance to take them into account or not.

We will review some basic facts regarding electrical analogues of mechanical systems. Electrical circuit theory makes use of four types of basic passive elements: resistors, capacitors, inductors and transformers. Although transformers can be useful as analogues in mechanical problems, which contain levers, we will not discuss this type of element. The drop across a circuit element, the flow through it, and its magnitude are related and thus describe the performance of a circuit element. Although other choices are possible, we will restrict our analogues to those where in the electrical case the drop over the element is equivalent to voltage and the flow through it to current. The relations in the electrical case are:

$$\text{(a) } E = IR, \quad \text{(b) } E = L\frac{dI}{dt}, \quad \text{(c) } E = \frac{1}{C}\int Idt \qquad (2.1)$$

where:

E = voltage	L = inductance
I = current	C = capacitance
R = resistance	t = time

In most practical cases R, L and C are supposed to be independent of E and/or I. Apart from these passive elements there are also generators of energy. They are represented by either a current or a voltage source. Symbols for electrical circuit elements and generators are given in Table 2.1.

If sinusoidal voltages and currents are assumed (AC, alternating current), and choosing the usual representation, then all the relations (2.1) can be represented by:

$$E = IZ \qquad (2.2)$$

where Z is the impedance of the circuit element in question. The magnitude of the impedance of an inductance L is ωL and the phase angle between the voltage and the current is $\pi/2$ radians. For the other elements the reader may easily find out the relevant relations.

In hydrodynamic systems flow and pressure drop are the generator functions. Given a Poiseuille flow the relation between pressure over a tube and the flow through it is given by

$$\Delta P = \frac{8\eta l}{\pi r^4}\dot{Q} \qquad (2.3)$$

Table 2.1 Symbols for electric circuit elements and generators.

Name	"Voltage" source	Current source	Resistance	Inductance	Capacity
Symbol					

where

ΔP = pressure drop l = length of the tube
$\dot{Q}$ = volume current $\equiv$ flow r = radius of the tube
η = viscosity of the fluid

If we introduce

$$R_H = \frac{8\eta l}{\pi r^4}$$ (2.4)

the relation between pressure and flow is given by

$$\Delta P = R_H \cdot Q$$ (2.5)

We will call R_H the hydrodynamic resistance. Comparison with equation 2.1a shows that these relations are equivalent if pressure is analogous to voltage and volume flow to current. Viscous resistance is analogous to electrical resistance.

A vessel with an elastic wall and filled to its resting volume can be expanded by introducing more fluid. The volume increase ΔQ is given by $\int \dot{Q} \cdot dt$. On the other hand this volume increase ΔQ gives rise to an increase of the pressure ΔP in the fluid. By definition $\Delta Q / \Delta P$ is the compliance C_H of the vessel. Equating one finds

$$\Delta P = \frac{\int \dot{Q} \cdot dt}{C_H}$$ (2.6)

If this is compared with equation 2.1c it is clear that the electrical analogue of compliance is capacitance.

A flow which is a function of time gives rise to acceleration or deceleration of fluid particles. The forces necessary for these accelerations summed over a volume can be expressed as a pressure drop. From Newton's law this can be expressed as

$$\Delta P = M_H \frac{d\dot{Q}}{dt}$$ (2.7)

in which M_H is the inertance of the fluid in observation. If a uniform velocity is assumed over the diameter of the tube (viscosity zero) one can derive that

$$M_H = \frac{\pi l}{\pi r^2} \tag{2.8}$$

One should notice that the total mass in the considered length of tube does *not* enter in this expression. By comparison of equations 2.7 with 2.1b it follows that the analogue of inertance is inductance.

Impedance is now defined analogous to electric impedance as $Z_H = (\Delta P/\dot{Q})$. In mechanical and acoustical problems using analogue reasoning, a similar analysis is possible. If a mass m is accelerated, the force F on it is given by Newton's law

$$F = m \cdot a = m\frac{dv}{dt}$$

with a = acceleration and v = velocity. For a spring with a stiffness S the well-known relation between force and displacement x is: $F = S \cdot x$ which can be written as:

$$F = S \int v \cdot dt \tag{2.10}$$

Comparison with the electrical equations 2.1 indicates that if voltage is chosen as an analogue for force and current for the velocity, the electrical analogue for a mass is an inductance while for the spring the analogue is a capacitance with a magnitude equivalent to $1/S$. A mechanical resistance is defined as a device with friction, for which holds the relation

$$F = R_m \cdot v \tag{2.11}$$

As in the electrical case one can now define the quotient of force and velocity as a mechanical impedance

$$Z_m \triangleq \frac{F}{v} \tag{2.12}$$

Table 2.2 gives the relations and symbols for impedance analogues.

As an example we give here a mass-spring with mechanical resistance (damping). A mass upon which a force F is exerted is suspended by a spring paralleled by a mechanical resistance (Fig. 2.5a). The resulting velocity can be found if this scheme is transformed into its electrical

Table 2.2   Relations and symbols for impedance analogues.

Name	Mechanical symbol	Electric analogue symbol	Magnitude	Mechanical impedance
Mass	▭	 〰〰〰	m	$i\omega m$
Spring	–〰–	–|�Ⓒ |├–	$1/S$	$\dfrac{S}{i\omega}$
Mechanical resistance	 ⊐–	 –▭–	R_m	R_m

analogue (Fig. 2.5b). As the velocity of all three elements is the same and the analogue of velocity is current, the series circuit (Fig. 2.5b) represents indeed the mechanical system. Resonance occurs for a circular frequency $\omega = \sqrt{(S/m)}$. In the same way as for electrical circuits, bandwidths and quality factors can be defined.

Using an impedance analogue it is obvious that *parallel elements* are transformed into *series elements* and this sometimes makes the use of such analogues difficult when the system consists of more elements. Very often therefore use is made of another concept: the mobility analogue. For this approach Newton's law is expressed in an impulsive form as $\int F \cdot dt = m$ and the analogue with equation 2.1c is seen. In this case the analogue of force is current and that of velocity is voltage; a mass is transformed into a capacitance with a magnitude m. A spring is now transformed into an inductance with magnitude $1/S$ and a mechanical resistance (dashpot) R_m into a resistance with a magnitude $1/R_m$. It is thus obvious that mechanical impedance and mobility type analogues are each other's reciprocals.

If we again look to the mechanical scheme of Fig. 2.5a this is now transformed into the electrical circuit of Fig. 2.6. In the mobility analogue

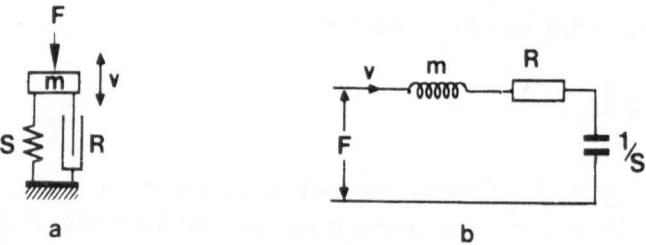

Fig. 2.5   Mass-spring system with mechanical resistance. *a)* Schematic drawing. *b)* Mechanical impedance analogue.

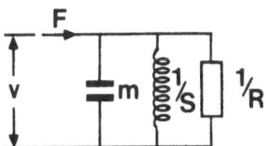

Fig. 2.6 Mass-spring system with mechanical resistance: Mechanical mobility analogue.

obviously parallel elements are transformed in parallel electric elements.

In electric systems voltage can be measured without breaking into the circuit, whereas a current is measured by introducing a current meter into the electrical circuit. In mechanical systems, velocity may be measured without influencing the system but force can only be measured by introducing a force measuring device into the mechanical system. The mobility analogue thus seems in this respect more advantageous. Which analogue to use is a question of taste. Both analogues are used throughout the literature. The two types must not be confused. Table 2.3 lists both types of analogues and their relations.

In acoustic problems the parameter which can be measured without breaking the circuit is the pressure and here an impedance type analogy is more advantageous. A voltage is chosen as analogue for the pressure. For the current one should choose a quantity proportional to the velocity. As in hydrodynamics the volume velocity U is a good choice. The acoustic impedance is then defined as

$$Z_A = \frac{P}{U}$$

In view of the compressibility of air the acoustic mass is defined as the mass of the fluidum *accelerated* by a net force *without appreciable com-*

Table 2.3 Relation between mechanical impedance and mobility analogues.

Mechanical quantity		Electrical analogues			
		Impedance analogue		Mobility analogue	
Name	Value	Name	Value	Name	Value
mass	m	inductance	m	capacitance	m
compliance	C_m	capacitance	C_m	inductance	C_m
stiffness	S	capacitance	1/S	inductance	1/S
mechanical resistance	R_m	resistance	R_m	resistance	$1/R_m$
force	F	voltage	F	current	F
velocity	v	current	v	voltage	v
displacement	x	charge	x	flux	x

Table 2.4 Survey of different types of analogues

Electrical			Hydrodynamical			Mechanical			Acoustical		
Name	*Symbol*	*Unity*	*Name*	*Symbol*	*Unity*	*Name*	*Symbol*	*Unity*	*Name*	*Symbol*	*Unity*
voltage	E	Volt	pressure	P	Pascal	force	F	Newton	pressure	P	Pascal
current	I	Ampère	volume-current	Q_H	$m^3 . s^{-1}$	velocity	v	$m . s^{-1}$	volume velocity	U	$m^3 . s^{-1}$
charge	q	Coulomb	volume	Q_H	m^3	displacement	x	m	volume	Q_A	m^3
resistance	R	Ohm	hydrodynamic resistance	R_H	$N. s. m^{-5}$	mechanical resistance	R_M	$N. s. m^{-1}$	acoustic resistance	R_A	$N. s. m^{-5}$
capacitance	C	Farad	compliance	C_H	$m^5 . N^{-1}$	mechanical compliance	C_M	$m. N^{-1}$	acoustic compliance	C_A	$m^5 . N^{-1}$
inductance	L	Henry	inertance	M_H	$kg. m^{-4}$	mass	m	kg	acoustic inertance	M_A	$kg. m^{-4}$
impedance	Z	Ohm	impedance	Z_H	$N. s. m^{-5}$	mechanical impedance	Z_M	$N. s. m^{-1}$	acoustic impedance	Z_A	$N. s. m^{-5}$

pression. The volume of the fluidum *compressed* by a net force without an appreciable displacement of its center of gravity defines an acoustic compliance. Acoustic resistance is associated with a viscous loss or a dissipative loss through radiation.

From the definitions it is easily seen which are the dimensions of the distinct elements. Although a mechanical impedance is defined theoretically in a point, it is measured in practice on a finite surface with an area A. As acoustic impedance is defined as the pressure on a surface element divided by the volume velocity through this element, we can relate mechanical and acoustical impedances. Since $P = F/A$ and $U = A \cdot v$ it follows that $Z_m = A^2 \cdot Z_A$. Table 2.4 gives a survey of the analogues so far discussed and their dimensions.

Very often mechanical and acoustic systems are coupled, for instance a radiating loudspeaker can be seen as a mechanical vibrating system coupled to the air. Depending on the mounting of the speaker, for instance in baffles or boxes, the mechanical system "sees" acoustic mass, compliance and/or resistance. An element which couples two systems is called a transducer. If the total system is to be analysed, it is useful to insert in the scheme an "ideal" transformer with the appropriate properties. For the mechano-acoustic transducing element the relations, using an impedance type analogue for both sides, are as in Fig. 2.7a. For the electrical-mechanical transducing a similar procedure can be followed. Fig. 2.7b given the scheme for an electro-magnetic transducing device using an impedance type analogue. Using the example of a loudspeaker one can in this way define the reaction of acoustic loading upon the mechanical behaviour of the cone-suspension system or inversely evaluate the acoustic efficiency of different cone-suspension systems. Microphones can also be analysed in this way.

2.3 Hemodynamics

2.3.1 *A simplified model of blood pressure ("Windkessel" model)*

A drastically simplified analogue of the left ventricle, the arterial system and the periphery is drawn in Fig. 2.8. In this analogue the left ventricle is during systole represented by a current source which repeatedly injects a current. To a certain extent a heart behaves like a current source; in experimental situations a dog heart, for instance, can still eject blood against pressures of more than 30 kPa ($\approx$250 mm Hg).

In the model of Fig. 2.8 no valves are needed since a current source has an

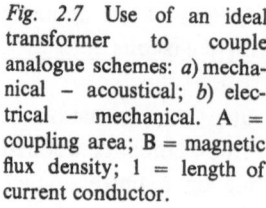

Fig. 2.7 Use of an ideal transformer to couple analogue schemes: *a)* mechanical – acoustical; *b)* electrical – mechanical. A = coupling area; B = magnetic flux density; l = length of current conductor.

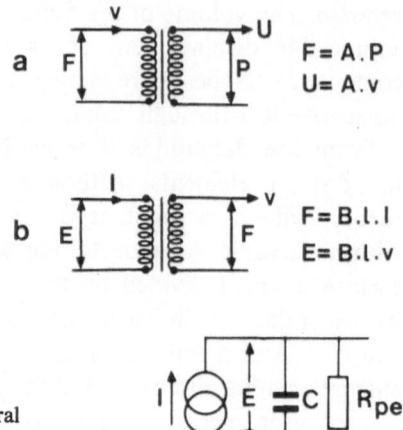

$F = A.P$
$U = A.v$

$F = B.l.I$
$E = B.l.v$

Fig. 2.8 Simplified analogue of the left ventricle, the arterial system and the peripheral vessels.

infinite impedance and no charge can flow back. A further simplification is that the elasticity of the aorta and arteries can be expressed by a pure compliance, translated into a capacitor C in the electric case. The peripheral vessels (arterioles and capillaries) are assumed to have only a viscous resistance; in the analogue this is represented by a single resistor R_p, the peripheral resistance.

With each beat a volume of blood is ejected into the aorta which is equivalent to a charge. For reasons of simplicity the duration of the ejection phase is neglected. The charge q causes a potential difference on the capacitor $E = q/C$. Further is $E(t) = E_s \exp(-t/R_p C)$, E_s being

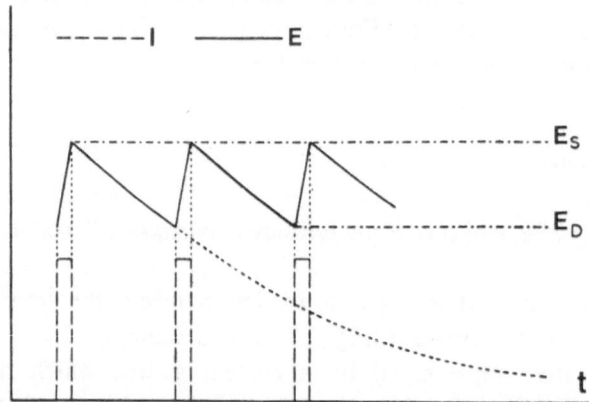

Fig. 2.9 Time course of the voltage in the system of Fig. 2.8, if regularly current pulses are injected in the circuit. A current pulse of short duration quickly charges the condensor. Discharge through the peripheral resistance R_{per} is exponential. E_S and E_D are the electrical analogues of systolic and diastolic pressures.

the electrical analogue of the systolic pressure. After a time t_H a new ejection takes place. Just before this ejection the potential (blood pressure) has fallen to the diastolic value (Fig. 2.9):

$$E_d = E_s \exp(-t_H/R_p C) \tag{2.13}$$

The difference between the diastolic and systolic potentials will be equal to $E = q/C$ so that: $E_s(1 - \exp(-t_H/R_p C)) = q/C$ from which follows:

$$E_s = \frac{q}{C(1 - \exp(-t_H/R_p C))} \tag{2.14}$$

and

$$E_d = E_s - q/C \tag{2.15}$$

As q = analogue to the stroke volume, it follows, providing that the stroke volume is kept constant, that an increase of the resistance R_p or increasing of the heart rate (decrease of t_H) have the same quantitative effect on systolic pressure P_s and diastolic pressure P_d because E, the analogue of $P_s - P_d$, is then constant and the average pressure is rising in both cases. If the walls of the arterial system are stiffer (less compliance), expressed by a smaller capacitor C in the analogue, the difference between P_s and P_d increases, given a constant stroke volume. Since the average flow (current) does not change, the average pressure does not change but the systolic pressure will rise under these conditions, whereas the diastolic pressure will fall. Because of the exponential course of the pressure fall during diastole P_s will rise more than P_d will fall.

Table 2.5 gives an impression of the quantitative influence for different values of various parameters. Only a change in stroke volume or in rigidity of the wall will give a change in pulse pressure. The model clearly indicates the importance of the time constant $R_p C$ which is of the order of 3 seconds in normal subjects. This means that if the heart stops beating, as is possible in stress situations, after approximately 9 seconds blood pressure has dropped to 900 Pa (≈ 7 mm Hg), a value incompatible with consciousness. With such a simple model various physiological (pathological) conditions like changing peripheral resistance or stiffening of the arteries due to senile sclerosis can be evaluated. It should be realised, of course, that in the intact circulation especially in the long range an extremely complicated multi-loop regulatory mechanism is active (see

Table 2.5　Systolic and diastolic pressure as a function of different characteristics.

	E_S	E_D
Normal	120	80
$(R_p) \rightarrow 2R_p$ $(t_H) \rightarrow \frac{1}{2}t_H$	215	175
$(C) \rightarrow \frac{1}{2}C$	144	64
$(C) \rightarrow \frac{1}{4}C$	215	55

The change for the systolic pressure (E_S) and the diastolic pressure (E_D) is given when one of the parameters is changed. When peripheral resistance increases or the frequency increases both pressures rise, but the stroke volume remains a constant. When the capacity (volume rigidity) decreases (as in sclerotic cases) the systolic pressure rises and the diastolic pressure falls.

for example Guyton, 7). There is no simple way to include this in the fundamental equations.

It was already stated that the model is based on many simplifications.

a) The duration of the systole is neglected. This will have a considerable influence at high heart rates but inclusion of the necessary mathematics makes the model less lucid.

b) The peripheral resistance is assumed to be independent of flow and pressure. In the introduction it has already been pointed out that this is not true. Nevertheless the general concept remains remarkably useful.

c) The non-linear elastic properties of the arteries also form a relatively important problem for the propagation of the pulse wave. For young persons there is a tendency for decreasing compliance at increasing pressure, a property well-known for purely elastic tubes. It is, however, soon more than compensated for by increasing age so that according to Langewouters et al. (11) Young's modulus of a segment of the thoracic aorta of a 70-years-old male is tripled (from 1.5 to $4.5 \times 10^6 \, \text{N/m}^{-2}$ when blood pressure rises from 12.5–25 kPa $\approx$ 100–200 mm Hg).

d) Complete neglect of the transmission properties of the system. In other words propagation velocity is considered to be infinite. In reality, however, the whole aorta is expanded only at the end of the systole. The related phenomena will be treated in the next section.

2.3.2 *The pulse wave*

The physics of propagation of the pulse wave in its simple form was early recognised (4, 15, 23) but also the difficult problem of the interrelation of acceleration forces and friction was already conceived by Euler in 1755 (19) whose work was for long forgotten.

It should be realized that a pressure pulse which propagates in a (visco-) elastic tube is a totally different thing from the particle transport which is the true "raison d'être" of the circulation. In its most simplified form pulse propagation is deduced from linearised hydrodynamic laws which are highly analogous with those of an electric cable. As soon as more factors governed by fluid dynamics and arterial wall properties are included complexity greatly increases. Some of the factors have a direct electrical counterpart, but in a number of cases the questions can be solved only with the parameters of the system itself and up to date a solution has not been found in a closed mathematical form for every situation. We will, however, mainly concentrate on the electrical analogue, i.e. the propagation of an electric disturbance through a cable. Our aim is to show that here, for a general use and certainly for demonstrations, an elementary model can also be a useful tool. The propagation velocity V of an electrical disturbance is $\sqrt{1/lc}$ where l and c are respectively the inductance and capacitance per unit length; it is independent of frequency. An important property is expressed by the characteristic impedance which in case of a loss-free infinitely long cable, and under certain conditions also in a dissipative cable, is a pure ohmic resistance. Its value is $Z\,char = \sqrt{l/c}$.

To obtain a good understanding of the characteristic impedance, we perform the following thought experiment. At the beginning of the cable we suddenly apply a constant potential. This potential step will propagate uniformly, since the transmission velocity is independent of frequency. The total charge will increase proportionally to time, which means that a constant current is drawn by the cable. An infinitely long cable thus acts as a pure ohmic resistance, the value of which is defined as the characteristic impedance.

From the formula for the propagation velocity it follows that the potential disturbance has passed in one second into a length of cable of $\sqrt{1/lc}$ meter. The capacity of that piece of cable occupied by the step potential is $c \cdot \sqrt{1/lc} = \sqrt{c/l}$. Since a constant current is drawn by the cable the charge transferred in one second $q = \sqrt{c/l}$. E is of course equal to the constant current $I = E/Z_{char}$ and so we arrive at $Z_{char} = \sqrt{l/c}$.

From the electrical-hydrodynamic analogues the propagation velocity of a pressure disturbance and the characteristic impedance of an elastic tube follow now directly. If the volume compliance per unit of length is C_T and if for the inertance per unit of length is taken $\rho/\pi r^2$,* the pro-

---

*If one takes for inertance $\rho/\pi r^2$ it is assumed that conditions prevail in which the flow profile is flat. There is considerable literature on this problem but all aggree that the inertance will be at highest $\frac{4}{3}$ of that for a flat profile. Actually the flat profile is energetically to be

pagation velocity

$$V = \sqrt{\frac{\pi r^2}{\rho C_T}} \text{ and } Z_{char} = \sqrt{\frac{\rho}{\pi r^2 C_T}}. \tag{2.16}$$

It should be noted that there is a fundamental difference between the Poiseuille resistance and the characteristic impedance. The former determines the constant flow through a tube when a constant pressure is applied and depends on the viscosity of the fluid and the geometrics of the tube. A rubber and a steel tube with the same radii have the same Poiseuille resistance but certainly not the same characteristic impedance since this latter depends on the elastic properties of the walls.

A further observation is that the propagation velocity in a rigid tube is determined by the compressibility of the fluid, i.e. approximately 1500 m/s in a water filled tube. It is thus seen that since the pulse wave velocity rarely exceeds 10 m/sec the effect of the compressibility of the fluid can be neglected.

When an electric disturbance is introduced into a cable, which is not terminated by its characteristic impedance, it follows from the boundary conditions that a reflection will occur at the end of the cable. The reflected wave will travel back to the beginning of the line with the same velocity as that of the original wave and it once again can be reflected, unless the source has an impedance equal to the characteristic impedance. The amplitudes of the reflections will depend on the ratios of the characteristic impedance Z_{char} and the terminal impedances Z. The reflected voltage will be equal to:

$$V_r = \frac{Z - Z_{char}}{Z + Z_{char}} \cdot V \tag{2.17}$$

If a wavefront arrives at the end of an open cable (Z = infinite) the amplitude of the reflected wave will add to the incoming wave and will double the voltage. If the cable with a length l is fed from a current source, the source impedance at the beginning is by definition also infinite; so if a wave, reflected at the end, arrives after $2\ l/v$ second at the beginning of the cable it is once again reflected with the same amplitude and a doubling of the

preferred as long as no viscosity is introduced in the fluid and the tube is rigid. The reason is, that disregarding to losses due to friction, which are proportional to velocity the determining factor is the kinetic energy, which is proportional to the square of the velocity.

voltage will also occur at the beginning. In Fig. 2.10 the voltages at the beginning, and at the end of a cable are schematically drawn for a loss-free cable for an impulse like input signal.

If we now apply this model to the human aorta, it is interesting to obtain an estimate of its characteristic impedance Z_{char} and to compare this with the peripheral resistance R_p. The situation, however, is extremely complex, even if one abstracts from nonlinearities of different kinds. The branching of the aorta and the different properties of the branches and the resistances they lead into, present problems very difficult to resolve, yet the lumping of all elements remains a sensible procedure if the limitations are well enough realised.

From $V \cong 7$ m s^{-1}, $\pi r^2 \cong 7$ cm^2 and $\rho \cong 1000$ kg m^{-3} follows $C_T \cong 1.5 \times 10^{-8}$ m^{4}N^{-1} and from this value is calculated $Z \cong 10^7$ Nsm^{-5}. It is obvious that whatever corrections should be applied the system is totally mismatched by the peripheral resistance $R_p \cong 17 \times 10^7$ Nsm^{-5}.

In reality the mismatch between Z_{char} and R_p may be estimated to be less because of the physiological property of the arterial system to increase its rigidity towards the periphery; according to Gow (6) a factor 5 may be expected. This would mean an increase of Z_{char} with a factor $\sqrt{5}$ to about 2.5×10^7 N s m^{-5}. If we accept this value in the ideal case we can thus expect a reflected pressure wave P_r with an amplitude

$$P_r = \frac{R_p - Z_{char}}{R_p + Z_{char}} \cdot P_p \cong \tfrac{3}{4} P_p \qquad\qquad (2.18)$$

where P_p = primary pressure pulse. In any case this would still be a near doubling of the pressure wave at the transition to the periphery and a considerable reflected pulse wave there. The importance of the reflected wave decreases, however, on its return owing to damping and reversed tapering. Some partly conjectural considerations may follow. A matched system where R_p and Z_{char} are equal would behave as an infinite cable and the flow and pressure pulse of the heart would be directly noticeable in all peripheral vessels and even in the venous system.*

No continuous flow would be present in the capillaries and this would give the blood less time for exchange of oxygen and nutrients with the surrounding tissues. The mismatch, however, gives, as we saw, rise to reflections which we can indeed find in reality (Fig. 2.2). The reflections show a large decrement, apparently due to much more damping in the

---

*There is a disease (acute beriberi) in which R_{per} is lowered dramatically and indeed a venous pulse of considerable magnitude is then present.

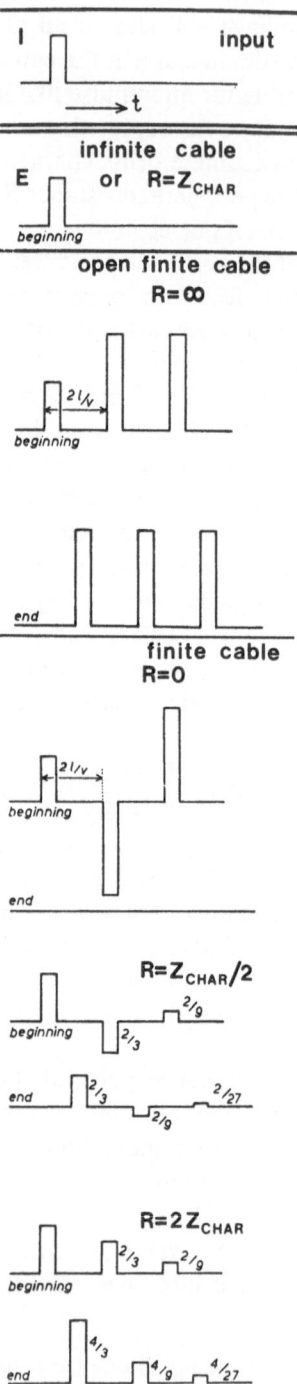

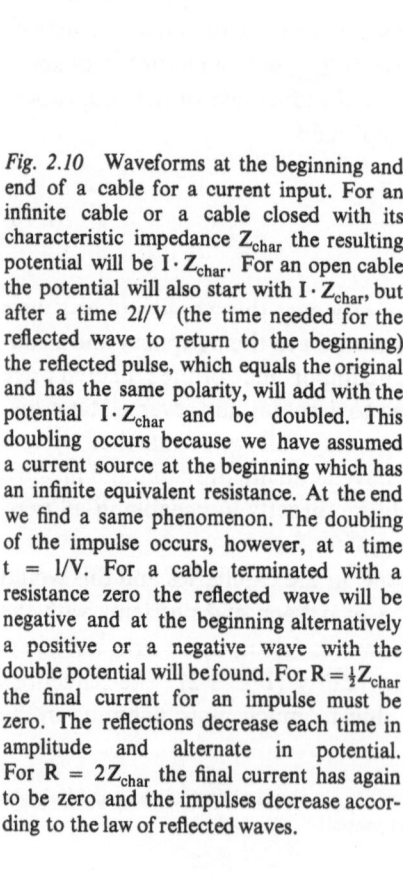

Fig. 2.10 Waveforms at the beginning and end of a cable for a current input. For an infinite cable or a cable closed with its characteristic impedance Z_{char} the resulting potential will be $I \cdot Z_{char}$. For an open cable the potential will also start with $I \cdot Z_{char}$, but after a time $2l/V$ (the time needed for the reflected wave to return to the beginning) the reflected pulse, which equals the original and has the same polarity, will add with the potential $I \cdot Z_{char}$ and be doubled. This doubling occurs because we have assumed a current source at the beginning which has an infinite equivalent resistance. At the end we find a same phenomenon. The doubling of the impulse occurs, however, at a time $t = l/V$. For a cable terminated with a resistance zero the reflected wave will be negative and at the beginning alternatively a positive or a negative wave with the double potential will be found. For $R = \frac{1}{2}Z_{char}$ the final current for an impulse must be zero. The reflections decrease each time in amplitude and alternate in potential. For $R = 2Z_{char}$ the final current has again to be zero and the impulses decrease according to the law of reflected waves.

system than can be expected from the Poiseuille resistance which in the aorta would only amount to $2 \times 10^4 \, \text{N s m}^{-5}$. The high losses mainly due to visco-elastic properties of the walls, however, make sure that under normal circumstances within one beat the waves are damped and pressure equality is practically reached in the larger arteries. The damping acts also as a high frequency attenuating factor and causes the development of an imaginary term in Z, which also means that the wave velocity becomes frequency dependent (dispersion). In practice such effects are difficult to evaluate. The net result of this all is that a continuous pressure is indeed built up on which the pulse wave is superimposed and it is seen that the mismatch at the periphery, the high damping and the high velocity of the pulse wave have together a very comparable effect as the bulk elasticity of the "Windkessel". To get an impression of the quantitative factors involved we can give the following estimate. In our representation the maximal pulse pressure is determined by $P_p = Z \times \dot{Q}_{max}$, Z being the impedance seen at the entrance of the aorta having a value of approximately $10^7 \, \text{N s m}^{-5}$ and $\dot{Q}_{max}$ the maximal volume current (flow). With $\dot{Q}_{max} \cong 500 \, \text{cm}^3 \, \text{s}^{-1}$ in a rest situation a maximal primary pulse pressure in the order of 5 kPa ($\cong 40$ mm Hg) can be expected. As was already stated in the introduction the systolic pressure is by no means a fixed value throughout the arterial system and one of the reasons is that towards the periphery the aorta's diameter decreases and the arteries have in general a relatively thicker wall. Geometric tapering adds to the elastic one and this leads the more to an increase of the characteristic impedances towards the periphery. From this we can understand the increasing size of the primary pulse wave as seen in Fig. 2.2.

Another phenomenon which is seen in the pressure curves is the sign of the closure of the aortic valves, the incisura. When the pressure of the left ventricle falls below that of the aorta, the valves will be closed by a back current of blood. After closure a certain column of blood, the length depending on the propagation velocity of waves in the aorta, is stopped abruptly. This will produce a positive pressure wave resulting from the liberated kinetic energy. Owing to its comparatively high frequency content it will be strongly attenuated along the aorta. This "waterhammer" effect coincides with a typical excursion in the phonocardiogram, ascribed to aortic phenomena. This suggests that the described mechanisms may contribute to the genesis of the aortic sound.

The mechanism that is described as "tapering" also causes the incisura to occur at a pressure higher than the diastolic. A discussion whether this is due to true tapering or to early reflections is academic, since the length occupied by the leading edge of the pulse wave is not negligible and the earliest reflections will occur long before the heart has ceased its contraction.

Strangely enough a number of the above phenomena can be closely demonstrated by a lumped model of only 10 elements. Fig. 2.11 shows the construction of a lumped delay line and the waveforms met at the entrance and the exit. The damping is a natural consequence of the resistances of the inductances employed. For instance the femoral pulse is reasonably well mimicked in this way.

It is simple to add a certain amount of tapering in the model by decreasing the condensors in this example by 20% per section. This makes the simulated waveforms more likely and in this way it is not difficult to find similarities in actual pressure curves in man and animal. Of course this is no proof for any theory about the pulse wave but it is given here for two purposes. The first is to show that conceptual physical reasoning with the use of analogues, can be fruitful, and the second to give a warning that the information in the pulse wave probably does not allow too strict conclusions about physiological and even physical processes in the circulatory system.

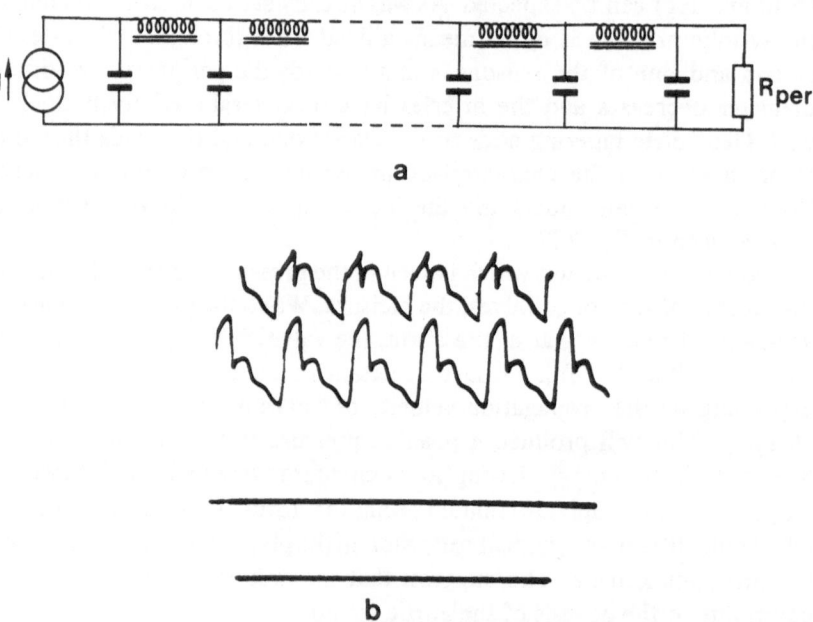

Fig. 2.11 a) Electrical lumped model analogue of an aorta. The inductances have a value of 1 Henry and a resistance of 550 ohm. The first capacitor has a value of 10 nF. In each section this value is decreased with 20% to imitate elastic and geometric tapering. The peripheral resistance has a value of about 70 k ohm. b) Wave forms obtained by this model. Input: rectangular current pulse with a length of 1.5 ms and·a repetition period of 5 ms.

Nonlinearity in the elasticity, the effects of multiple branching and reflections all give enough variables to match different experimental data.

2.4 Pressure recording with catheters

Catheters are special, rather stiff, tubings with a small lumen which are introduced into vessels and heart. They serve several purposes among which is the sampling of blood but they are mostly connected to blood pressure measuring devices. It was noticed at an early state (8) that the physical properties of the system will determine the faithfulness of the recorded pressure wave in the dynamic case. Although it is possible to describe approximately the system as a whole with a set of mathematical equations this does not give much direct insight into the factors involved and even less into deviations to be expected from varying factors. In fact they are too many for an exhaustive description. What will interest us here is the transfer characteristics of the system with regard to the time course of the pressure pulse including movement artefacts. It is useful to approach the system by three separate analogue circuits, where different elements dominate the response. In reality there is an overlap of the properties of all of them but it will be shown that the treatment can be successful enough for practice (21). Let us first shortly describe the three models.

In the first model we describe the Poiseuille resistance of the stiff catheter and the non-infinitely stiff pressure transducer as a low pass filter (Fig. 2.12). This kind of model will play a role especially if air bubbles (even small ones) are introduced in the pressure chamber, a not always avoidable condition and this may be the limiting factor with thin catheters as used for children. A 6 db/octave high frequency cutoff can be expected from the electrical diagram, which, if the cutoff frequency is low enough, can lead to smoother curves. The nomogram of Fig. 2.13 gives the cutoff frequencies for a given volume elasticity (reciprocal of compliance) of the pressure transducer respectively coupled to 3 catheters with different lumens (drawn

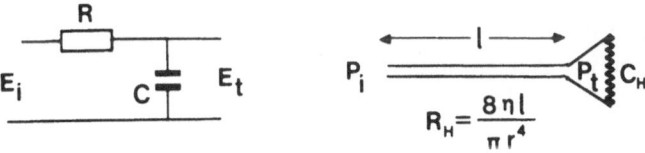

Fig. 2.12 Catheter and pressure transducer as a low-pass filter. Electrical analogue and schematic drawing. R_H: Poisseuille resistance of catheter, C_H volume compliance of pressure transducer.

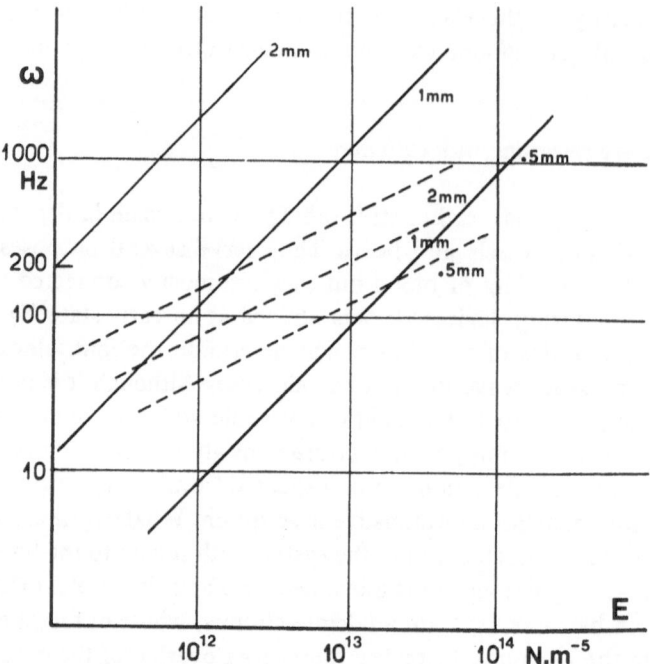

Fig. 2.13 Nomogram for the frequency limitation of catheters with a length of 80 cm and three different lumens are given against the volume elasticity E of the manometer system. Drawn lines for the low-pass filter approximation, (ω 70%), dashed lines for the second order approximation (ω_{res}). From the figure immediately an impression can be obtained as to which restriction is the stronger one. (Note: ω is circular frequency = $2\pi f$.)

lines). Introducing an air bubble decreases the volume elasticity and a lower cutoff frequency will be the result.

In reality as we also noticed in section 2.3 for accelerated flow the relation between acceleration forces and friction losses determines the flow profile. At higher frequencies the velocity profile broadens more and more towards the wall of the tube. This is a nonlinear effect comparable to the skin effect in electric high frequency transmission through copper wires. It leads to an increase of friction losses at high frequencies and also to dispersion. Because frequency is explicit in the underlying nonlinear equations, for each waveshape the effects must be calculated separately. Fourier analysis is not simply applicable.

In the second model (Fig. 2.14) the catheter is assumed to be frictionless, it has an inertance $M = \rho l/\pi r^2$. Together with the volume elasticity $1/C_H$ this forms a second order system which has a resonance frequency $\omega_0 = \sqrt{\pi r^2/C_H \rho l}$. In reality there is of course friction and this will make the resonance less pronounced. The resonance frequency determines the cut

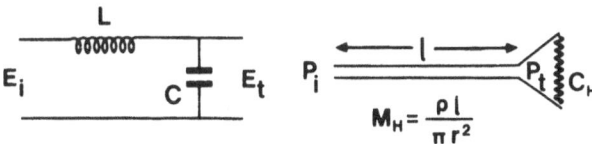

Fig. 2.14 Catheter and pressure transducer as a second order system. Electrical analogue and shematic drawing. M_H: inertance of the catheter filled with fluid, C_H as in Fig. 2.12.

off frequency of this model and the response above this frequency will decrease with 12 dB/octave (amplitude decrease with a factor 4 for a frequency doubling). Deviation of a flat flow profile in the catheters will give a somewhat larger inertance, which will not exceed a factor of $\frac{4}{3}$. In the nomogram of Fig. 2.13 the performance of this model is schematically drawn for the same parameters as for the friction limitations (dashed lines).

There is a strong tendency to judge the contractile quality of the heart by parameters as $(dP/dt)/P_{max}$ in which P is pressure in one of the ventricles. This means that the high frequency reproduction should be better than estimated from the pressure curve because of the differentiation which favours high frequencies. It looks safe to have a cutoff frequency approximately 10 times the average heart rate for such precision work. For children this means up to 20–30 Hz. Because of the inherent variance of the heart rhythm the term harmonics which could lead, and has led, to confusion is to be avoided. The manufacturers of pressure transducers strive of course towards constructions that limit the influence of the transducers on the performance of the whole system. Modern transducers have compliances that are small enough to give no major limitations. For example: for a Statham P23Db or a Bell and Howell 4-32I transducer a compliance of 0.04 mm^3/100 mm Hg ($\cong 0.3 \times 10^{-14}$ m^5 N^{-1} or a volume elasticity of 3.3 × 10^{14} N m^{-5}) is claimed. As can be seen from Fig. 2.13 this does not any more mean a limitation to the desired frequency response. It underlines, however, the importance of avoiding air bubbles. A bubble with a diameter of 3 mm has a volume elasticity of 7×10^{12} N m^{-5} and again from Fig. 2.13 we see that this may limit the performance of the system. In practice the values given by the manufacturers for the volume displacement are very often based on the displacement of the membrane only, but the volume displacement of the dome may not be negligible. The elasticity of stop-cocks used in the system for flushing etc. may also play a role.

In the third model we consider the catheter as a transmission line. In Fig. 2.15a the response of a catheter to a step function of a pressure source is given. Its shape is that of a damped oscillation, as follows directly from Fig. 2.11. The pressure step is transmitted and reflected against the high impedance of the transducer and reflected negatively against the zero im-

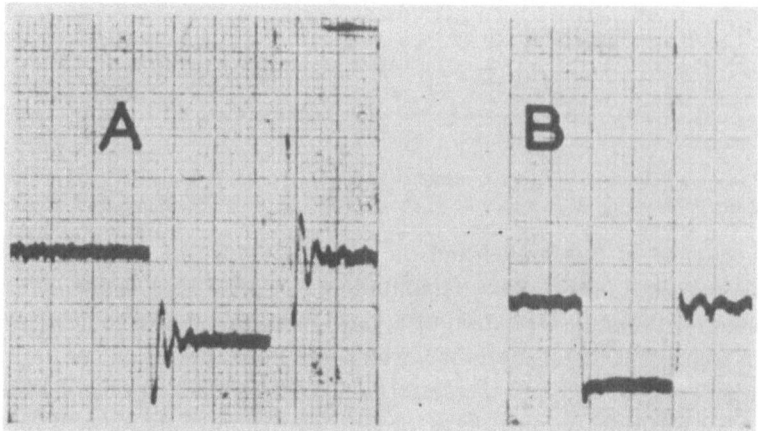

Fig. 2.15 A pressure step is made by opening a valve and allowing water under hydro-static pressure to flow into free space. By stopping the flow one gets a step function plus an impulse function. *A* is without any parallel damping; *B* with a needle parallel to the catheter and coming out into the open air.

pedance of the pressure source. (Whereas, for the pulse wave we considered the heart as a volume current source acting as a high impedance for this wave, for pressure measurements the arteries act to a marked degree as a high compliance pressure source, i.e. a low impedance.) Since the damping in the catheter can be low, artefacts and steep fronts can bring about strong oscillations.

When we neglect for a moment the fluid resistance and other damping factors as the friction in the catheter wall we can calculate the transmission velocity and the characteristic impedance in the same way as for arteries. Typical values for a catheter with a diameter of 1.2×10^{-3} m are a propagation velocity of 200 m/s and a characteristic impedance of 1.7×10^{11} $N s m^{-5}$.

Latimer (12, 13) gives a recommendable and very thorough discussion of the problem in which he critically discusses the literature and presents solutions to the problem. These are based on the same physical considerations as in this chapter. In his case a needle with the same characteristic resistance as that of the catheter is placed between the pressure source and the catheter itself (Fig. 2.16a). The pressure transmitted is now half of that at the entrance of the needle; it is doubled at reflection against the allegedly high impedance of the chamber of the pressure transducer and after return to the entrance of the needle is not reflected again. It has the advantage of being a closed system, but the disadvantage that it presents a high resistance to blood sampling etc. and it also requires calibration. Our own comparatively simple solution (21) makes use of a fine needle adjusted to the charac-

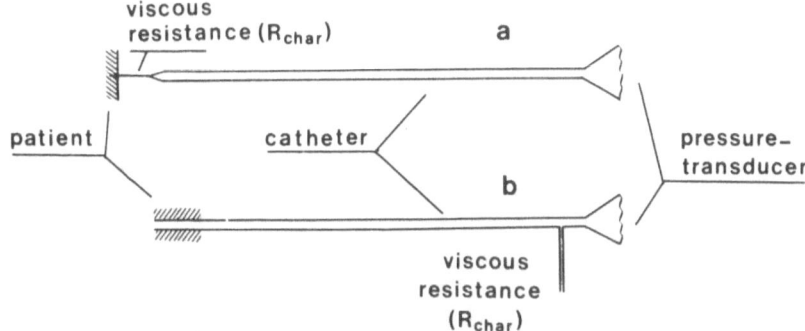

Fig. 2.16 a) Matched series damping of a catheter at the entrance of a catheter. *b)* Matched parallel damping at the end of a catheter.

teristic impedance that is applied parallel to the transducer (Fig. 2.16b). The bore of the needle has to be such that at the frequencies of interest the inertance term is still smaller than the resistance. If this should form a problem, a parallel circuit of fine capillaries may be advantageous to ensure that the resistance term will remain dominating. Of course the system requires calibration before starting and after termination of the measurements, but the correction for low frequencies being $Z/(R + Z)$ is in the order of 10% for the typical values presented. It is also obvious that a solution as is sometimes proposed, in which a fine needle is used as a method to decrease the effects of resonance artefacts but now between catheter and transducer, is less preferable than the impedance matching which in principle does not attenuate higher frequencies.

In Fig. 2.15b the effectiveness in praxis of the parallel damping system is demonstrated. The frequency response is flat to a high degree which can be seen from the undisturbed waveforms recorded. Fig. 2.17 gives an example from a patient recording.

2.5 Cardio-acoustics

2.5.1 *General*

It will be clear from the introduction that the genesis of heart sounds is a relatively complex problem, which has not been solved completely.

In recent years several investigators have tried to make a model of the sources and the transmission of the vibrations. Verburg (22) concludes from his investigations of accelerograms on the chest wall that a dipole source (a vibrating sphere along one axis (16)) seems appropriate for a number of

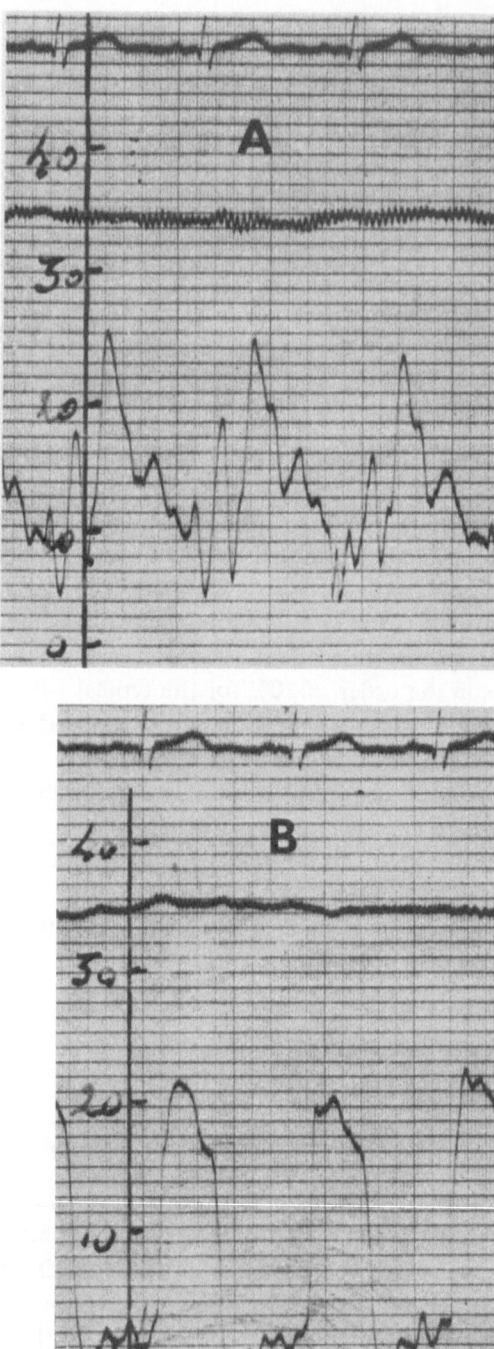

Fig. 2.17 A pressure re-
gistration in the right ventricle with a no. 7 catheter.
A without B with parallel
damping needle.

phenomena. Cheung (2) adopts several spheres which are allowed to vibrate along three different axes, in fact a multiple dipole model. Kozmann (10) investigates a combination of monopoles, dipoles, and quadrupoles. With respect to the transmission one may think in terms of wave transport of the vibrations. Every source having a more or less complicated movement will produce compression as well as shear waves in an elastic medium. An example may also be found in geological vibrations. Human soft tissue is nearly incompressible due to the high water content. This leads to a much smaller propagation velocity for shear waves than for compression waves. Unlike the conditions encountered with geological vibrations the distance between the source and the measuring site is in the human torso comparatively small so that both wave types cannot be separated and they will interact strongly. In view of this strong coupling one can calculate that the propagation of vibrations is much slower than of compression waves. The viscous properties of tissue also introduce dispersion and shear waves especially will be damped rather quickly. At low frequencies the propagation velocity of the waves is such (at 10 Hz compression waves velocity $\cong$ 1500 m s^{-1}, shear wave velocity $\cong$ 10 m s^{-1}) that owing to the relatively large wavelength almost no phase differences will exist, which means that the propagation phenomena (time differences, damping) need not be taken into account and displacements at the outside of the body due to forces and/or displacements on the inside can be described as being quasi-static.

At high frequencies (> 500 Hz) the propagation velocity has become so large that time lapses due to transmission are negligible; the damping, however, is not and should be calculated using the wave concepts. In the frequency range between 10–500 Hz not negligable time lapses of 1–10 ms due to the transmission can be expected.

It will be clear that analogue models, describing this complex situation with a certain validity, are practically impossible. For some problems like the coupling of mechano-electric transducers (microphones) to the chestwall, it seems to be permissable however to use a very simple mechanical impedance analogue model for the chest wall structures. As the transducers can also be described as mechanical structures it will be a useful concept to solve the coupling problem in terms of mechanical impedances.

2.5.2 *Transducers on the chest wall*

After the transmission to the chest wall the vibrations due to the heart action are either listened to by means of a stethoscope or are recorded using suitable transducers. We thus deal with the problem of transfer of vibra-

tions from one mechanical system (the chest wall) to another (the stethoscope, i.e. finally the ear, or a transducer).

At the chest wall the vibrations from the source are transmitted through
the elastic tissues between the source and surface. We assume the source
to be moving independently of any external manipulation. Locally we may
thus say that the recording site is in first order connected by a spring to a
velocity or displacement source. As the tissue also has viscous properties,
we postulate a mechanical resistance (dashpot) parallel to the spring. At
the surface the spot in contact with the transducer behaves as a mass
suspended on the already mentioned spring but also connected to a spring
representing the stiffness of the skin. A mechanical scheme of this situation
is drawn in Fig. 2.18a. We abstract in this model from all further mutual
coupling.

Keeping in mind that, in using an electric analogy, parallel elements
translate to series elements, we can draw the electric analogue as in Fig
2.18b.

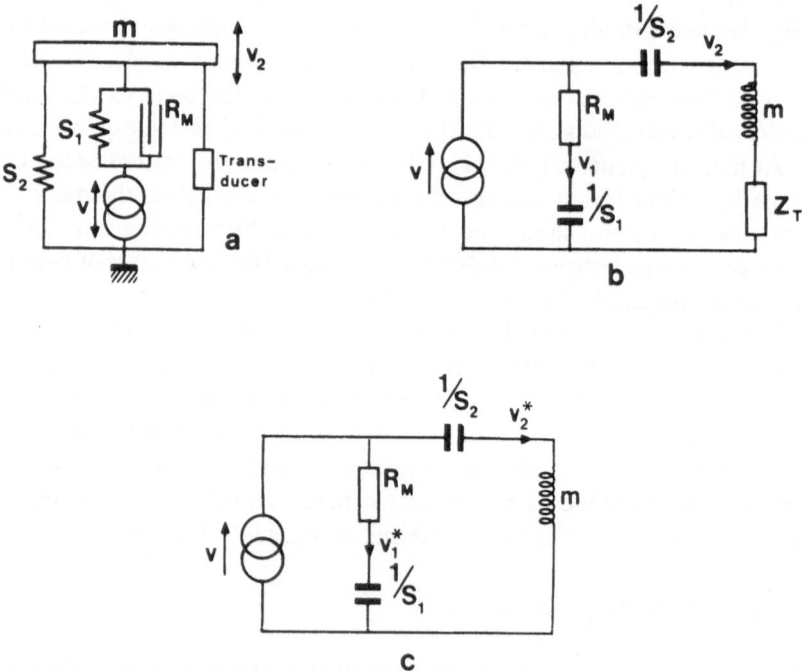

Fig. 2.18 Coupling of the chest wall to a transducer: *a*) mechanical scheme; *b*) electrical
analogue with applied transducer; *c*) electrical analogue without transducer. m = mass
below contact area; S_1 = stiffness of the tissue below contact area; R_M = mechanical
resistance of the tissue below contact area; S_a = stiffness of skin surrounding contact area.

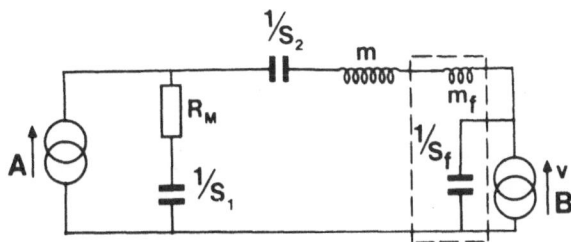

Fig. 2.19 Electrical analogue for a chest wall impedance measuring system. m_f = mass of force gauge; S_F = Stiffness of force gauge; A = heart vibrations source; B = external vibration source.

In the case where no transducer is applied to the chest wall the electric analogue is as in Fig. 2.18c.

If the aim of the measurement is to approach the velocity of the chest wall under no load conditions one can analyse from these schemes what properties the transducer should have to achieve this aim as nearly as possible: the quotient of v_2 and v_2* should be near one in the frequency range of interest. As the reader may easily verify:

$$\frac{v_2}{v_2{}^*} = \frac{1}{1 + \dfrac{Z_T}{Z_B}} \qquad (2.19)$$

where $Z_B = D + i\omega M - i(S_1 + S_2)/\omega$, the mechanical impedance of the chest wall.

As long as $Z_T \ll Z_B$ the measured velocity (displacement; acceleration) is practically the undisturbed velocity (displacement; acceleration) of the chest wall. That Z_B is the chest wall impedance is seen by considering that if a known vibration with a constant velocity is forced upon the chest wall at the spot of interest, the force F necessary to do so is a measure of the mechanical impedance before the probe. From Fig. 2.19 which gives the electrical analogue for such a situation, it can be deduced that this is Z_B in series with the mass m_f of the force-sensitive device and parallelled by the compliance $1/S_f$ thereof. In the frequency range of interest this compliance proves not to be of influence so that a correction has to be made only for the mass of the device to obtain the proper value of the chest wall impedance. A problem when measuring chest wall impedance may be interference with the original heart vibrations (source A), but by proper dimensioning of the impressed velocity and filtering of the force signal this interference may be reduced to an acceptable level. In practice a circular

* Refers to unloaded situations.

piston is connected to a suitable electromechanical vibration source through a so-called impedance head in which the functions of measuring velocity and force are combined and where the mass of the force gauge is known. The electric input signal to the source is regulated in such a way that a constant velocity is produced through the frequency range of interest. It is necessary to apply this installation to the skin with a certain static force.

It will soon be found that, if a measurement is made of the chest wall mechanical impedance, its value throughout the frequency domain is very much dependent on this static force applied to the chest wall at the measuring site (Fig. 2.20). An appreciable increase in impedance is found with increasing static force. This demands correction of the model.

Although this is an essentially nonlinear problem we may approach this through a linear model in view of the fact that the forces due to the heart vibrations are small compared with the static forces. The same applies for the forces necessary to measure the mechanical impedance. From Fig. 2.20 we find that the general shapes of the curves in first approximation can be considered as that of a mass-spring system with damping in accordance

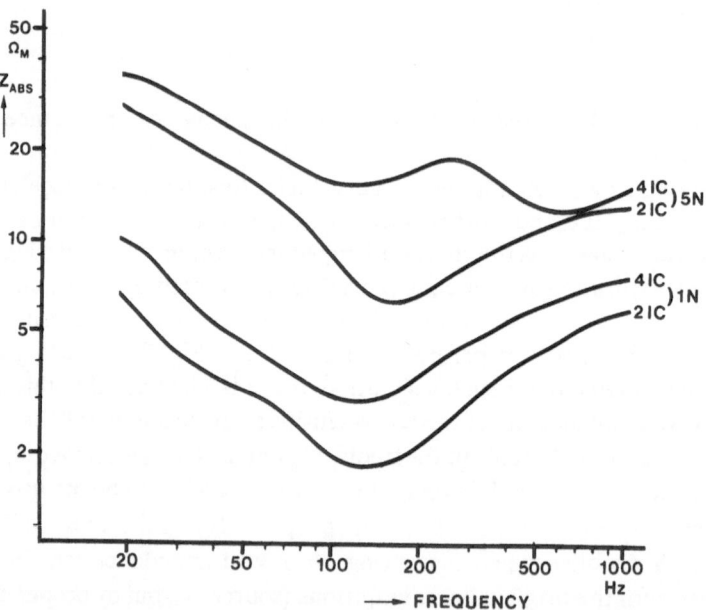

Fig. 2.20 Some typical examples of measurements of the absolute value of the mechanical impedance of the chest wall at two different places and two different static loads; diameter contaat area 15 mm. IC: intercostal space. Measurements were made on left side of the thorax, mid-clavicular.

with the model from Fig. 2.18. In the higher frequency range beyond that of the figure where the impedance is mainly defined by the term due to mass, it is found that the impedance is nearly independent of the static load. We can therefore conclude that the mass term is not influenced by the static load. Further it can be observed that both the stiffness and the mechanical resistance increase with larger static loads. From the model one cannot deduce whether this is due to an increase of either S_1 or S_2. It does not seem very likely, however, that due to a load perpendicular to the surface, the spring S_2, representing the stiffness of the skin beside the contact area, will change much in first approximation. So we may assume that the increases in stiffness with a larger static load can be attributed to the spring S_1.

As the velocities in the two branches (Fig. 2.18b) in the electrical analogue are inversely proportional to the impedances of these branches and the source by definition is a velocity source, an increase of the stiffness S_1 and the mechanical resistance R_M due to a larger static load should give a larger velocity v_2 through the branch with the transducer. With many transducers it can easily be observed that a larger static load gives a larger output (Fig. 2.21).

If the measured velocity v_2 under a certain static load is corrected to the undisturbed velocity v_2^* of the chest wall we have to keep in mind that part of the stiffness and damping of the chest wall will change its value with a changing static load. The relation between the measured and undisturbed velocities is now given by:

$$\frac{V_2}{V_2^*} = \frac{1 + Z_2/Z_1^*}{1 + (Z_2 + Z_T)/Z_1}$$

where

$$Z_2 = j\omega M + \frac{S_2}{j\omega}$$

$$Z_1 = R_M + \frac{S_1}{j\omega}; Z_1^* = R_M^* + \frac{S_1^*}{j\omega}$$

As all given impedances are frequency dependent, the correction will be a complex one. However, as long as the weight and the impedance of the transducer is small, we may estimate that the values of Z_1 and Z_1^* will not be very different and the simple formula (2.19) again applies.

The choice of small transducers with a small mechanical impedance seems of importance for research into the transmission of heart sounds

---

* Refers to the unloaded situations.

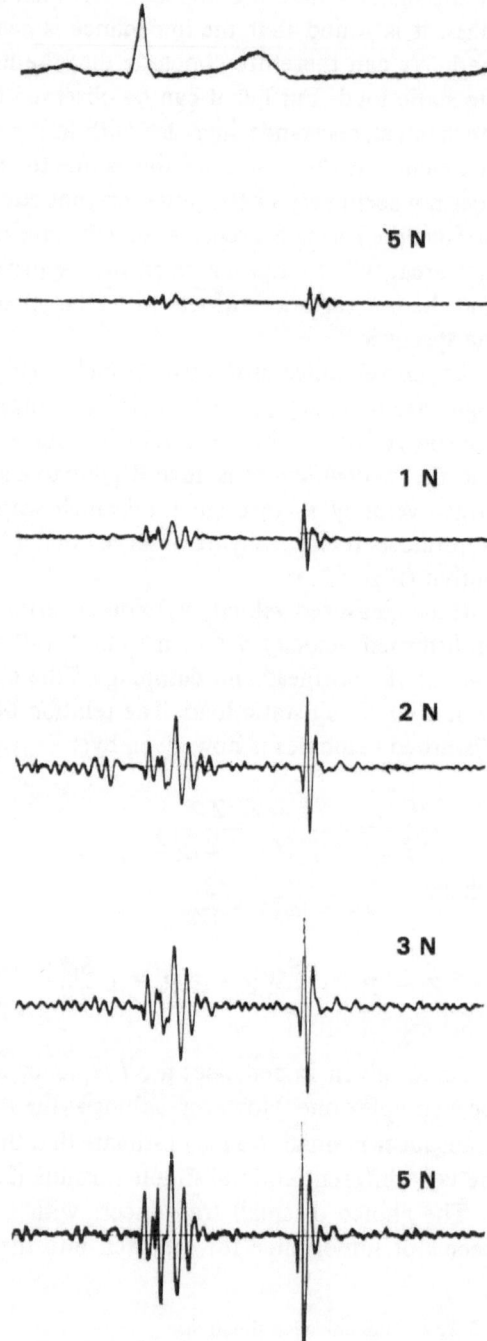

Fig. 2.21 Phonocardiograms of a normal subject measured under different static loads.

and the modelling of its sources; they form the base of the instrumentation of Kozmann (10), Verburg (22), and Cheung (2).

2.5.3 *The mechanical impedance of the chest wall*

The impedance of the chest wall and in general other tissues is related to its density, various elastic constants and viscous properties. It is also dependent on the area of the contact site. We may thus use the measured mechanical impedance to get an estimate of these mechanical properties. One then needs a model which relates the mechanical impedance and the mechanical properties. The mechanical problem can be stated very simply: a circular piston vibrating on a half infinite visco-elastic medium. The boundary conditions are: a prescribed velocity at the contact area and no tractions anywhere at the surface except at the contact area. The solution, however, is not at all simple and to our knowledge has not been reached.

Von Gierke et al. (5) approach the problem through the use of a solution given for the field and impedance of a sphere vibrating along one axis in a visco-elastic medium by Oestreicher (16). They demonstrated by their experiments that over a large frequency range the only difference is given by a factor depending on the geometry of the contact area.

From these analyses it follows that in a frequency range from 0–2000 Hz, assuming the human tissues to be nearly incompressible, the mechanical impedance may indeed be written as that of a mass-spring system with damping, although both the term due to mass and that due to damping are frequency dependent. Here we have an example of how a lumped model can be deduced from a solution of an essentially distributed system. From this theory it follows that the mechanical impedance, besides being determined by the mechanical properties, is also dependent on the radius of the contact area. In its most simplified form one should expect that the virtual mass should have a cubic relation with the diameter of the contact area, the measured stiffness a linear and the mechanical resistance a quadratic relation. From Fig. 2.22, which summarises the result of several investigators, it is seen that except for the mass term these relations are not easy to verify. One explanation could be the strong dependency on static load.

2.5.4 *Mechanical impedances of phonocardiographic microphones*

Many types of microphones are in use in phonocardiography. A main distinction can be made between contact microphones and air-microphones. Although the latter are also in contact with the chest wall, the transfer

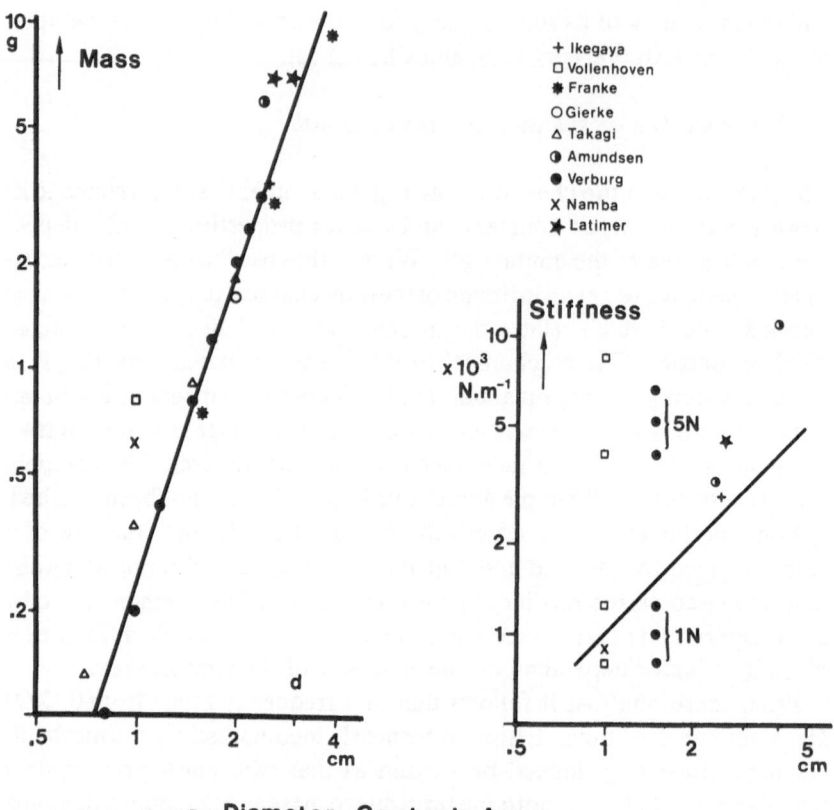

Fig. 2.22 Mass and rigidity terms of the mechanical impedance of human soft tissues in relation to the diameter of the contact area collected from several authors (after Latimer). In the mass diagram the line gives a third power relation between mass and diameter of the contact area. In the stiffness diagram a line is drawn based on a theoretical linear relation for a shear modulus of 25,000 N. m^{-2} and Poison's constant 0.5. Vollenhoven used static loads between 1 and 3 Newton, Takagi used 5 N, Ikegaya 0.5 − 1.5 N, and Amundsen from 1.8 to 9 N. The lower values of the rigidity always correspond with lower static loads. The values for the rigidity from Verburg are for different intercostal spaces.

of vibrations to the mechano-electric transducer is in this case via air, in contrast with the other types where a direct mechanical transfer takes place.

We will deal with two types of contact microphones, the heavy seismic microphone and the lightweight accelerometer; further we give a general simplified treatment of the air microphone.

Fig. 2.23a, b gives sketches of the contact microphones. From the mechanical construction it can be seen that they are both built up from two masses with a spring between although the ratio between the two masses

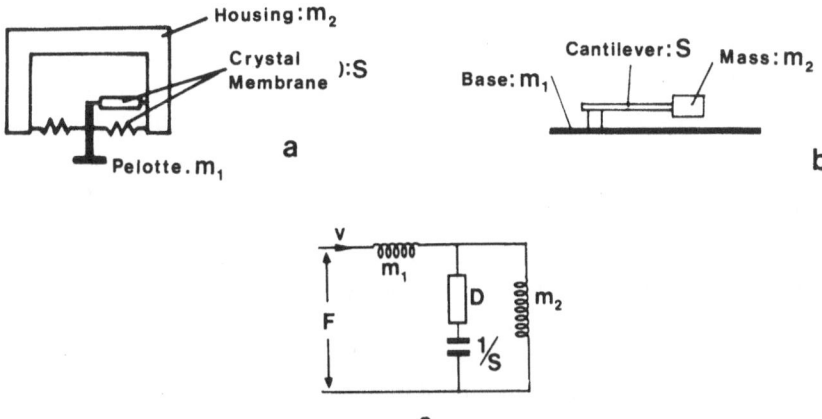

Fig. 2.23 a) Simplified drawing of a heavy-weight contact microphone. *b)* Simplified drawing of an accelerometer-type contact microphone. *c)* Electrical analogue scheme.

is quite different in each case. This means that by using mechanical impedance analogy we arrive at the same electrical scheme (Fig. 2.23c).

Assuming that in both cases a part of the spring is formed mainly by a piezo-electric crystal or a piezo-resistive element, the important variable in the electrical analogue is the charge on the capacitor (analogue of the distortion of the crystal). The distortion of the crystal develops an electric voltage proportional to this distortion and for simplicity we assume that this voltage is measured exactly.

From the electrical analogue scheme one can calculate for instance the output voltage for a constant input velocity, or one can estimate from experimental curves some of the microphone parameters. For the coupling with the chest wall the mechanical impedance Z_t of the device is of importance. Its magnitude can again be calculated from the electrical analogue scheme. To get an impression of the impedances involved the following simplified formula (damping neglected) will be of use:

$$Z_t = i\omega \left(m_1 + m_2 \frac{1}{1 - \frac{\omega^2}{\omega_c^2}} \right) \tag{2.21}$$

with ω the circular frequency and ω_0 a resonance frequecy ($\omega_0^2 = S/m_2$).

For a heavy-weight microphone $m_1 \ll m_2$. Since m_2 is rather large ($\cong 0.6-1$ kg) ω_0 will be small (in practice about 40–60 Hz). Above this frequency the second term in (2.21) is mainly due to the spring S. Together with the mass m_1 (order of magnitude 10 grams) it forms a series resonance circuit and we may expect a minimum of impedance. Due to a relatively small mechanical resistance this resonance is rather pronounced and only in a very small frequency band will the impedance be much smaller than the chest wall impedance, as was seen in 2.5.2 a desirable condition for quantitative measurements. This type of microphone does not seem such a sensible choice for such work if laborious calculations to correct the data are to be avoided.

For small accelerometers where m_2 is much smaller than m_1 one can choose the resonance circular frequency ω_0 above the frequency range of interest at the cost of sensitivity. Below this resonance the mechanical impedance is mainly given by ωm. This means that as long as $(\omega m_1) \ll$ (chest wall impedance) the output of the device represents the true acceleration of the chest wall. Even if this condition is not completely fulfilled the corrections are more feasible. Accelerometers are therefore advantageous for quantitative measurements of chest wall vibrations. For instance accelerometers can be made with a total mass of about 2 grams. At low frequencies the impedance of the device is much smaller than that of the chest wall, which is in that region mainly due to the rigidity. At higher frequencies the impedance is indeed comparable with that of the chest wall, but here the chest wall impedance is mainly due to a mass and correction for the transfer characteristics of vibrations can easily be made.

A simplified drawing of an air microphone applied to the chest wall is drawn in Fig. 2.24. Pressure changes in the air in the cavity are transformed into displacement variations of the membrane of the microphone. The membrane itself is either coupled to a moving coil (dynamic microphones) or forms one plate of a capacitor (condensor microphones).

The air cavity between the membrane and skin influences the behaviour of the microphone in practice. Different types of cavities have been analyzed by Suzumara and Ikegaya (20). In a cylindrical cavity resonance may occur for plane waves in the axial direction and for cylindrical waves perpendicular to the axis. However, as long as the wave lengths of the frequencies of interest are much larger than the diameter of the air cavity, which is the case for phonocardiographic applications, cylindrical

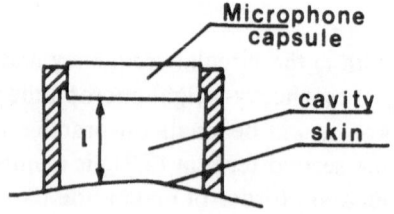

Fig. 2.24 Simplified drawing of an air microphone applied to the chest wall.

waves need not to be considered. The cavity can be seen as a kind of horn
between the membrane and the skin. For a cylindrical cavity with a diameter of the microphone membrane one can calculate by solving the wave
equations that the impedance seen at the skin side of the cavity is as given
by Olson (17).

In this simple situation it can be shown that as long as the acoustic im-
·pedance of the transducer is large (stiff membrane) the impedance presented at the chest wall in the frequency range of interest in phonocardiography
is mainly due to the compliance of the cavity. For instance, from the data
of Suzumara and Ikegaya (20) it can be shown that, for a length of the
cavity of 5 cm and a diameter of 3 cm, this holds to about 500 Hz. The
compliance has then a value of about 2.6×10^{-10} m^5 N^{-1}. Resonance
phenomena are in this situation found at a frequency of 1800 Hz. Decreasing the length of the cavity to 1 cm will decrease the compliance to

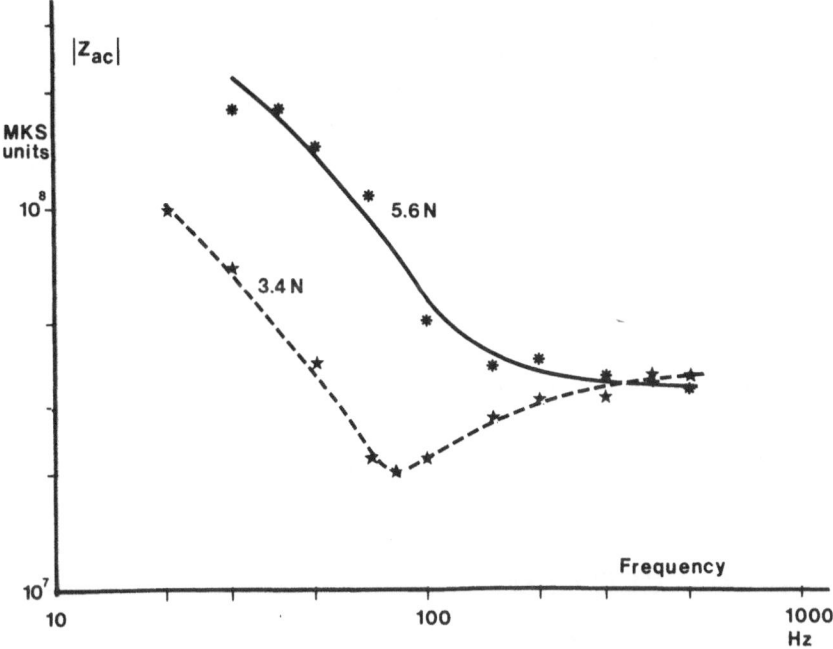

Fig. 2.25 Absolute values of the acoustic impedance of the chest wall, measured by air
vibration under different loads, for an area with a radius of 1.2 cm. (Data from Amundsen,
1.) To compare these values with those in Fig. 2.20 it should be kept in mind that acoustical
impedances and mechanical impedance are related through the contact area (see section
2.2). In view of the different experimental situations the values can be considered to be in
the same order of magnitudes.

$5.2 \times 10^{-11} \, m^5 N^{-1}$ and this compliance now mainly controls the impedance up to about 2500 Hz; resonance occurs first at 9000 Hz.

Comparison can now be made with values of the acoustic impedance of the chest wall given by Amundsen et al. (1). These values are measured by air vibration and seem here more representative (Fig. 2.25).

For an area with a diameter of 2.5 cm acoustic impedances of the chest wall are found of the order of $10^8 \, N \, s \, m^{-5}$ (Fig. 2.25). The main determining factor is a compliance below a frequency of 50 Hz and an acoustic resistance above 300 Hz. These frequencies may change, however, with applied static forces. A general conclusion can be that at low frequencies the transfer is defined by the ratio of the compliance of the chest wall and that of the cavity of the air-microphone, which are of the same order. At frequencies around 300 Hz the cavity impedance is in general still controlled by its acoustic compliance and has now become much smaller than the impedance of the chest wall, so here the load on the chest wall is negligeable. In the intermediate frequency range the situation is much more complicated and careful measurements are imperative. If the cavity is not cylindrical but is narrowed at certain points, acoustic inertance and/or resistance will be added to its compliance and this may alter the performance of the microphone completely. Although one may reach good clinical results with air microphones especially in the high frequency range, this type of microphone also has marked disadvantages for quantitative analysis.

References

1. Amundsen, O., K. Gjaeveness, and Langeland, T. (1971): The acoustical impedance at the surface of the human body in relation to auscultation. Acoustica, 25, 89–94.
2. Cheung, Y.S. (1977): A feasibility study of the spatio-temporal analysis of cardiac precordial vibrations. Thesis, University of London.
3. Einthoven, W. and Geluk, M.A.J. (1894): Die Registrierung der Herztöne. Pflügers Arch. ges. Physiol., 57, 617–639.
4. Franke, O. (1899): Die Grundform des arteriellen Pulses. Z. Biologie, 87, 483–526.
5. Gierke, H.E. von, Oestreicher, H.L., Franké, E.K., Parrack, H.O., and von Wittern, W.W. (1952):Physics of vibration in living tissues. J. Appl. Physiol., 4, 886–900.
6. Gow, B.S. (1972): The influence of vascular muscle on the visco-elastic properties of blood vessels. Ch. 12, Cardiovascular fluid dynamics (ed. D.H. Bergel); Academic Press, London, p. 66–110.
7. Guyton, A.C. and Coleman, T.G. (1967): Long term regulation of the cir-

culation. Ch. 11, Physical bases of circulatory transport (eds. E.B. Reeve, A.C. Guyton); W.B. Saunders, Philadelphia. p. 179–201.

8. Hansen, A.T. (1949): Pressure measurements in the human organism. Acta physiol. scand., *19*, suppl. 68, 1–227.

9. Huerthle, K. (1895): Ueber die mechanische Registrierung der Herztöne. Pflügers Arch. ges. Physiol., *60*, 263–290.

10. Kozmann, G. and Kenedi, P. (1976): Surface acceleration mapping, a new method for heart sound investigation. Report KFKI 76-35, Hungarian Academy of Sciences, Budapest.

11. Langewouters, G.J., van Dieren, A., Erens, H., Goedhard, W.J.A., and Wesseling, K.H. (1977): Visco-elastic properties of human thoracic aorta, theoretical discussion and some results. Report FUPREP 7703, Physiological Laboratory, Free University of Amsterdam.

12. Latimer, K.E. (1968): The transmission of sound waves in liquid-filled catheter tubes used for intra-vascular blood pressure recording. Med. & Biol. Eng., *6*, 29–41.

13. Latimer, K.E. and Latimer, R.D. (1969): Measurements of pressure wave transmission in liquid-filled tubes used for intra-vascular blood pressure recording. Med. & Biol. Eng., *7*, 143–168.

14. Leon, F.L. and Shaver, J.A. (eds.) (1975): Physiological principles of heart sounds and murmurs. Am. Heart Ass. Monograph 46; Am. Heart Ass., New York.

15. Moens, A.I. (1878):"Die Pulskurve." Brill, Leiden.

16. Oestreicher, H.L. (1951): Field and impedance of an oscillatory sphere in a visco-elastic medium with an application to biophysics. J. Acoust. Soc. Amer., *23*, 707–714.

17. Olson, H.F. (1957): Acoustical engineering. Van Nostrand, Princeton.

18. Rushmer, R.F. (1976): Cardiovascular dynamics (4th edition). W.B. Saunders, Philadelphia.

19. Skalak, R. (1972): Synthesis of a complete circulation. Ch. 19, Cardio-vascular fluid dynamics (ed. D.H. Bergel). Academic Press, London. p. 341–376.

20. Suzumura, N. and Ikegaya, K. (1977): Characteristics of air cavities of phonocardiographic microphones and the effects of vibration and room noise. Med. & Biol. Eng. & Comp., *15*, 240–247.

21. Tweel, L.H. van der (1957): Some physical aspects of blood pressure, pulse wave and blood pressure measurements. Am. Heart J., *53*, 4–17.

22. Verburg, J. (1977): Transmission of heart sounds, a new model. Ch. 29, Biomedical computing (ed. W.J. Perkins); Pitman Medical, Tunbridge Wells, p. 263–268.

23. Young T. (1809): On the functions of the heart and arteries. Philos. Trans., *99*, 1–31.

3
THE ELECTRONIC APPROACH TO MEASUREMENTS
B. Veltman

3.1 Introduction

The introduction provides a general outline of an electronic measuring and data-handling procedure so that the reader may become acquainted with the vocabulary.

The tools for observing nature and performing experiments can be modelled on the straightforward manner of Fig. 3.1a. It is useful to realize that the measurement problem fits into the more general context of *system analysis and synthesis*. The main points of the theory and many practical procedures are common to a variety of problems in measurement theory, filter theory, control theory, parameter estimation and simulation. In Fig. 3.1b, M represents a system and n an *equivalent output noise* (all noise sources in input and system are supposed to be translated into their effects on the output signal).

In the measurement problem the system M is given and the input x, i.e. the message is converted by M into a signal y. The observation consists of y and noise n and the problem is to reconstruct x from y + n. In the filter problem, however, x is given, y is specified and M has to be defined. A special example of this problem is the control problem where only a part of M (the controller) can be modified.

In parameter estimation (see chapter 9) x is chosen, (n + y) is measured and M has to be reconstructed.

In simulation (of a stochastic behavior) x and M are given and y has to be determined from the disturbed output (n + y).

Although the concepts behind the model presentation in Fig. 3.1a are of a much wider application than the world of electronic measurements only, we will concentrate on measuring, so that we can use a consistent terminology.

According to Fig. 3.1a, a physical variable is brought into contact with a *measuring pick-up* or transducer at the *front-end* side of a measuring chain. The transducer produces an electrical output signal (voltage, current, charge, waveform change, impedance change), which represents the physical variable. It will be clear that the presence of the transducer

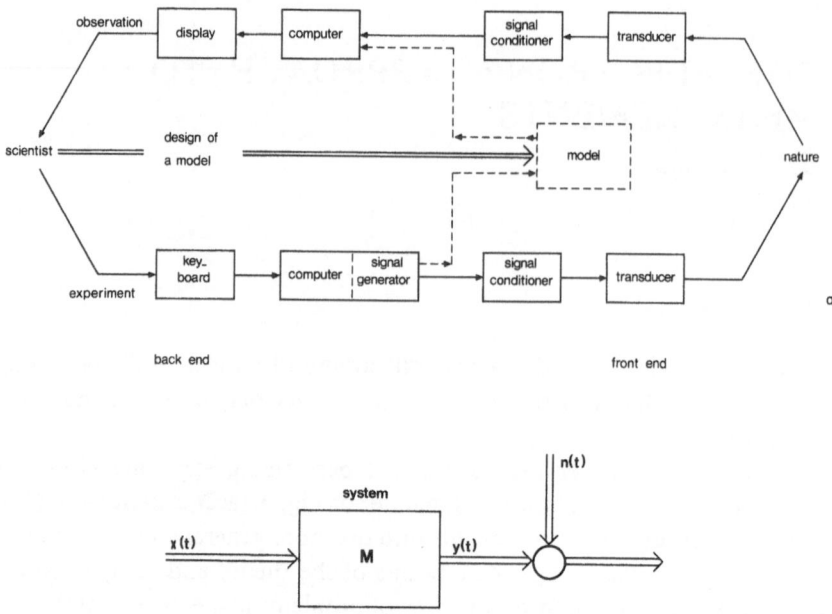

Fig. 3.1a In natural sciences, the process of making observations, carrying out experiments and performing simulations with model building, relies heavily on electronic instrumentation.

  b Imbedding the measurement problem in system analysis and synthesis. For measurement the input (vector) x has to be reconstructed from the disturbed output (equivalent noise n added to the output) and the given system properties of M.

preferably has to leave the conditions at the measuring site undisturbed: the transducer has *to match* its measuring task.

If the output signal is proportional to the physical quantity the transducer is said to be *linear*. Linearity of a transducer is a prerequisite for legal metrology; for scientific measurements it is often very helpful. A non-linear characteristic can of course be compensated afterwards with the aid of computers. With fluctuating phenomena, however, this compensation can be an awkward, or even impossible, task. Think, for instance of a pulsating flow, measured by an instrument where, according to Bernoulli's law, the fluid velocity v is proportional to the square-root of the pressure difference Δp. The average velocity $\bar{v}$ is then proportional to the average of the square root: $\overline{\sqrt{\Delta p}}$. Usually however, $\sqrt{\overline{\Delta p}}$ is measured. Linearizing afterwards means taking the square-root of the average pressure difference $\sqrt{\overline{\Delta p}}$. For the proper correction of $\sqrt{\overline{\Delta p}}$ to $\overline{\sqrt{\Delta p}}$ one has to know the waveform of the fluctuation in advance.

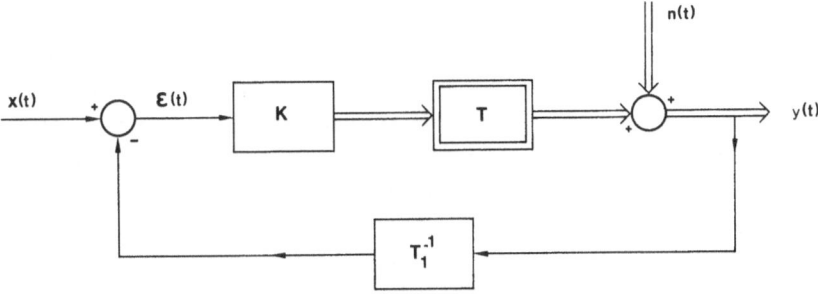

Fig. 3.2 Feedback is a favourable way of overcoming undesired behaviour of a given measuring device T and to diminish the influence of noise n. In series with T, a high-gain factor K is introduced. The desired overall behaviour T_1 is brought into the feedback loop as T_1^{-1}.

$$y = \frac{KT}{1 + KTT_1^{-1}}x + \frac{1}{1 + KTT^{-1}}n.$$

For $KTT_1^{-1} \gg 1$: $y \approx T_1 x$, independent of linearity, dynamic behaviour, drift etc. in T and K. In most measuring systems $\underline{T_1 = 1}$.

This example introduces the importance of the *dynamic behavior* of a measuring device. It can be described as the ability of the system to reproduce rapidly changing incoming waveforms (signals as instantaneous functions of time) at its output.

The dynamic behavior should not be confused with a totally different concept: the *dynamic range*. The dynamic range is the span of the input signal, being the distance from the maximum input value, with proper operating of the instrument, to the smallest detectable input.

A general way to improve the linearity as well as the dynamic behavior is to make use of the principle of *feedback*. The effect of linear feedback is shown in Fig. 3.2. The transducer T no longer measures the physical variable x directly, but measures the difference between its own output y and the variable to be measured, x. As this difference will be much smaller than the variable itself, the influence of the non-linearity is reduced. A greater sensitivity is of course required for the transducer, as the smaller input signal has to produce the same output value. The desired increase in sensitivity can be obtained with the gain K of the transducer. The feedback-effect can then be described as follows (see Fig. 3.2):

$$y = \frac{KT}{1 + KTT_1^{-1}}x + \frac{1}{1 + KTT_1^{-1}}n$$

For $KTT_1^{-1} \gg 1$ one has $y = T_1x$. Taking $T_1 = 1$ we have $y/x \approx 1$, which is independent of the properties of T! Not only is the relation between y and x a linear one, the independence of T also implies that the *frequency response* of the measuring system is improved: the system is better able to follow high frequencies in the input signal. A slow dynamic behaviour of a measuring instrument makes it unsuitable for the observation of rapid changes (*large bandwidth* signals) of the physical variable. Although feedback seems to be one of the possibilities for overcoming this constraint, it is by no means a panacea. It has to be possible, in one way or another, to compare the physical variable with the transducer output. The measuring method has to be suitable for that purpose. With a hot-wire anemometer, for measuring the velocity of gases, feedback is a straightforward procedure: the gasflow cooles off the hot-wire. The resulting resistance change in the hot-wire is converted into an electric current, being the transducer output. This current can easily be used to heat the hot-wire. So within the hot-wire the amount of heat taken off by the gasflow and the amount of heat brought in by the output current are compared. The resistance change is proportional to the difference between these two heat flows. As the introduction of feedback into the measuring circuit is not an easy or even impossible task, the undesired properties of the transducer can sometimes be *compensated* in the way as presented in Fig. 3.3: in mathematical terms

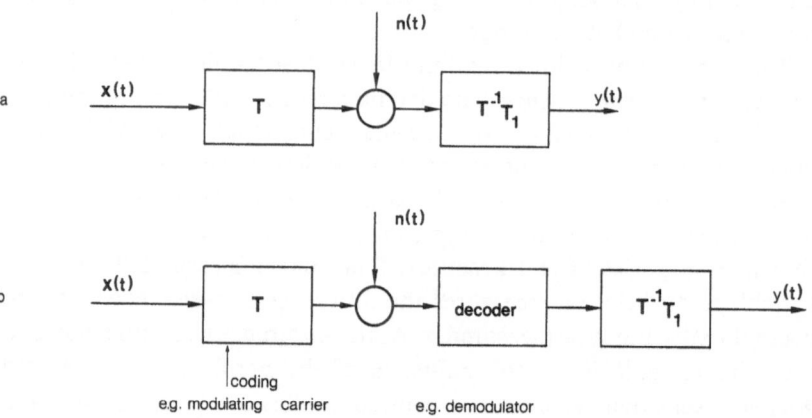

Fig. 3.3a If feedback is not possible, series compensation may improve the overall behaviour.

$$y = T_1x + T^{-1}T_1n.$$

T_1 eliminates noise and brings in the desired relation between y and x.

  b Time varying behaviour in T makes compensation doubtful. Labelling an input by modulation is a powerful tool to diminish noise influences.

one says: the operator T has to be followed by the inverse operator T^{-1} in order to make TT^{-1} equal to 1. The problem then is, whether T^{-1} represents a *physically realizable* behaviour. If not, the problem of designing the best, realizable, approximation arises.

Two important aspects of dynamic behaviour will be dealt with later in this chapter, i.e.:

– feedback may give rise to *instability* (e.g. spontaneous oscillations in the system, even without any input);

– for an undisturbed reproduction of a waveform by a measuring system a *linear phase characteristic* is required.

Gradually the second block in Fig. 3.1a, the signal conditioner, has taken part in the discussion. A variety of tasks can be attributed to this conditioner. If the transducer is of the *interrogative* type, the signal conditioner has to generate the interrogating signal to the transducer as for instance in the hot-wire anemometer. An active transducer, e.g. a thermocouple, *generates* a signal. A passive transducer, e.g. a resistance thermometer, modifies the interrogating signal. Interrogating procedures can often be used in a favourable way to give the measured signal a special label, to distinguish it later on from disturbing influences (e.g. modulation of a carrier). The signal conditioner has to receive and to amplify the transducer answer. It also has to contribute to the elimination of noise and has to avoid other influences than the desired signals in its output. For the sake of simplicity we also include the measuring act itself in the task of the signal conditioner: the comparison of the measured quantity with a built-in scale. This latter operation provides a tracing of the magnitude of the measured variable to an international standard. One can describe the measurement procedure, including all calibration tracing behind it, in a much finer detailed model. This may be helpful in the analysis of the total absolute error in a measurement. For our purpose, a survey of terminology and component properties will enable critical appraisal of the specifications of measuring instruments, the conceptual scheme in Fig. 3.1a is sufficient. The final result of the signal conditioner is a specified output signal, which obeys certain current or voltage level requirements and is ready for use in a computer, either as an *analogue voltage* or, after analogue-to-digital conversion, as a sequence of numbers, a *time series*.

A variety of entries to the computer is possible. Sometimes the signals are first recorded and played back for computer analysis. This procedure is called *off-line* data-handling. The tendency is to have measuring sites permanently connected to a computer, this is called *on-line* data-handling. If the calculated results and/or precalculated actions, are immediately used for control of the experiments, it is called the *in-line* use of a computer.

For off-line recording, analogue and digital magnetic recorders are available. The 32-channels *analogue recorder* still gives the highest density in data-packaging.

Punched paper-tape registration is becoming obsolete, being slow and expensive in equipment maintenance and still having a high error-rate. It is substituted by the more rapid, reliable and quiescent *cassette recording*. The economic *floppy-disc*, with its fast access, is also becoming an attractive tool for high packaging and easy retrieval of data. *Data-buffers* are necessary to adapt these mono-speed devices to the variation in speed requirements of measurements. It is needless to say that the *data-format* is an important specification in signal registration and computer analysis. Format conversions between the measuring apparatus, the transmission channel, the recording medium and the computer brand, are expensive and time-consuming manipulations. Standardizing, e.g. the IEC (International European Committee) format, is an important decision!

The linking of measuring instruments to computing devices and their peripherals may still require *interfaces*. Potentially they provide an adaptation of three aspects of an interconnection: the *voltage and/or current levels* representing the data, the *format* of the data and the *synchronization* of the mechanisms for transmitting and accepting data (including buffering and the control protocols).

The cable connection between measuring stations, computing devices and peripherals is usually organized in a *bus-structure*. The specifications of a bus contain among other things the *format of the data* to be transported, the *control protocol* for controlling and identifying data streams along the cable, the *access time* of data, the allowable *bus length* (important for proper synchronization of high speed devices) and the *reliability provisions* for an error-free communication (e.g. repeated transmission or handshaking).

The availability of low-priced *micro-processors* (see also chapter 8) has proved an important contribution in facilitating the computational work that has to be done with a measurement. Commercially available measuring instruments are increasingly provided with built-in micro-processors for performing control tasks within the instruments and for standardizing data-formats and transmission protocols. For very fast calculations (results within microseconds), local micro-computers can be connected directly to these instruments. For extensive further calculations and permanent recording, these micro-computers are connected via a *concentrator* to a large *background machine* (number cruncher). The latter has the advantage of *powerful compilers* for easy programming. A *large instruction set* and many *hardware operators* with a *large working memory* for efficient computation, an *economic background memory* for dumping of data and sophisticated peripherals for *high quality displays*. Fig. 3.4 shows such a hierarchical scheme, where the concentrator not only takes care of a proper connection with the background computer, but also substitutes local micro-processors in modest speed calculations. The final remarks in this introductory section are devoted to the conversion of analogue signals into time series: the analogue-to-digital conversion (A/D) and its inverse, the D/A conversion.

The A/D converter transforms a continuous time function into a time series, i.e. a discrete series of numbers coded in electrical voltages or currents. Several concepts of measurement (system)-theory are needed to describe the specifications of the conversion. Fig. 3.5a gives a general outline of the A/D conversion. The corresponding time functions are presented in Fig. 3.5b. The *sample-and-hold* circuit takes a sample from the continuous time function and keeps it at the same value for a short period of time to allow for a unique conversion. The *sample width* is an important parameter: it is either small, in order to adopt a "needle" function, or it is broad to average out noise effects (see Fig. 3.6). With a wide sampling interval certain filtering properties can be obtained. Much care is necessary in choosing the proper *sampling rate*. With too low a rate serious errors occur with the reconstruction of the original signal or in dealing with its

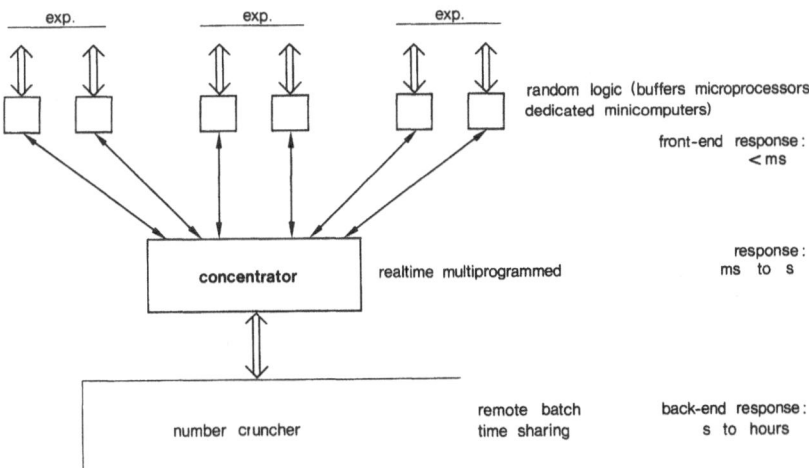

Fig. 3.4 A three-level hierarchical computing system, often used in laboratory automation. Front-end processors for direct control, display and μs-calculation; the concentrator for durable recordings, loading, transmission to a giant machine, programme development, disc memory and low-speed calculation. The number cruncher for extensive, non time-critical calculation.

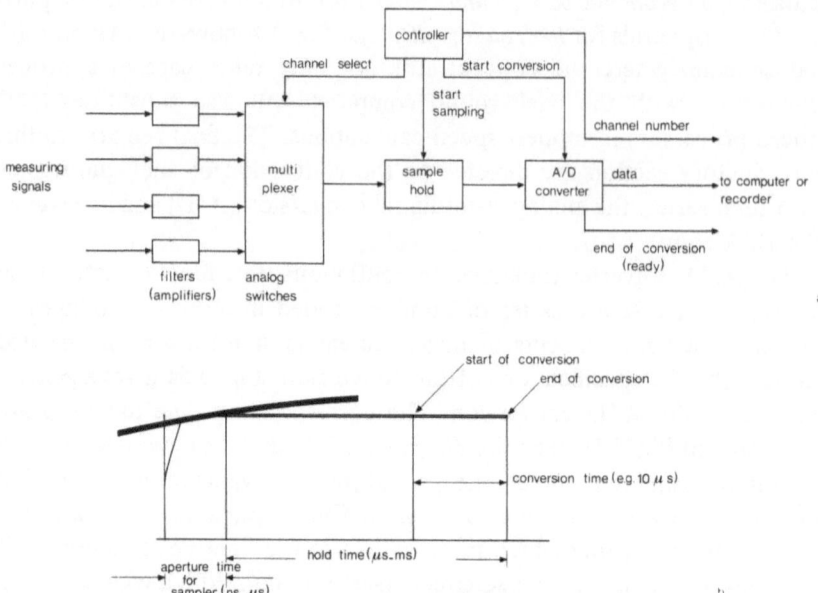

Fig. *3.5a* Analogue-to-digital conversion with multiplexing. Filters are necessary to limit the signal bandwidth for an adequate sampling rate.

   b   Timing diagram for sampling, hold and conversion.

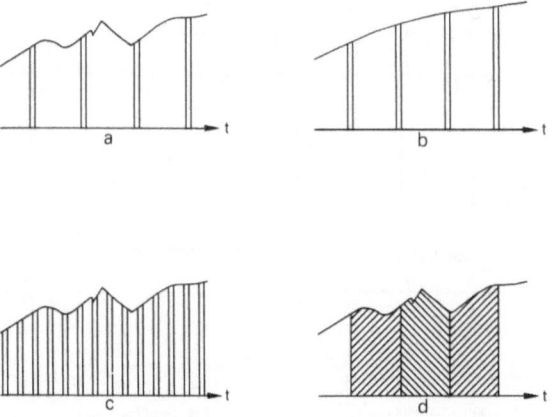

Fig. *3.6* Sampling rates have to be matched with signal *and* noise bandwidth.

   a   Too low a sampling rate; original waveform can never be reconstructed

   b   Filtering the noise permits a low sampling rate

   c   Without filtering a high sampling rate is necessary

   d   Certain conversion methods include filtering by taking an area *interval* as the measured quantity (integrating is low-pass filtering).

frequency properties (spectrum folding, see section 3.2). On the other hand, too high a rate implies much computational work. The mathematically based guide-line is the *Nyquist-rate* (or Shannon-theorem): all properties of the signal are preserved if the sampling rate is larger than twice the highest frequency present in the time function (signal + noise!). To satisfy the a priori needed knowledge of the highest frequency, it is advantageous to filter the signal with noise, prior to conversion, to limit the highest frequency deliberately. The effect of too low a sampling rate can *never* be corrected later! (see also chapter 7 and 9). If one has to restore a continuous waveform use of a much higher sampling rate is recommended. This will allow of an easier interpolation scheme (see Fig. 3.7).

Before sampling, the signal is usually amplified (and *multiplexed* if only one A/D converter has to convert more than one signal) before it is converted. The conversion has a specified *input range*, e.g. 0 to 10 volt or −10 to +10 volt. The width of that range is called the *input span* (also called dynamic range). The outputs of A/D converters are numbers. In most cases these numbers are represented in a binary form (if the output is displayed in decimal numbers, the instrument is called a *digital voltmeter* or DVM). The output span is determined by the *resolution* of the conversion (resolution is the smallest possible change in the output) and the output

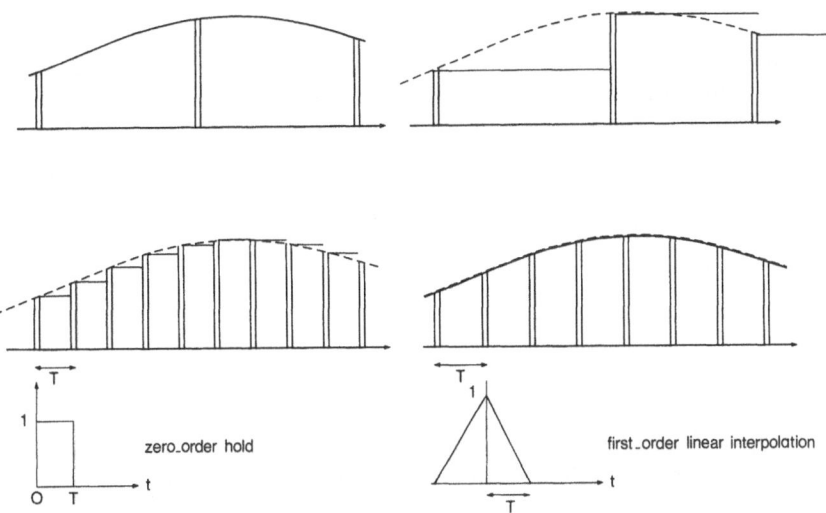

Fig. 3.7 Reconstruction of waveforms from sampled data requires complicated interpolation if Nyquist sampling rate is used. With simpler interpolations appreciably higher sampling rates become necessary. E.g. for a maximum deviation of 0.01% in a sinusoidal signal zero-order hold extrapolation requires 62,800 samples per period! With first order linear interpolation 223 samples are necessary.

range by the coding of the conversion (the coding determines which numbers will be indicated). A 12-bit converter can produce 2^{12} different output values; it has a lower resolution than a 14-bit converter with 2^{14} different output values for the same span. However, the 12-bit converter may still have a higher *precision*. This is the ability of the instrument to produce the same number for the same input signal, under the specified environmental conditions.

Precision is a lower quality aspect of a measuring system than *accuracy*. Accuracy determines the distance of the measuring result from the *true value*, as it is specified by an international standard. Superior measuring instruments are accompanied by *calibration* reports which are *traceable* to an international standard (from a primary international standard, to a secondary national standard to working standards in authorized calibration laboratories).

It is self-evident that only a few institutes are able to trace the *accuracy* of a 16-bit A/D converter, working at 50,000 samples per second on a time varying signal.

3.2 Digital circuits

A modest amount of knowledge about binary logic and the design of digital circuits greatly benefits the insight in the specifications and limitations of digital data-handling equipment for measurements.

A short survey of the principles and the terminology is given for a better understanding of such operations as calculation, coding, signal generation and digital filtering with either micro-processors or "random-logic".

The success of digital electronic technology is rooted in the wide availability of economic components for the registration of binary states: a relay is open or closed, a ferromagnetic core is magnetized in either one or the reversed direction, a transistor is conducting or non-conducting. Indeed the switching *speed* of semi-conducting gates (within 10^{-9} s), their low *power consumption* (μ-watt level), their small *size* (a few mm^2) and above all their *low cost* have caused a significant shift from traditional analogue techniques to digital implementations.

In measurement instrumentation the progress in component technology, from single-chip (a few gates per chip) to medium-scale-integration (MSI) and large-scale-integration (LSI, over 5000 gates per chip) is culminating in the systematic use of micro-processors for measurement purposes. Micro-processors not only substitute analogue procedures (often with better specifications) but also add new features to the instrument e.g. automatic calibration.

The most widely known use of digital circuits is in *calculation*. However, in measurement instrumentation other applications are equally important: *coding*, *sequential switching* (programmable switch), *digital filtering* and even *noise generation*. It is important to recognize that different codes are required for calculation, for communication and for numerical displays to simplify the operations, or to give maximum protection against disturbances.

Calculations are performed in a weighted binary code: each position of a *binary digit* or *bit* represents a specified weight. This has the advantage of simple algorithms for addition and multiplication, corresponding to the rules for weighted decimals in daily calculus. For a more convenient subtraction algorithm most computers perform their calculations in a pure binary *two's complement* code (see Fig. 3.8).

The transmission of digital data is called *pulse-code modulation* (see chapter 7) and offers the possibility of a highly reliable data transport; the communication of analogue signals over long distances is commonly performed in *amplitude modulation* or *frequency modulation*. For pulse transmission sophisticated coding techniques are used to detect and even correct error(s) in the received data. A simple and commonly used version is the *parity bit*; in order to make the total number of "1"'s in a word – a string of "1" and "0" – even, either an extra "1" or a "0" is added. This is of course only *error detecting*. With *handshaking* the receiver sends the message back to the transmitter for verification. It is clear that the latter is a time consuming procedure. In between these two extremes a variety of coding schemes exist to improve the reliability of the message (Fig. 3.9). In general a longer word length is necessary for a higher performance, thus limiting the bandwidth of the signal, in real-time situations, for a given channel capacity.

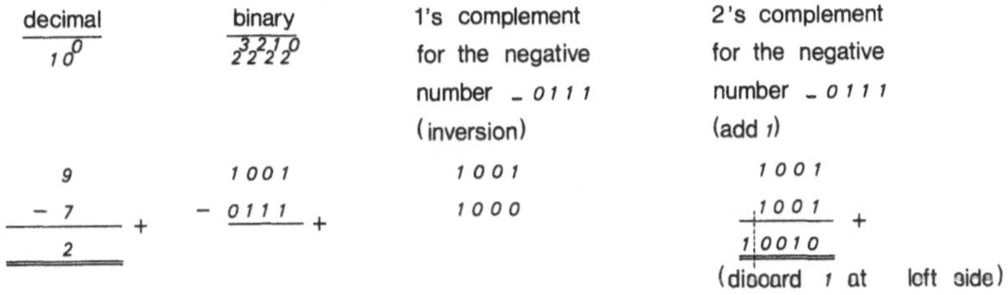

decimal	binary	1's complement for the negative number -0111 (inversion)	2's complement for the negative number -0111 (add 1)
10^0	$2^3 2^2 2^1 2^0$		
9	1001	1001	1001
-7 $+$	-0111 $+$	1000	$,1001$ $+$
2			$1|0010$
			(discard 1 at left side)

Fig. 3.8 Proper coding is important for easy data handling. Arithmetic circuits often use sophisticated algorithms to simplify hardware. The examples show how the "2's complement" code enables a subtraction to be treated as an addition.

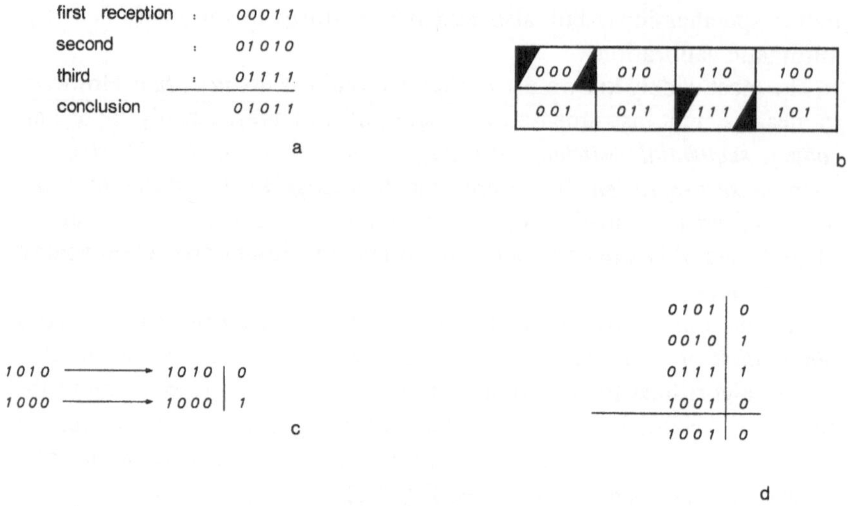

first reception : 00011
second : 01010
third : 01111
conclusion 01011

a

000	010	110	100
001	011	111	101

b

1010 ——————→ 1010 | 0
1000 ——————→ 1000 | 1

c

0101 | 0
0010 | 1
0111 | 1
1001 | 0
1001 | 0

d

Fig. 3.9 Improving the reliability of transmitted data by redundant coding.
a repeated transmission, majority vote
b creating distance between allowed symbols: 010 should be 000
110 should be 111
(the example is 2-error detecting or one-error correcting)
c adding a parity bit: one-error detecting
d grouping data in blocks; adding row and column parity: one-error correcting, 2-error detecting.

With twisted pairs of wires and push-pull amplifiers at the transmitting station and differential amplifiers at the receiving station, the pulses themselves can be transmitted directly over reasonable cable distances (km's) at relatively high speeds, say 10^6 pulses per second. However, one has to be aware of the propagation delay (1 μs for 300 m) if the data transmission has to be synchronized with computer actions (flag and control). In instrumentation for measurements one has to be careful with "earth" connections to remote stations: *optical couplers* offer a satisfactory solution for galvanic isolation!

Computer manufacturers specify the maximum distance, requested for a fast communication, between the data source and the computer. Direct memory access to the computer (*DMA*) provides a rapid entrance (clock speed of the computer) of data blocks. Often an extra data buffer is necessary to exploit the potential capacity of a DMA facility. An important aspect in the design of a data-net for measurements is the choice between a *bit-parallel* or a *bit-serial* method of transmission. Parallel is fast, serial is economic in copper wiring. For laboratory purposes (relatively short distance) parallel transmission is often advantageous. For plant/process control a series connection is preferred (bus structure). Normalization in the measurement area has led to the *IEC-standard*: *the byte parallel series transmission* (Fig. 3.10); one byte corresponds to 8 bits.

Instead of direct transmission of pulses, a modulation of a "1" or a "0" on a carrier is often used. Along cables this is an acoustic frequency (used with frequency-keying or with phase-shift-keying to distinguish "1" and "0"), corresponding to telephone or telegraph traffic. With aerial connections, this acoustic frequency is again modulated on a radio frequency (MHz band). The commercially available *MODEMS* (modulator/demodulator) provide a series transport of data at speeds up to 9600 baud (this is 10 bits/sec; 7 bits are used for information, 1 for parity, 1 start bit and 1 stop bit). An increasing number of computer brands have standard input-output connections adapted to the MODEM protocols.

For display purposes a decimal representation is usually requested. Each decimal is separately coded in a binary way: the *binary-coded-decimal* or *BCD*. Most common is the 8,4,2,1-code, but others, e.g. 4,2,2,1-code or "excess"-codes might have certain advantages for additional error detection. For the display of the numerical symbol itself another conversion is necessary: from the 4 bit BCD into e.g. 7 bar positions, or into a dot-matrix or whatever the character generator needs (Fig. 3.11). Although, with to-day's MSI-circuits almost all conversions can be per-

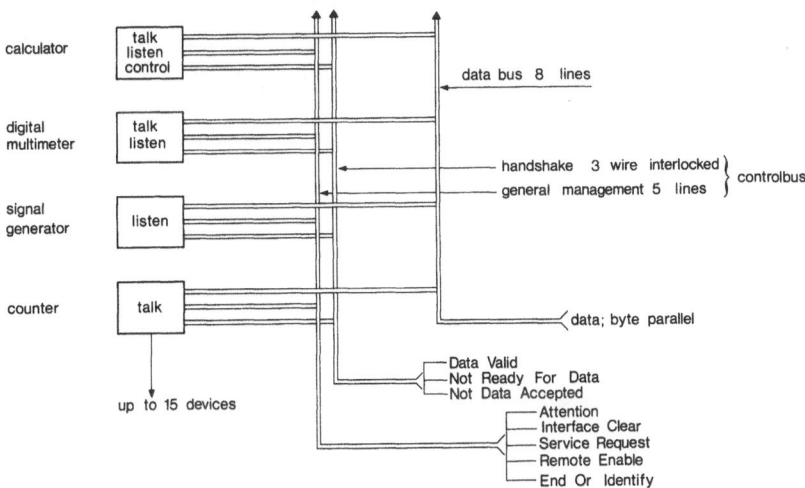

Fig. 3.10 Operating a multi-device instrumentation system for data transfer includes a variety of coding aspects. This drawing describes the IEC-protocol. Mnemonic codes for commanding and checking. The data are transferred in a bit-parallel way. The data bits may be coded in ASCII (ISO-7 bit code).

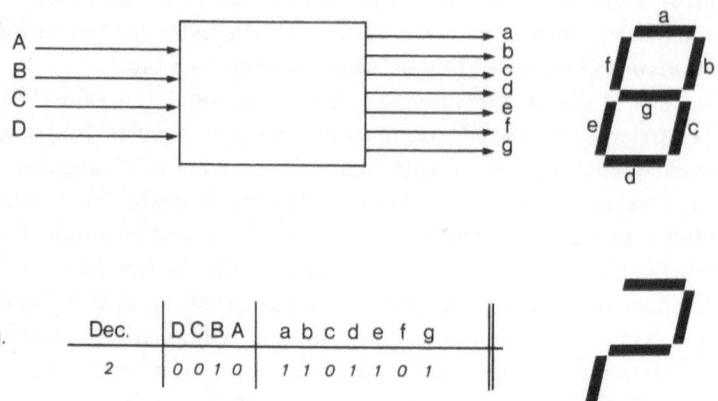

Fig. 3.11 Example of a combinatory circuit (direct input-output relation, no memory): translation of a binary coded decimal (BCD) into a decimal character for display.

formed on one chip, careful consideration is necessary to match computer input/output, data transport net, analogue-to-digital output and display devices as matrix printers, TV monitors, oscilloscopes, light emitting numericals and teletypes, for a minimum complexity. Coding is a matter of instantaneous *combinatory logic*; it uses only gates and thus very high speeds are possible. The majority of applications of digital circuits also use memory elements, this is called *sequential logic*.

An example is the generation of a sinusoidal test signal for measuring purposes. A number of successive samples of a sine wave can be stored in a memory and repeatedly reproduced to form a "continuous" sinusoidal signal as is shown in Fig. 3.12a. The advantages over the analogue genera-

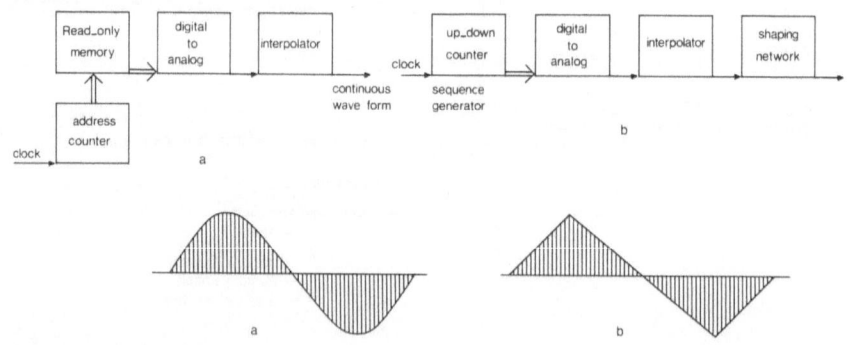

Fig. 3.12a Sine-wave generation with a ROM. Accurate waveforms demand an extensive memory (see also Fig. 3.7).

b With a proper sequence generator the number of memory elements is drastically reduced (10 flip-flop produce 1024 different states).

tion of such a signal are obvious: by changing the clock-pulse the frequency is changed; one can also start instantaneously at any desired phase. Moreover, several perfectly synchronized sinusoids can be produced simultaneously with a proper multiplexing scheme. In the past such waveforms were stored in *shift registers*. As this is a *volatile* memory element (it loses the data when the power is switched-off) it has first to be loaded, e.g. by hand switches, by punched-tape or by a computer. To-day the *programmable read-only memory* or *PROM* allows a non-volatile storage of data. In a PROM each storage location can be addressed individually. One has to provide an address counter to obtain the proper sequencing of data; this is in contrast to the (volatile), *first-in, first-out register* or *FIFO* where only the ingoing sequence is reproduced at the output (higher densities and lower costs than PROM's). With the reproduction of standard sequences, either for a continuous waveform or for a prescribed switching pattern, there is always the choice between storage of the complete sequence or of an algorithm, making use of feedback loops, for generating the sequence. For a triangular waveform one might store 1024 samples (see Fig. 3.12b), but it is much simpler to use a binary counter to make the 1024 steps with only 10 flip-flop elements ($1024 = 2^{10}$).

It is worthwhile emphasizing that not only for a linearly increasing sequence, but for any sequence, a "counting" circuit can be designed in the form of a *feedback shift register*. One has to specify the desired transition from former output to next output in a truth-table. This then can be simplified in Karnough diagrams and is followed by an implementation of the desired output variables as a function of inputs and the preceding outputs with flip-flops and gates.

In a straightforward approach one uses a pure binary counter with as many positions as the sequence length with an added combinatory logic circuit to provide the desired output from the counter output. With measurement instrumentation the use of stored generation with PROM's, or loading FIFO's from micro-computers, is increasing because of its flexibility at yet decreasing costs.

A special kind of sequence, which is often used with response measurements, is the *pseudo-random binary sequence* or *PRBS*. Randomness in a test-signal is necessary if one prefers to eliminate the anticipatory aspects of a system under test. In the past a random signal was also chosen for its wide-bandwidth properties: a sequence of *random or aselect* numbers is called *white noise*, as all frequency components are equally present. A PRBS signal is a good approximation to a wide-bandwidth signal, with the fundamental difference that succeeding samples are *not independent*; for all practical purposes, however, they can be taken as *uncorrelated*. Although a PRBS is a deterministic signal, one needs a very long memory

to be able to predict the next value; as such its pseudo-randomness satisfies most requirements for an "aselect" input. The advantages of a PRBS are:
- it is an on-off signal, which means a high energy level compared to the maximum amplitude; it allows simple calculations (e.g. with correlation one multiplies only with +1 and −1);
- it is periodic, which means that the correlation is (practically) zero *over a finite interval* (the period length); this in contrast to white noise which is uncorrelated only over an infinite interval;
- it is simple to generate and to reproduce. Fig. 3.13 summarizes the properties of PRBS signals.

It must be mentioned that, although PRBS is the most widely spread on-off signal, there exist other test signals which have an even better performance for measuring the parameters of a certain system in a noisy environment. It is for instance possible to construct periodic on-off signals with their energy concentrated in a priori specified frequencies; the so-called *multi-frequency signals*. In general one can design a random test signal with both a specified amplitude distribution and a specified power density spectrum!

Increasing attention in measurement instrumentation is paid to digital filters (see also chapter 9). Filters modify the frequency spectra of signals

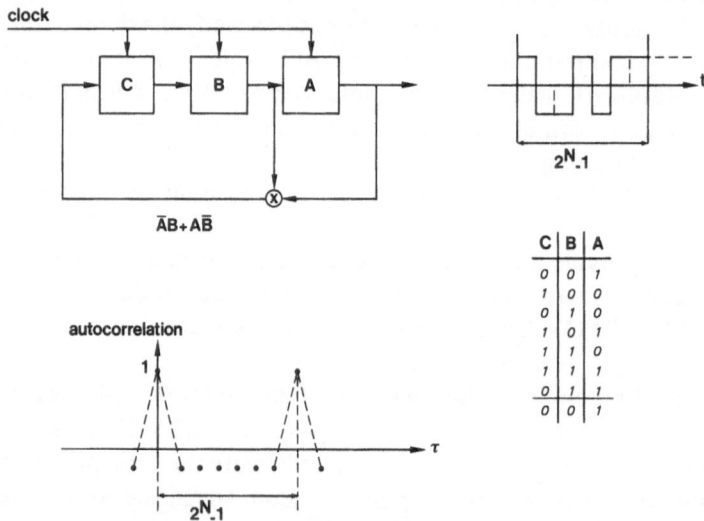

Fig. 3.13 Simple example (N = 3) of a linear feedback shift register, generating a maximum length chain code upon incoming clock pulses. The sequence of 0's and 1's at the output is called pseudo-random binary noise, a wide-band signal. The circuit goes in a "random" way (programmed) through all possible positions. With 128 flip-flops (and a clock of 1 MHz) the sequence repeats after $2^{128}\,\mu s \approx 10^{23}$ years!

e.g. for relative attenuation of the noise spectrum or for compensation of frequency characteristics of a transducer. They are also used for determining the frequency content of a signal. The basic form of a digital filter consists of feedback and feedforward coefficient loops along a digital delay line (a clocked shift-register; see Fig. 3.14). Digital filters have many advantages over the traditional active analogue filters:

- any accuracy in the coefficients can be obtained (by increasing the length of the coefficient registers) with perfect reproduction
- the time scalè of the filter behaviour is very precise, as it is coupled to the clock frequency; crystal stabilized clocks have a relative uncertainty smaller than 10^{-6}
- The time scale is flexible: a simple change in the clock frequency and the filter works in another frequency region (the memory elements in the filter also allow a use at ultra low frequencies!)
- the filter can be simulated perfectly on a digital computer
- with non-recurrent filters independent shaping of amplitude and phase characteristics is possible
- multiplexing (using the same filter for several signals) is easy, as only the memory elements containing the samples belonging to each signal have to be changed
- time varying filters are easy to implement (change in the coefficient register)

There are also drawbacks. In real-time applications the frequency region

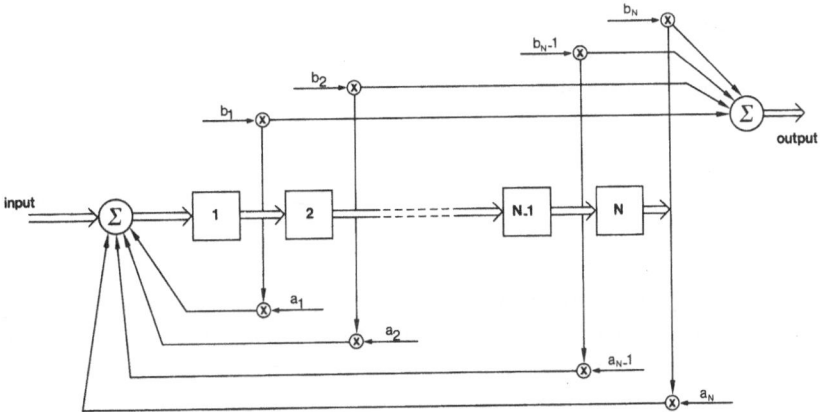

Fig. 3.14 Canonical form of an N-points digital filter. The feedback (recurrent loops) coefficients a_i are related to the "poles" of the filter, the zero's of the denumerator; the feed-forward coefficients b_i are related to the "zero's", the zero's of the numerator.

of the signals is limited to a few MHz and there are no "passive" digital filters, and they are more complicated and more expensive than analogue filters. These drawbacks have only a relative meaning in most measurement instrumentation. In particular through LSI and the increased use of micro-processors, the cost barrier is shifting considerably.

To illustrate the various problems in the design of digital filters, the *running average* will be treated. A running average is formed by adding a number of succeeding samples of a signal, see Fig. 3.15. The conventional analogue approach is to approximate the *uniform weighting function*, i.e. the impulse response. As this is rather complicated, one often uses simple RC-low pass filters. With digital techniques the ideal weighting function can be obtained in the straightforward way of Fig. 3.16a. In this *non-recurrent solution* one has to carry out N + 1 additions for each signal step.

It is easy to recognize from the system of Fig. 3.16b that a *recurrent filter* needs only one addition and one substraction, because the operation takes away the earliest signal value and adds the latest value, leaving the greater part of the sum unchanged.

In this example it is quite obvious how the non-recurrent filter can be substituted by a more economic recurrent solution. Unfortunately, it is seldom that simple. Several methods have been developed for a systematic design of recurrent filters (which of course have to be stable in spite of their feedback loop). A straightforward procedure is to start from a well-established analogue filter (e.g. a Butterworth or a Chebyshev filter) and to use the *impulse-invariance* method for the design of a digital filter with the same impulse response as the sampled version of the analogue filter. This simple approach neglects the potential features of digital filters for sharper frequency bands, for binary coefficients or for lower sampling rates at the filter output.

An important aspect in the design of digital filters are the effects of *rounding-off* or *quantisation* errors.

A multiplication of two 10-bit numbers causes a 20-bit output. Due to word length limitation this may be rounded-off to 16-bits. The error, the difference between the 20-bit and the 16-bit output, may diverge in a feedback loop or in further calculations and needs careful consideration.

The example of Fig. 3.16b illustrates this. Suppose that 10-bit numbers are added and that the result is again rounded-off to 10-bits. It will be clear that averaging over more than 1024 samples will be completely irrelevant. Early samples will disappear in the quantization error. Such situations do indeed happen with average response devices!

A final remark has to be made about the hardware implementation:

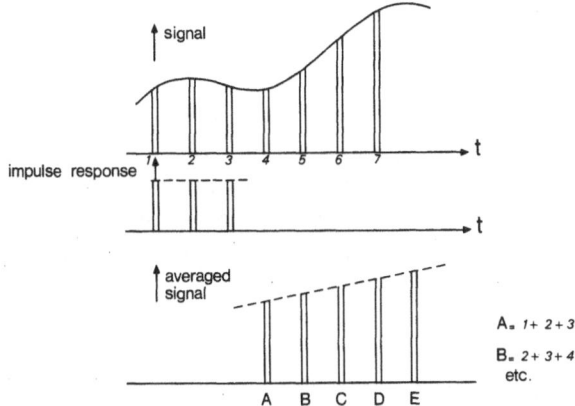

Fig. 3.15 A 3-points running average filter; the signal is convoluted with the filter's pulse response.

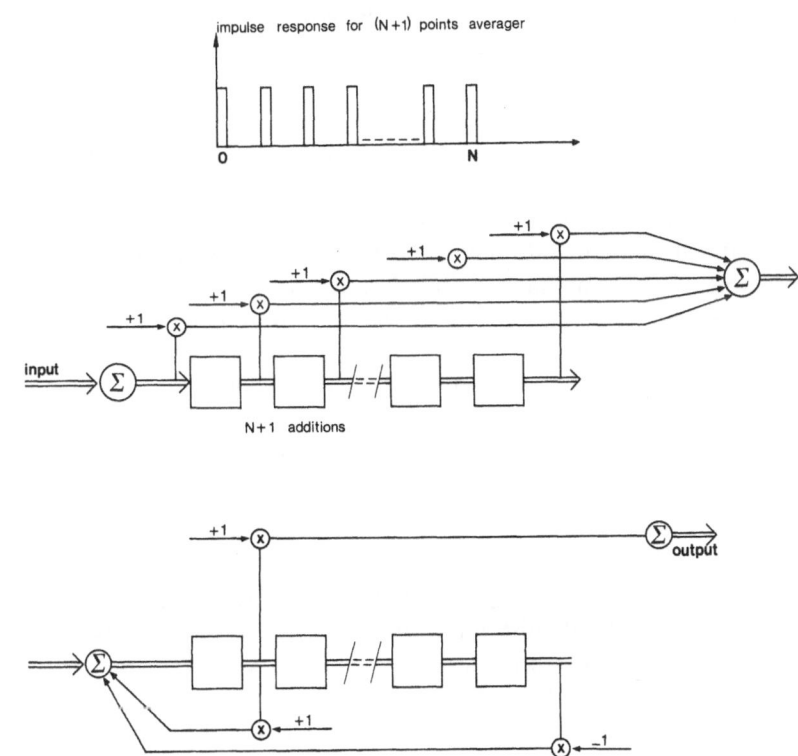

Fig. 3.16 Carrying out the summation for averaging in a straightforward way (a); doing the same in a more sophisticated way (b) makes a large difference in equipment.

micro-processor or random logic? With micro-processors one is limited to a given instruction set: relatively simple operations have often to be performed through software. Together with the speed limitation caused by the one-chip design and the fact that usually only one computing register is used, micro-processors work relatively slowly and require much programming. On the other hand, they are highly universal and their economic price allows the use of more micro-processors in one apparatus. An important innovation is that micro-processors have brought computing devices into the hands of the vast group of electronic technicians. This intensive use may lead to a rapid development of design rules and tricks.

With random logic one designs its own instruction set, tailored to the problem. Many operations can be performed in parallel or pipe-lined. One can use the latest and fastest components. All this implies that random logic allows of a much more rapid operation. On the other hand, the bread-boarding, the worst-case design, the manufacturing, the testing and, above all, the more specialized servicing are sufficiently unattractive to make the micro-processor preferable in many cases.

3.3 Analogue circuits

In spite of the powerful new techniques with digital circuitry, the very nature of most measurements (*continuous* time functions) keep the interest in analogue components alive. Partly for such unique applications as instrumentation amplifiers, partly because in certain devices analogue techniques with operational amplifiers still surpass their digital equivalents in speed and economy. Hybrids (mixed analogue/digital integrated components) form an important new field, covering such items as comparators, voltage dependent oscillators, phase-locked loops, sample-and-hold circuits, analogue switches, etc. In hybrids the analogue part limits the quality specifications in most cases.

The majority of analogue circuits in measurement instrumentation use the *operational amplifier* as the basic building brick. The ideal operational amplifier is a voltage amplifier with an infinite gain, no current drain at its input (infinite input impedance), and an unlimited current supply at its output (zero output impedance). Fig. 3.17 summarizes the properties and illustrates how, with feedback circuits, a variety of operations as *sign inversion, summation, and integration* can be obtained.

Although in practice the operational amplifier is far from ideal (the gain not being infinite and not even constant), the feedback circuit stabilizes the behaviour according to the model of Fig. 3.2. Although the input

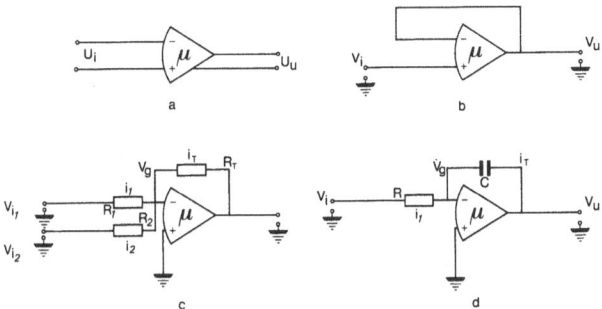

Fig. 3.17   *a* The ideal operational amplifier. No input current and a gain equal to infinity

      b The voltage follower: $V_m = \mu(V_i - V_u)$; if $\mu \to \infty$ then $V_u = V_i$.

      c The summator: $i_1 + i_2 - i_T = 0$. With $\mu \to \infty$ V_g has to be about zero for a finite output V_u.
Thus

$$V_u = -\frac{R_T}{R_1}V_{i_1} - \frac{R_T}{R_1}V_{i_2}$$

includes sign inversion and coefficient multiplication.

      d The integrator $i_1 - i_T = 0$. $V_u = -\dfrac{1}{RC}\displaystyle\int V_i dt$ + initial constant or $- RC\, dV_u/dt = V_i$.

impedance is not infinite, a high input impedance can be obtained by using an operational amplifier in a *voltage-follower* stage.

In literature on measuring instruments the operational amplifier is often used as a symbol for modelling the functional structure of a device. This means that the same *symbol* may represent a simple one-stage transistor voltage follower or a very sophisticated low-drift, high-gain, wide-bandwidth instrumentation amplifier.

The *instrumentation amplifier* itself is a good example of such a modelling. One can distinguish three sections: the preamplifier, the voltage amplifier, and the current amplifier. The *preamplifier* is a low-drift circuit (selected components) with a high input impedance and high *common-mode rejection ratio*. The latter property suppresses noise and disturbances induced as equal potentials on both wires of the transducer (this is usually only effective in the low-frequency region!). The *voltage amplifier* raises the level of the signal from $\mu V/mV$ to mV/V, a level where active filtering can also be applied. The output stage uses the operational amplifier in a *current-source* circuit for driving e.g. a high frequency galvanometer.

An important application field of operational amplifiers are *active filters.* From a comparison of Fig. 3.18 with Fig. 3.14 it is clear that the canonical structure of an analogue filter and a digital filter have much in common. The main difference being the time behaviour of the basic component: an infinite impulse response for an integrator vs. a finite impulse for a shift-register element. It is also clear that with active filtering the same configuration can be used equally for *low-pass, high-pass* and *bandfiltering*. Simply by taking another output terminal in the same circuit, another frequency behaviour is obtained. It is important to emphasize the requirement of a *linear phase* characteristic in the majority of applications with measurements. A linear phase behaviour of a circuit leaves the incoming signal waveform undistorted. Due to the linear phase shift the output waveform is shifted in time compared to the input (see Fig. 3.19a). Compensation of phase characteristics can be obtained with *phase-shifters* or *equalizers* (see Fig. 3.19b).

With external feedback circuits connected to amplifiers it is important to verify the *stability* of the total circuit. With the arrival of operational amplifiers in integrated circuit form, the intrinsic stability properties for all feedback circuits, were omitted. One reason for this is the difficulty in manufacturing sufficiently large capacitors in integrated technology. The other reason is that matching the amplifier behaviour to a particular feedback circuit for a marginal stability gives a higher circuit performance (*bandwidth gain product*) than the crude worst-case design for stability.

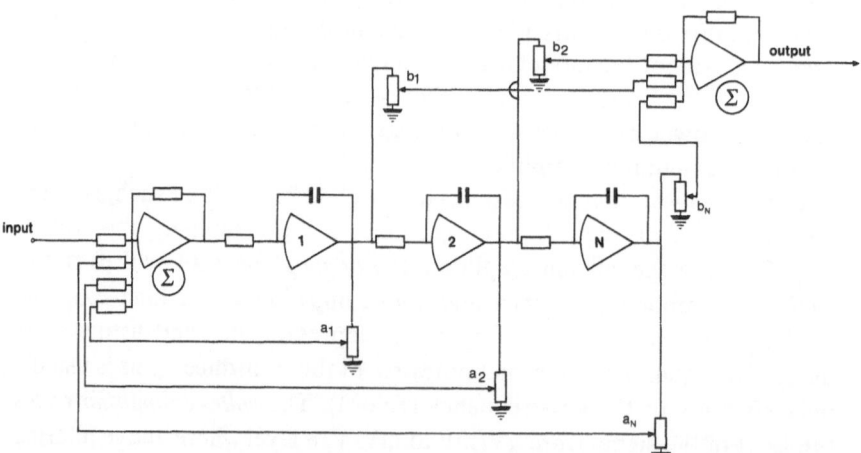

Fig. 3.18 Canonical form of an Nth order analogue filter with operational amplifiers. The structure is similar to the digital filter in Fig. 3.14. For the proper signs extra sign-inverters may be necessary in the feedback and feedforward loops.

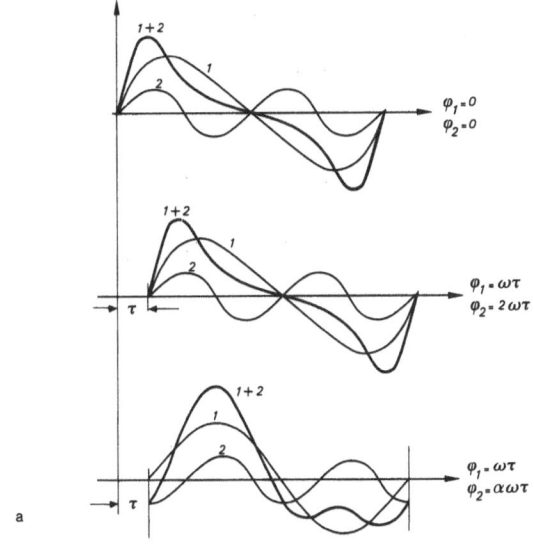

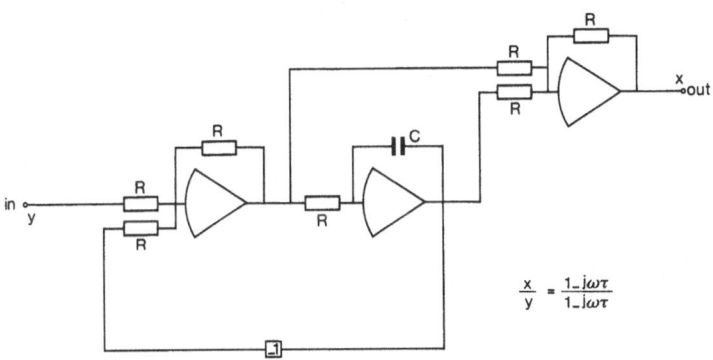

Fig. 3.19a The importance of a linear phase shift in a measuring system to avoid distortion of a waveform. Only an equal time shift of the harmonics reproduces the input wave form. Equal time shifts correspond to proportional phase shifts.

b. $\varphi_1 = \dfrac{\tau}{T} \cdot 2\pi = \omega\tau, \quad \varphi_2 = \dfrac{\tau}{T/2} \cdot 2\pi = 2\omega\tau,$

Phase shift circuit: $\tau \dot{x} + x = -\tau \dot{y} + y; \tau = RC.$

The feedback stability problem in general can be analyzed by considering the phase and/or gain margin in the frequency response characteristics. From Fig. 3.20 it can be seen that a sinusoidal wave in a feedback system with gain 1 and phase shift $-180°$ forms a sustained oscillation, inde-

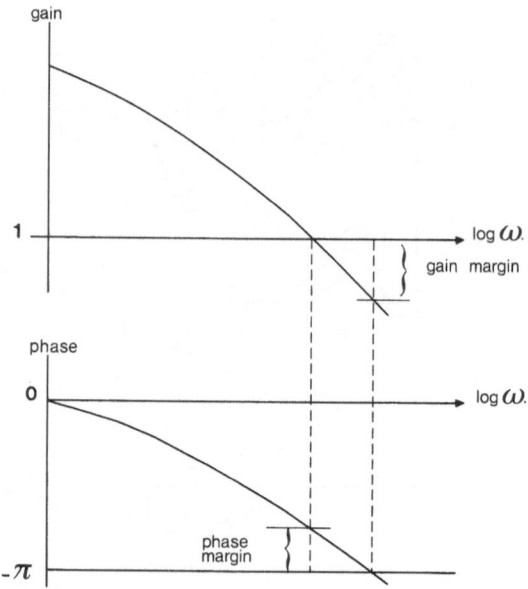

Fig. 3.20 Stability is checked by gain and/or phase margin in the frequency response. The smaller these margins are, the more oscillatory is the system response.

pendent of the input. If the gain is appreciably lower than 1 at the $-180°$ point, additional gain may be added until a prescribed *gain margin*, but still less than 1, is obtained (one may also look for a certain *phase-margin* at the point where the gain equals 1).

The gain and phase margins specify the damped oscillatory behaviour of the feedback circuit: a certain distance to the state of continuous oscillation (Fig. 3.21a). If a system can be described by a 2nd-order differential equation (as is always the case with mechanical systems, due to the presence of inertia) it is possible to obtain a satisfying transient behaviour (one overshoot) together with a linear phase characteristic (Fig. 3.21d1).

In a more sophisticated approach the differential equation, describing the dynamic relation between input and output, is modified into an algebraic equation by the *Fourier* or the *Laplace transform*. The roots of the denominator of this algebraic equation quantify the various modes of oscillation of the system. With a topological display of these roots (called a *poles and zero's plot*), it is possible to predict the stability of the closed-loop behavior from the open-loop properties. Roughly spoken the poles determine the eigen frequencies of the system, and the zero's their weight in the total time domain behaviour. This *root-locus* technique enables the design of *compensating networks* to improve the open-loop

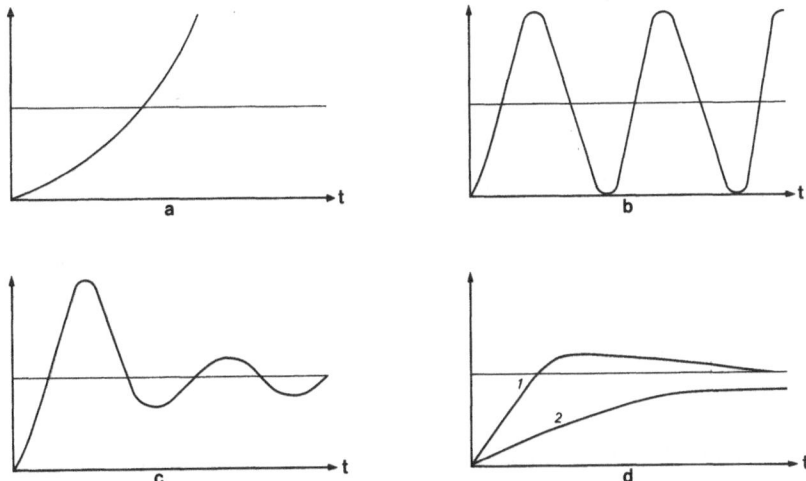

Fig. 3.21 Different ways of system behavior on a step-wave input.
   a instable system
   b sustained oscillation (boundary between instable and stable region)
   c stable oscillatory response
   d *1* damped response
          2 overdamped response

behaviour in such a way that a high gain can be permitted in the feedback configuration.

For the *generation of periodic signals* sustained oscillators are obtained from the very same circuits. E.g. for a sinusoidal signal one may use a 2nd-order system with a negative damping. In order to limit the ever increasing amplitude of the oscillation an *automatic volume control* (AVC) is necessary.

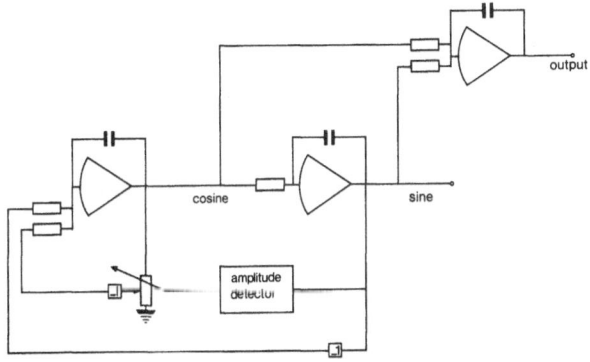

Fig. 3.22 Quadrature sine-wave generator. The output delivers any phase between sine and cosine. For a stable sustained oscillation the amplitude has to be measured and used for control of the damping in the circuit.

A method of realizing this is to control the damping with a voltage-dependent-resistor (VDR) or a field-effect-transistor (FET). While approaching the desired amplitude in building up the oscillation, the damping is brought to zero. This amplitude control is a feedback circuit in itself, causing *transients* if the frequency is switched over to another value. With the circuit of Fig. 3.22 any phase-relation·between two sinusoids can be obtained by weighted adding of sine and cosine of the *quadrature oscillator*. The specification of a sine-wave generator should contain data about the harmonic content (in the above circuit the AVC causes distortion), the amplitude- and frequency stability and the frequency range within the span of a continuous adjustment (this may be very important with biological experiments).

With a *clipping* circuit a sinusoidal signal can be shaped into a *block-signal* and from this a *triangular signal* is obtained by integration or a pulse waveform by differentiation.

4
RADIOACTIVITY*

Richard B. King and James B. Bassingthwaighte

4.1 Introduction

Tracers are substances not naturally present in a system but whose be-havior is identical to that of the "mother substance" about which infor-mation is desired; they should be detectable in amounts or concentrations that are very small compared to that of the traced mother substace. Their great utility is that they can be added to a system which is in steady state without disturbing it. In this chapter we are concerned with the use of radioactive tracers, beginning with the physics of radioactive emission and giving some hint as to the mechanisms for detection and calibration tech-niques used in tracer studies in human applications. The possible applica-tions of tracers cover the whole field of medicine, and we will touch on a few particular applications which illustrate the general principles under which they may be used to provide insight into biological systems. In particular, they are useful for measuring the mass of some component of a system, for example the amount of sodium in the body, or for measur-ing the kinetics of the exchange and regulatory processes of an ion, a substrate, or a metabolite.

4.2 An overview of the physics of radioactive emissions

4.2.1 *Atomic structure*

Though there are many subatomic particles, the atom is basically a nuc-leus with a surrounding cloud of orbiting electrons. The nucleus is com-posed of protons, which have a positive charge, and neutrons, which have no charge. Neutrons and protons have approximately the same mass. These particles exist in the nucleus at discrete energy levels. Transition of a neutron or proton from one level to another results in the release or ab-

*Supported in part by research grants HL 19135 and HL 19139 from the National Institutes of Health, U.S.A.

sorption of energy. Atoms of the various elements are defined by the number of protons existing in the nucleus, the atomic number. It is these protons which give each element its unique chemical properties. Addition or removal of protons from a nucleus transmutes that atom into an atom of a different element. The number of neutrons, on the other hand, may vary without changing the chemical properties of the atom. Nuclei which contain the same number of protons but which have different numbers of neutrons are called isotopes.

Surrounding the nucleus are the orbiting electrons. These are much less massive than neutrons or protons and have a negative charge which is equal in magnitude to the charge of proton. Like nuclear particles, electrons exist in discrete energy levels which are called shells. When an electrom moves from one shell to another, there is either release or absorption of energy.

The measure of atomic mass is the atomic mass unit (amu). Neutrons and protons each have a mass of approximately 1 amu, while that of the electron is only 0.00054 amu. The atomic mass of any atom is, therefore, approximately equal to the number of protons plus the number of neutrons in its nucleus.

Not all combinations of neutrons and protons from stable nuclei. There exists an optimum neutron to proton ratio. As shown in Fig. 4.1, the optimum ratio at low mass numbers is one; as nuclei progress to higher mass numbers, this optimum ratio becomes greater than one. If the neutron to proton ratio varies too far from the optimum, the nucleus is unstable and will undergo radioactive decay.

4.2.2 *Modes of decay*

Unstable nuclei reach stability either by emitting a particle or by the process of electron capture. Among the several particles that can be emitted is the beta particle. Beta particles have the same mass as an electron, and can have either a positive or negative charge. If the neutron to proton ratio is greater than the optimum, a neutron will decay to become a proton plus a negative beta particle. This beta particle will be emitted from the nucleus (see Fig. 4.1). If the neutron to proton ratio is less than the optimum, a proton will decay to a neutron plus a positive beta particle, which will be emitted. Positively charged beta particles are called positrons. When either a negative beta particle or a position is emitted, a neutrino or an antineutrino is emitted simultaneously. The total energy of the emission (i.e., the energy of the beta particle plus the energy of the neutrino) is con-

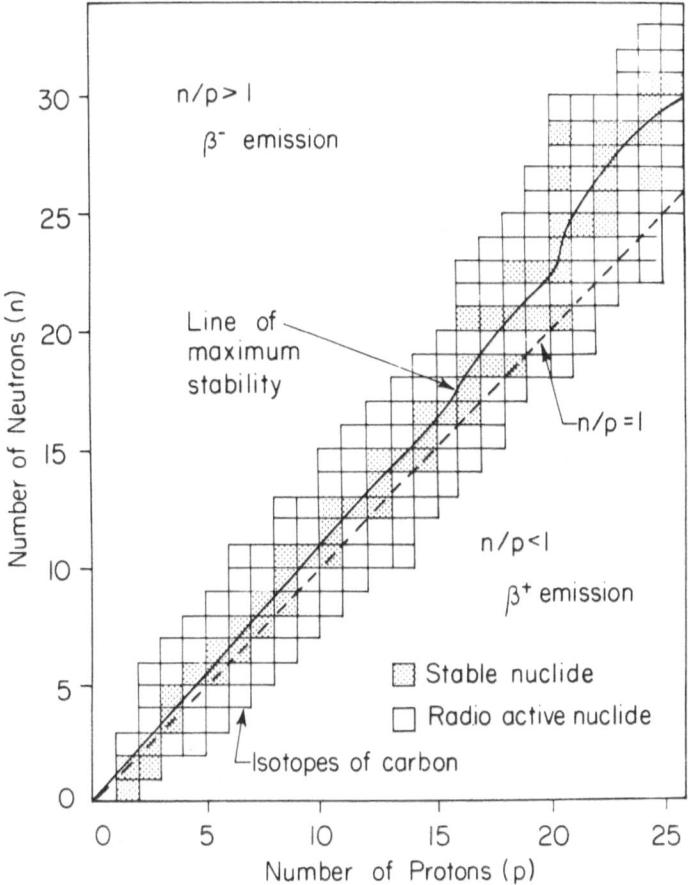

Fig. 4.1 A chart of the nuclides showing the relation of nuclear composition to stability and the areas of β^+ and β^- emission.

stant, but the distribution of energy between the two particles varies. Thus the beta particles emitted by a given radionuclide do not have discrete energies, but have rather a spectrum of energies which can be characterized by a maximum and a mean energy. The maximum and mean beta particle energies of several commonly used radionuclei are given in Table 4.1. The range of a beta particle in any given medium (such as air, or human tissue) is proportional to its energy. The beta particles emitted by carbon-14 and tritium (hydrogen-3) are referred to as "soft" betas because of their low maximum energy; they are absorbed by a few mm thickness of aluminum.

Table 4.1 Energies of β particles and γ rays emitted by some radionuclides commonly used as radioactive tracers.

Nuclide	Half-life	β Energy (KeV)		γ Energies (KeV)
		Maximum	Mean	
^{3}H	12.26y	0.0186	0.0057	–
^{14}C	5730y	0.156	0.045	–
^{24}Na	14.96h	1.389		1.369, 2.754
^{32}P	14.28d	1.710	0.69	–
^{35}S	87.9d	0.167	0.0488	–
^{36}Cl	3.08×10^{5}y	0.714		0.511
^{42}K	12.36h	3.52		1.524
^{45}Ca	165d	0.252	0.075	–
^{46}Sc	83.9d	0.357		0.889, 1.120
^{51}Cr	27.8d		0.315*	0.320
^{59}Fe	45.6d	0.475		0.192, 1.095, 1.292
^{65}Ni	2.56h	2.13		0.368, 1.115, 1.481
^{65}Zn	245d		1.106*	0.511, 1.115
^{85}Sr	64.0d		0.499*	0.514
^{90}Sr	27.7y	0.546		–
^{95}Nb	35.0d	0.160		0.765
^{125}I	60.2d		0.030*	0.035
^{131}I	8.05d	0.606	0.19	0.080, 0.284, 0.364 0.637, 0.723
^{141}Ce	32.5d	0.518		0.145
^{201}Tl	73h		0.084*	0.135, 0.167

*These nuclides undergo internal conversion. The energy given is the energy of the conversion electrons.

The large alpha particle, consisting of two neutrons and two protons, is also produced by radioactive decay. Since it has two protons its charge is twice and its mass approximately 7500 times that of a beta particle; therefore alpha particles have a very short range. They are, for example, entirely absorbed by mylar which is only a few microns thick. Because of this short range, radionuclides which emit alpha particles are of limited use as radioactive tracers. They are, however, of use in radioautography.

Neutrons may also be emitted as a result of radioactive decay. Since they are uncharged they have an extremely long range, but this lack of charge makes them difficult to detect and of little practical use as radioactive tracers.

Another mode of radioactive decay is electron capture, in which the nucleus takes up one of the electrons from the inner electron shell. One of the protons in the nucleus combines with the electron to form a neutron. This gives an electron cloud that is now in an unstable energy configuration and must reorganize in order to stabilize.

4.2.3 *Stabilization after radioactive decay*

When atoms undergo radioactive decay, the result is a new atom, called a daughter, which is unstable either in its nucleus or its electron shell. If the electron shell is unstable, the electrons will quickly seek a stable configuration, usually within 10^{-10} seconds after the decay. The result of this transition to a stable configuration is the emission of X-rays, which are low-energy photons; the most energetic X-ray has an energy of approximately 0.012 MeV.

Most radioactive decays result in an unstable nucleus which may stabilize by either of two processes. The first is by internal conversion, in which energy is transferred from the nucleus to an electron which is then ejected from the atom. Unlike beta particles, these conversion electrons have discrete energies. The most common method whereby the nucleus attains stability is by the emission of a gamma ray. Emission of gamma rays accompanies the emission of most particles. They are emitted from the nucleus within approximately 10^{-12} seconds after decay. More energetic than X-rays, they also have discrete energies, ranging from about 0.02 MeV to greater than 7.0 MeV. The energies of gamma rays emitted by some commonly used radionuclei are given in Table 4.1. Gamma rays have extremely long ranges when compared to beta particles. In attaining stability, a nucleus may emit a cascade of several gamma rays. When an atom decays, it may decay to a metastable state. These states, which are stable for more than 10^{-12} seconds, are converted to stable states by the process of isometric transmission accompanied by the emission of gamma rays.

Positrons will eventually interact with an orbital electron of an atom. When this occurs the positron and the electron are annihilated and two gamma rays, each with an energy of 0.511 MeV, are emitted with an angle of 180° between them.

4.2.4 *Activity and half-life*

The rate at which radionuclei undergo decay is referred to as activity. The standard rate of activity, the Curie (c), is equal to 3.10×70^{10} decays per second. As this is a large amount of activity, the more familiar units are the millicurie (mc) and the microcurie (μc).

The number of radioactive nuclei which will decay in any given time is proportional to the number of atoms present at that time. The proportionality constant is called the decay constant and is usually referred to by the symbol λ. This is referred to as exponential decay, and can be expressed mathematically as:

$$A = A_0 e^{-\lambda t} \tag{4.1}$$

where A_0 is the amount of activity present at time zero, and A is the amount of activity present at time t. If the natural log of A is plotted against time, a straight line will result (see Fig. 4.2.) The slope of this line is λ. The curve can also be characterized by its half-life ($t_{1/2}$), defined as the amount of time it takes for the activity to fall to one half its original value. Numerically it is equal to:

$$t_{1/2} = \frac{\log_e 2}{\lambda} \tag{4.2}$$

Thus after one half-life, only half of the original activity will remain and after two half-lives only one quarter remains.

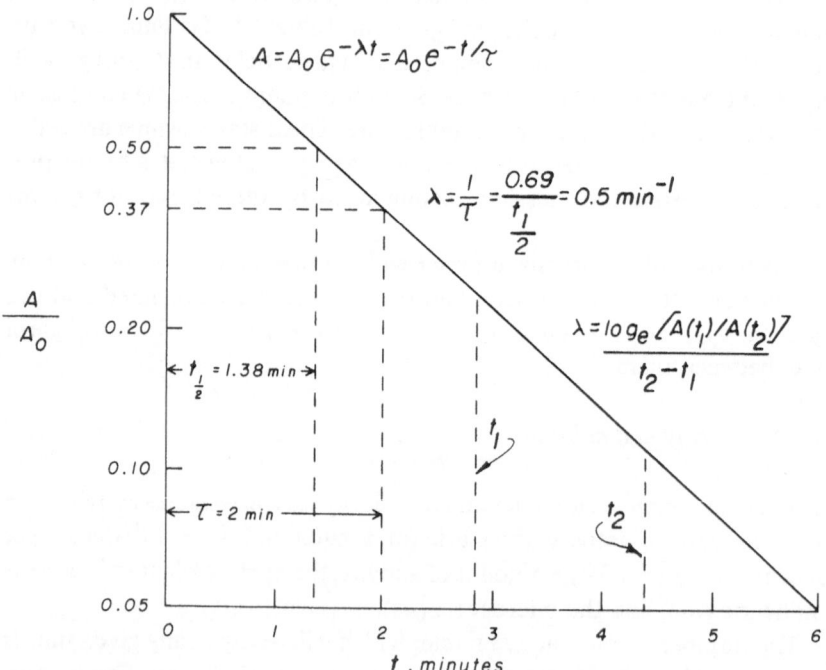

Fig. 4.2   An exponential decay curve.

4.3 Radiation detection systems

A block diagram showing the components of a typical radiation detection system is shown in Fig. 4.3. Some types of detectors and signal analyzers are discussed below.

4.3.1 Gamma detectors

A common type of gamma ray detector is the crystal scintillation detector consisting of crystals whose lattice structure has been doped with impurities. These impurities are atoms known as fluors. When a fluor atom interacts with a gamma ray (or beta particle) its electron shell is raised to an excited energy state. The atom soon returns to the unexcited state, emitting a photon that is of lower energy than the incident radiation and in the visible, or near visible, range.

An example of a scintillation detector is the sodium iodide (NaI) crystal. In the NaI crystal, the visible light from the fluor is transmitted to the photocathode of a photomultiplier tube where it interacts giving rise to primary photoelectrons. These primary electrons interact with a string of anodes with each interaction giving rise to many secondary electrons. Thus, the interaction of the gamma ray with the crystal results in a cascade of electrons. The amplitude of the resulting current pulse is proportional to the energy dissipated when the gamma ray interacted with the detector. Several interactions are required for an energetic gamma ray to transfer all its energy to the fluor, and the gamma ray may escape from the crystal after transferring only a fraction of its energy.

There are also solid state detectors, such as the GeLi detector, in which the interaction of the gamma ray with the detector gives rise directly to the production of current. As with scintillation detectors, the amplitude of the pulse is proportional to the energy dissipated by the interaction of the gamma ray with the detector. The active volume of GeLi detector is

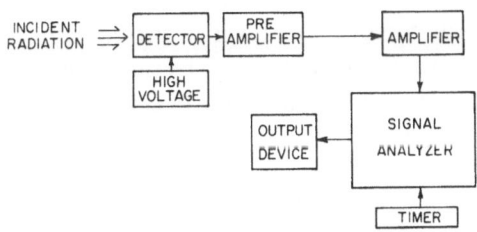

Fig. 4.3 A block diagram showing the components of a typical radiation detection system.

extremely small compared to that of NaI crystals. These solid state detectors operate at low temperatures and are usually cooled by liquid nitrogen.

Not every gamma ray emitted by a sample results in a current flow in the detector. The detector efficiency is the fraction of gamma rays incident on a detector which do so; it is influenced by several factors. One of these is the active volume of the detector. The larger the active volume, the higher the efficiency. Another factor is the energy of the gamma ray. If the energy of the gamma ray is low, it may be absorbed by the casing of the detector and not interact with the active volume; if it is high, the probability increases that the gamma ray will pass through the detector without any interaction. Not every gamma ray emitted by the sample will be incident upon the detector. The more the detector surrounds the sample, the higher the probability that the gamma ray emitted will be incident upon the detector. Therefore, well detectors have a higher efficiency than flat detectors.

4.3.2 Beta detectors

One type of beta detector is the gas ionization detector, in which the beta particle passes through a gas filled chamber, producing ion pairs. A potential gradient is maintained across the gas; thus the ion pairs migrate towards either the anode or the cathode, resulting in a current flow. The relationship between applied voltage and current response of the gas ionization detector is shown in Fig. 4.4. The detectors are classified as simple ionization, proportional, or Geiger-Müller counters depending upon the range in which they are operated, as shown in the figure. They are constructed in many ways, but all have in common a thin window through which the beta particle enters the ionization chamber. This window is necessary to minimize the absorption of beta particles before they enter the chamber.

Another type of beta detector is the liquid scintillation detector. In preparing samples for these detectors, the samples are dissolved in an organic solvent, commonly toluene, which also contains a fluorescent compound at a concentration of about 1%. The sample is thus in intimate contact with the fluor, and absorption of the beta particles without interaction with the fluor is minimized. The interaction may, however, be a two step process. Since the concentration of the fluor is low, the beta particles tend to interact predominantly with the solvent molecules, resulting in the emission of photons which also interact with the fluor to give rise to lower

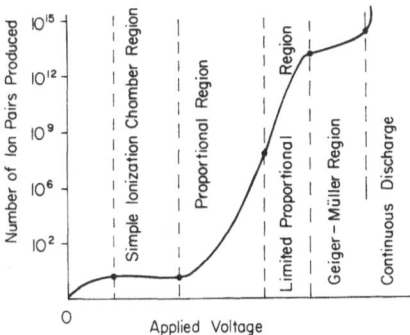

Fig. 4.4 The number of ion pairs produced by a single beta particle in a gas ionization detector as a function of the voltage applied across the detector.

energy photons. Photons from the fluor pass out of the vial where they are sensed by one or more photomultiplier tubes. Thus, the detection scheme for liquid scintillation counters is similar to that of the crystal scintillation detector.

4.3.3 *Signal processing*

The block diagram of a radiation detection system is shown in Fig. 4.3. Detectors have been discussed above. The high voltage source is necessary for current flow in the detector. The signal from the detector is input to a preamplifier, and the output of the preamplifier is routed to an amplifier, then to the signal analyzer. There are two main types of signal analyzers: the single channel analyzer (SCA) and the multichannel analyzer (MCA). The SCA has two discriminators which can be set to selected energy levels; thus there is one energy window which lies between the levels set by the upper and lower discriminators. Any signal which comes to the SCA and falls into this energy window will be passed and counted. In the multi-channel analyzer, the signal is input to an analog to digital converter (ADC) which digitizes it, and is then stored in a memory.

The signal analyzer is controlled by a timer. There are two types of timers, real-time and live-time. Whenever a pulse is sent from a detector to a signal processing system, it takes a finite amount of time for that signal to be processed. For this amount of time the processing system is dead; that is, it is unable to accept another pulse. As count rate increases, dead time also increases. In a given amount of clock time, therefore, the amount of time during which a signal processing system is able to accept signals is proportional to the sample activity, and this can result in errors in computation of activity. This difficulty can be obviated by the use of a live-time timer, in which the timer clock runs only for that period of time in which the signal processing system is able to accept data, making counting time independent of sample activity. The final block of the signal processing system is the input/output device, which may range from simple printers to complex computer-based data processing systems.

4.3.4 Processed gamma spectra

Fig. 4.5 shows the observed spectra for ^{85}Sr and ^{60}Co. The spectra as observed from NaI and GeLi detectors are illustrated, and are the summation of the counts due to background plus the counts due to sample activity. The background is comprised of counts due to cosmic rays, naturally occurring background radiation, and other extraneous sources such as other samples in the room. Background counts can be decreased by shielding the detector with lead. The components of the activity due to the

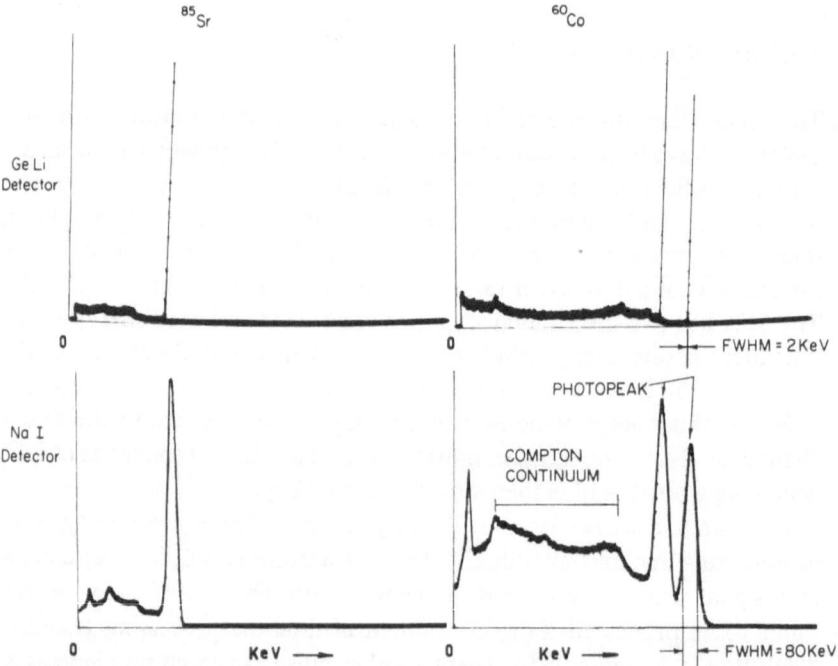

Fig. 4.5 The spectra of ^{85}Sr and ^{60}Co as observed on a crystal scintillation detector (3″ × 3″ NaI, Harshaw Chemical Co.) and on a solid state detector (30 cc GeLi, Nuclear Diodes, Inc.).

sample are the photopeak and the Compton continuum. These regions are shown in Fig. 4.5. The Compton continuum results from gamma rays which interact with the crystal but do not deposit their entire energy within it.

One measure of the detection system is its resolution, usually defined as the full-width-half-maximum (FWHM) of the photopeak. It is equal to the width, in energy units, of the peak at the point where it has fallen to one-half its maximum value. The FWHM is illustrated in Fig. 4.5. The resolution is a measure of how well a detector system can separate two photo-peaks which are nearly equal in energy. The FWHM is a function of several parameters, including gamma ray energy. The FWHM of any detector system increases as the energy of the incident radiation increases.

There are several factors that effect the appearance of the observed spectrum. The detector high voltage affects the energy calibration of the detector and its resolution. Resolution is increased as high voltage is increased. The gain and zero shift set on the amplifier will also affect the appearance of the observed spectrum; however, it will not effect the resolution. The resolution of the system is, however, severely affected by the dead time of the system. As dead time increases, resolution decreases. The severity of the effect of dead time upon the resolution varies with the type of detector and with the electronic characteristics of the detector system. Another factor which influences the spectrum is the sample geometry. With larger and larger samples, there is increased probability that gamma rays will be absorbed in the sample itself; thus, the observed efficiency will be decreased. Scattering of gamma rays in the sample also degrades resolution, but this effect is minor except for very low energy gamma rays. All these effects can be seen graphically on the output of a MCA. They affect the SCA in the same manner; however, the effects are not as severe if a wide energy window is used. When using a narrow energy window, care must be taken that the effects of such things as dead time and sample self-absorption are known.

4.3.5 *Processed beta spectra*

Typical spectra for tritium and carbon-14 as observed in a beta scintillation system are shown in Fig. 4.6. These are the most common tracer markers. Their spectra are seen to overlap considerably; assessment of their relative abundance in the same sample requires careful accounting of the amount of overlap of the spectra. Typically, spectra windows are set up so that each will contain mainly emissions from one or the other of the pair of

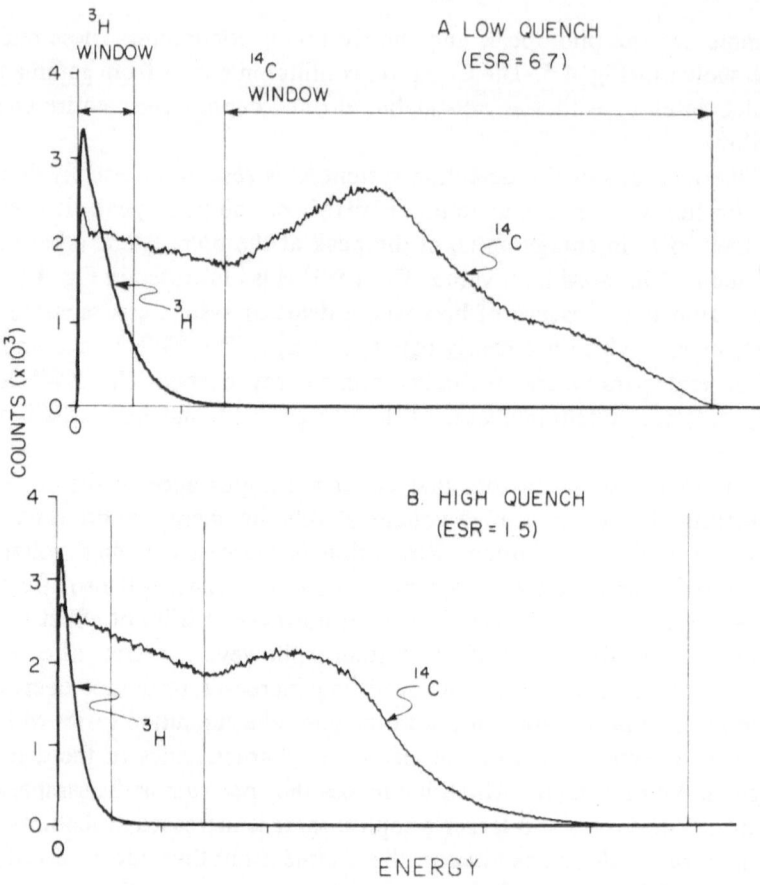

Fig. 4.6 The spectra of ^{3}H and ^{14}C as observed from a liquid scintillation detector illustrating the effect of quench on the shapes of the spectra.

isotopes. Quantitation requires knowing the exact percentage appearing in each window; that is, the relative efficiency of the detector for each tracer in the spectral windows.

Liquid scintillation counting has problems that practically do not exist for gamma scintillation counting, namely the effects of quenching agents or of fluorescence not due to the radioactive compound. Certain agents give changes in colour which cause light absorption and reduce the apparent emission energies (colour quenching). Chemical quenching, on the other hand, is due to components of the solution: either the scintillation solution itself or the substance being counted may absorb some of the energy associated with the beta particle and degrade it to a form which does not excite the fluor. Since water, a highly polar compound, is a quenching agent, minimally before putting it in with the scintillator fluid. Another problem may be the formation of an emulsion or fogging of the outside of the scintillation vial, both of which can produce optical quenching, simple light absorption.

The upper panel of Fig. 4.6 shows spectra with minimal quenching. The lower panel shows the effect of quenching on the shapes of the spectra. The spectra have a similar form, but are compressed into lower energy ranges. If the window settings for the pulse height analyzer are constant, as in the figure, then the overlap from one into the other is changed by differing degrees of quenching. Because of the high sensitivity of the form of the spectra and thus the overlaps to quenching, and because of the inherently large variation in quenching even within series of similarly prepared samples, it is necessary to measure the amount of quenching for each sample.

4.4 Tracer sample counting

4.4.1 *Gamma counting with a single tracer*

The simplest type of tracer experiment is one using a single radionuclide as a tracer. With a single tracer the single channel analyzer, which integrates all the counts in the selected energy window, can be used. The lower edge of the energy window is usually set to the lower edge of the photopeak and the upper window to its upper edge. The window should be selected so that the ratio of sample counts to background counts is maximized. (The count rate can be increased by opening the window on its lower side to include the Compton continuum. The advantage of this increase must be balanced against the increase in background counts.) The equation for the counts in the window is:

$$W = B + KC \tag{4.3}$$

W is the number of counts in the window, B the number due to background, K the efficiency of the tracer in the window, and C is the activity of the sample. The efficiency can be determined by counting a standard of known activity. K is then equal to the ratio of observed counts in the window to the activity of the standard.

4.4.2 *Gamma counting with multiple tracers*

The amount of information gained in a given experiment can be increased by using more than one tracer. Fig. 4.7 shows the spectrum obtained from a sample containing two radionuclei. In using two tracers, the signal must be processed by either two single channel analyzers (SCA) or a multi-channel analyzer (MCA). If two SCA's are used, the signal from the amplifier is routed to both analyzers. The windows are set so that most of the gamma rays from one tracer fall into the energy window of one SCA, and those from the second tracer as much as possible into the window of the other, as suggested by Fig. 4.7.

The equations for the observed counts of two radionuclides are given by:

$$W_1 = B_1 + K_{11}C_1 + K_{12}C_2 \qquad (4.4)$$

$$W_2 = B_2 + K_{21}C_1 + K_{22}C_2 \qquad (4.5)$$

W, B, and C are as described in equation 4.3. K is again efficiency; the first subscript refers to the window number, the second to the tracer number. Thus K_{11} is the efficiency of tracer 1 in window one, and K_{12} is the efficiency of tracer 2 in window 1. The efficiencies are again determined by counting standards of known activity. In using two tracers, two standards are required. The first standard must contain only isotope 1, and the second only isotope 2. The equations for the two tracers are linear simultaneous equations which can be solved by various methods.

Errors in the efficiencies, the K's, can be the major source of error in determination of the activities in Eq. 4.4 and 4.5. The efficiencies may be either absolute or relative. For relative efficiencies, K_{11} and K_{22} are defined as unity and the other efficiencies are computed as fractions. For example, when a sample containing only isotope 1 is counted, K_{21} can be computed as the ratio of W_2 to W_1. This removes the error due to uncertainty of the activity of the standard, but the result will be only relative numbers. When using relative efficiencies, known aliquots of the experimental tracer solution must be used as standards against which the samples can be compared for estimating absolute concentrations or masses. If absolute activities are required, then absolute efficiencies must be determined by counting samples of known standardized activity purchased from a reliable manufacturer.

This technique can be extended to more than two tracers. If SCA's are

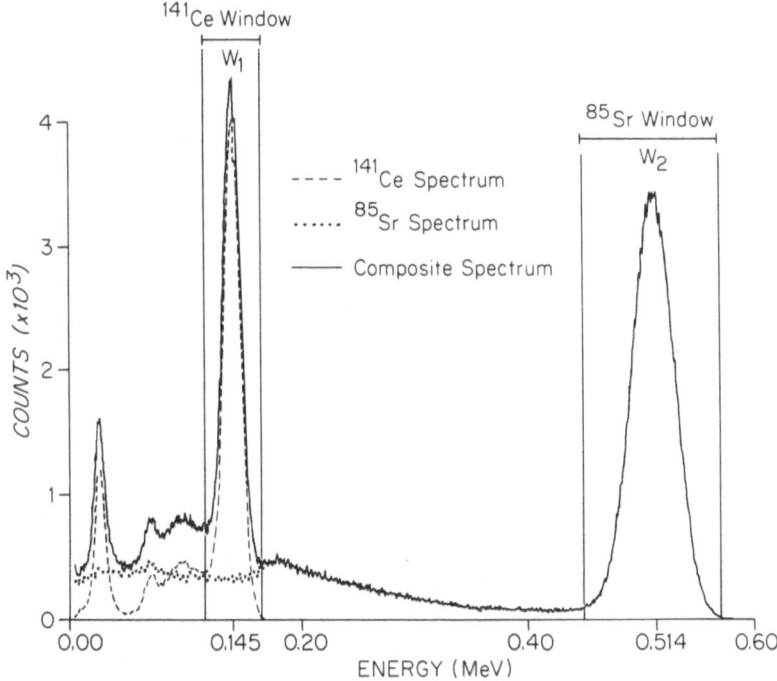

Fig. 4.7 The spectra of ^{141}Ce and ^{85}Sr as observed on a 1024 channel multichannel analyzer. The individual spectra and the composite spectra are shown.

used for multiple tracers, one analyzer is required for each tracer used. The factor limiting the number of tracers that can be used is the resolution of the detection system. As the number of tracers used increases, the effect of each tracer on windows other than its own is increased. Windows must be selected that minimize the efficiency of a given tracer in windows other than its own; efficiencies must be determined as precisely as possible in order to minimize the error made in separating the activities. When an MCA is used, a "single window" may be more complex, e.g. by being comprised of separate energy ranges each containing a gamma peak for its tracer.

The efficiency of a tracer in a window chosen for an energy range higher than around its highest energy photopeak is usually very low, and is often negligibly different from background. For example, ^{141}Ce spillover into the ^{85}Sr window in Fig. 4.7 is effectively zero. The converse is, however, usually not true, as the Compton continuum can contribute significantly to lower energy windows (see Figs. 4.5 and 4.7). For this reason, the relative amounts of the tracers used in a multiple tracer experiment must be judiciously chosen to balance counting efficiency against overlap into other windows. Ordinarily, one will therefore use a larger dose of the lower energy tracers, for example, one might choose the relative doses of ^{141}Ce and ^{85}Sr so that the ^{141}Ce activity is several times that of the ^{85}Sr activity in order to obtain the same accuracy of estimation of their activities.

4.4.3 *Beta counting with a single tracer*

The general system is diagrammed in Fig. 4.3. For beta counting there are analogous elements to those used for gamma counting.

For any single tracer, having a spectrum such as one of those shown in Fig. 4.6 the maximum counts are obtained by using a broad window. Ordinarily one has control over the energy window by selecting a lower level and an upper level in the discriminator of the pulse height analyzer, and counts the number of pulses of all energies lying between these two boundaries of the window. In most liquid scintillation counters, electronic noise will produce falsely high counts if the lower edge of the energy window is set too low, making it advisable to use a setting above the minimum. The practical level can be ascertained by looking at background counts while shifting the lower level setting. As the lower level is decreased, the background will increase almost linearly until the region of electronic noise is encountered. At this point the background will rise much more quickly for a given change in lower level setting.

When counting a sample, the counts obtained in the window are given by:

$$W = B(Q) + K(Q) \cdot C \tag{4.6}$$

W is the count rate in the window; B is the background count rate, which is indicated by the parenthetic Q to be a function of the quench factor; C is the true count rate or disintegrations per minute from the sample, and K is a counting efficiency factor which is, again, a function of the quench factor and of the window settings. To gain accuracy it is important to maximize the ratio of true counts C to background B. This is one reason for setting the lower level high enough to minimize electronic noise. The upper level window setting should not be higher than the highest energy of the emitted betas, in order to avoid accumulating background counts from a region yielding no true tracer counts.

Since one will ordinarily be counting a series of samples of somewhat different quench, the curve of B(Q) vs. quenching factor must be determined. This is done by taking a series of background counts from different liquid scintillation vials containing no tracer, but some small and variable amounts of quenching agent. Similarly, the efficiency $K(Q)$ as a function of quenching factor must be determined. This is usually done by having 20 to 100 liquid scintillation vials containing the same amount of tracer and a fixed amount of liquid scintillator fluid, but with varying amounts of a quenching agent added to the solution. (Chemical, colour, and optical quench all have different effects on the shape of the spectrum, so there is no perfect or universal quench correction curve. Nevertheless the influences of chemical and colour quench are fairly similar, and there are no standard methods for correcting for each separately. Insofar as it is possible, it is therefore best to obtain quench correction curves with the same mixture of quenching agents as occur in the sample.)

In most modern liquid scintillation counters, the degree of quenching is measured by using an external standard counted along with each of the samples. The technique is to use a gamma-emitting source (for example, barium-133, americium-241, or cesium-137) which is a long-lived radioisotope emitting gamma rays of moderate energy only. The liquid scintillator sample is shielded from the gamma source for the counting of beta emissions, then the gamma source is moved adjacent to the sample and it is recounted to obtain the estimate of the quench. The interaction of the gamma rays with the counting glass vial and the scintillation solution produces Compton electrons which have a spectrum resembling the shape of the beta spectra of carbon-14 and tritium, particularly the former. A separate pair of pulse height analyzers with two preset windows provides the count rates obtained in each of these windows. After subtraction of the counts in each of these special windows due to the beta emitter itself, the ratio of counts in one window to that in the other due to the gamma emitter is calculated. This is the so-called "automatic external standard ratio" (AES or ESR). This ratio of counts due to the gamma emitter we have simply called the quench factor, Q, and recognize it as a measure of quenching. The usual way of obtaining a quench plot is to take the counts from each sample and plot them as a function of Q, as shown in Fig. 4.8.

It should be noted that there can be no totally unquenched sample. In order to relate the quenching factor to the actual dose of tracer injected, an aliquot of the tracer dose must be counted in the system along with the samples, and similarly corrected for quenching. The counts in the samples can then be compared with the counts in the dose aliquots. In obtaining these relative numbers, errors due to uncertainty in the actual activity of the dose are removed. If the absolute activity of the samples is required, special techniques are needed which are not discussed here.

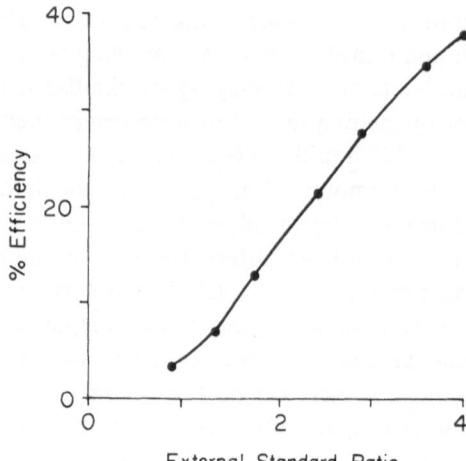

Fig. 4.8 A typical curve of relative efficiency as a function of ESR on a liquid scintillation detector.

4.4.4 *Beta counting with multiple tracers*

The principle of the method of multiple beta-emitters is essentially the same as that described for two gamma-emitting tracers. It can be extended fairly readily to the counting of three tracers, but only with some loss of accuracy. With standard liquid scintillation counters one obtains the counts in pre-chosen windows rather than the whole spectrum. For two tracers, two windows are used. The counts in each window are the sum of the counts due to background, and the counts due to tracers 1 and 2. These statements are summarized as:

$$W_1 = B_1(Q) + K_{11}(Q) \cdot C_1 + K_{12}(Q) \cdot C_2 \qquad (4.7)$$

$$W_2 = B_2(Q) + K_{21}(Q) \cdot C_1 + K_{22}(Q) \cdot C_2 \qquad (4.8)$$

W_1 is the counts in window 1, B_1 the background counts in the same window; C_1 and C_2 the actual concentrations of tracers 1 and 2, for example C_1 concentration of tritium, C_2 concentration of carbon-14. $K_{11}(Q)$ is the efficiency of counting tracer 1 in the band of the spectrum defined as window 1; the parenthetic Q indicates that K_{11} is a function of the quench as measured by the external standard ratio. Similarly, $K_{12}(Q)$ is the counts due to carbon-14 in window 1. Thus, in order to determine the actual concentration C_1 and C_2, one has to measure the counts in windows 1 and 2, the background counts in both windows with no tracer present, and the efficiencies for each tracer in each window across the range of different quench factors. This pair of equations must be used for each sample. There-fore, for determining the two concentrations C_1 and C_2 in a sample, eight items of information must be obtained: the W's, B's, and the four K's. There is statistical error in each of these measures, which inevitably reduces the accuracy, but with careful determination of each number, and counting for long enough so that the numbers are reasonably accurate, the counts due to each of the two isotopes can be distinguished accurately.

Using the data on the coefficients from Fig. 4.9 and the observed count rates in windows 1 and 2, the true counts due to the isotopes, C_1 and C_2, can be obtained by solution of equations 4.7 and 4.8. The solution for this set of two equations with two unknowns will ordinarily be done by straight-forward algebraic techniques. With three isotopes, necessitating three equations and three unknowns, and when there is some indeterminacy in the coefficient values, the B's and K's, somewhat more complex techniques are used to obtain the best estimates of the C's.

4.4.5 Decay corrections

If a single sample is counted a number of times, the observed count rate will decrease because of radioactive decay. This holds true when a series of samples is counted, each at a different time after the end of an experi-ment. In order to place all the observed counts on the same time basis, cor-

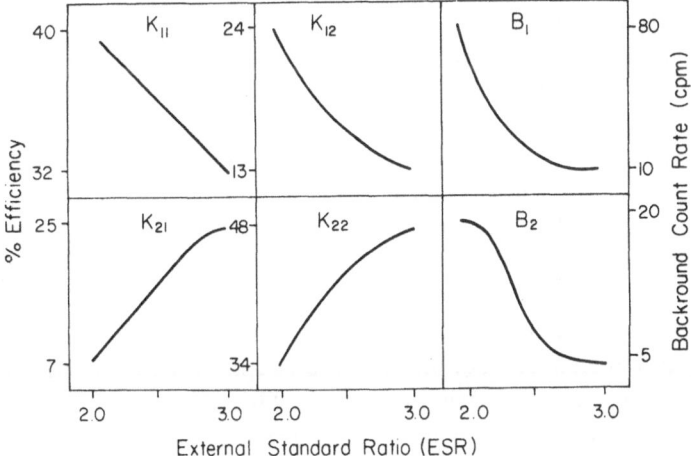

Fig. 4.9 Typical shapes of relative efficiencies and background counts as a function of ESR on a Nuclear Chicago Mark II liquid scintillation detector system using a ^{133}Ba gamma source. The quenching agent used was dog plasma. With no quenching agent present, the ESR was 3.6. These curves provide the coefficients used in equations 4.7 and 4.8.

rection must be made for decay. For long-lived isotopes such as tritium and carbon-14, with half-lives of 12.3 and 5730 years, the correction is small and may usually be ignored, but for many commonly used radioactive tracers the decay correction is significant and must be taken into account.

The decay correction is based on the exponential decay, equation 4.1. In this case what is desired is the initial activity A_0 so that all samples can be corrected back to the same time, usually the time of the experiment. The equation can be rewritten as:

$$A_0 = Ae^{\lambda t} \tag{4.9}$$

This equation assumes that the sample is not decaying while it is being counted. When the half-life of the radionuclide is not many times greater than the time period for counting, this assumption can lead to considerable error. When decay during counting is to be taken into consideration, then:

$$A_0 = A\lambda t_c \frac{e^{\lambda t}}{1 - e^{-\lambda t_c}} = Ae^{\lambda t} (1 + 0.5\lambda t_c) \tag{4.10}$$

t is the time from the experiment to the beginning of the count, and t_c is the duration of the counting period.

4.4.6 *Relative doses with reference to the experiment design*

The indeterminacy of the counting rate for radioactive decay processes is the square root of the number of counts. Thus if one desires 1% accuracy, that is, a standard deviation of 1% of the observation, then one must measure at least 10,000 counts ($\sqrt{n}/n = 0.01$). High doses of tracers are desirable to increase the accuracy of measurement while making reasonable counting times possible. When using multiple tracers, however, it must be remembered that increasing the dose of one of the tracers relative to the others will increase the counts overlapping into other windows. Looking at equation 4.4, if we had 100,000 counts of C_2 in its window and its overlap in window 1 were only 10%, this would be 10,000 counts, which would have a standard deviation of 100 counts. If tracer 1 were counted with 50% efficiency in its window we would have to see at least 200 counts in window 1 merely to recognize its presence. Additional noise due to background counts degrades the accuracy further. To compensate for this, one will ordinarily give a relatively large dose of the lower energy tracer so that it will not be swamped by the overlap of the higher energy tracer into the lower energy window.

When using tracers that overlap into each other's windows (and this is commonly the case because one wishes to have large windows to gain maximum efficiency), the ratio of doses of the two tracers may depend on the part of the experiment in which one is most interested. This problem is illustrated in Fig. 4.10.

As shown in Fig. 4.10, I^{131}-labeled albumin (an intravascular tracer) and C^{14}-labeled sucrose were injected simultaneously into the coronary arterial inflow, and a sequence of venous samples was collected from the coronary sinus outflow. The concentrations of the tracers were measured in each sample. In the first few seconds the fraction of the albumin dose appearing in the outflow exceeds that of the sucrose, a part of which is escaping from the capillary blood into the surrounding tissue, while during the tail of the curve the sucrose level is higher than the albumin. In fact the albumin curve quickly approaches zero, so that the sucrose concentration after 30 or 40 seconds may be two orders of magnitude greater than the albumin concentration. The front end of the curve, as we shall see in section 4.5.2 is useful for the measurement of an apparent instantaneous extraction of the sucrose relative to the nonpermeant albumin. If the primary interest is here, one can feel secure in making the sucrose dosage high relative to that for albumin, because the albumin spillover into the sucrose window is substantial but the reverse spillover is small. For determining the ratio of albumin to sucrose concentration with great accuracy in the tail of the curve, very much larger doses of albumin relative to sucrose would have to be injected.

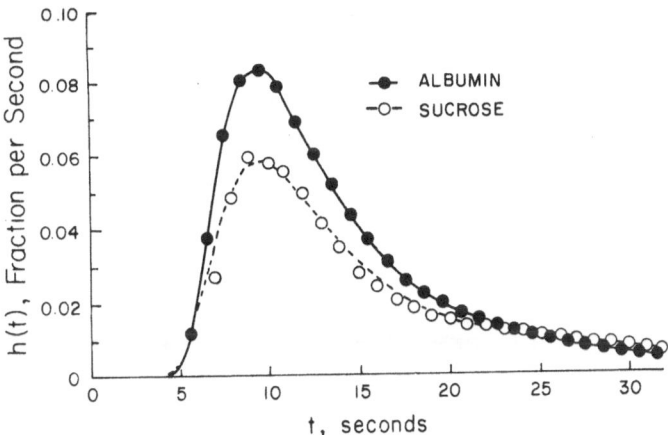

Fig. 4.10 The concentration as a function of time of albumin, an intravascular tracer, and sucrose, a diffusible tracer, measured in the coronary sinus of a dog heart following injection of the tracers into the inflow at t = 0.

4.5 Applications

The examples are chosen to illustrate fundamental principles. In these examples we will not provide all the fine detail that would be required for careful scientific application. For a general overview concerning cardio-vascular applications, see Bassingthwaighte (2, 3), Bassingthwaighte and Holloway (4), and Zierler (19, 20). Compartmental analysis is thoroughly presented by Jacquez (13).

4.5.1 *Application of tracer dilution techniques to the measurement of volume and flows*

The steady-state volume of distribution.   The method of volume determination simply involves addition of a dose, m (cpm), of a tracer to a constant volume, V (ml), mixing thoroughly and then measuring the final concentration, C (cpm/ml) (Fig. 4.11). From C the volume, V, is computed as:

$$V = \frac{m}{C} \tag{4.11}$$

This principle is applied commonly in the measurement of blood volume. A known amount of a tracer, such as radio-iodinated serum albumin which remains confined mainly to the vascular system, is added to the blood by intravenous injection. After allowing 10–15 minutes for complete mixing of the indicator with the blood, a sample is drawn and the concentration

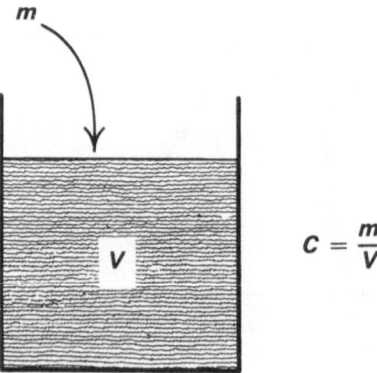

Fig. 4.11 The concentration (C) after mixing is the mass of the added tracer (m) divided by the volume of the whole system (V).

measured. Blood volume and its measurement have been well discussed by Lawson (17).

Gaudino and Levitt (9) used a similar technique for measuring the volume of distribution of inulin, "inulin space", by intravenous injection into animals whose kidneys had been removed to prevent the excretion of inulin through glomerular filtration. Inulin (MW ~5000) does not enter cells but permeates capillary membranes so that the plasma concentration equilibrates with the interstitial fluid space, the region between cells and outside of the blood stream. They estimated that inulin space, the sum of plasma space plus interstitial fluid space, comprises 26% of all body water. The identification of the space as extracellular fluid is supported by the fact that there are a number of other indicators that fill only slightly larger spaces, sucrose, sodium and chloride filling 30–35% of body water volume (7). These smaller molecules probably owe their larger volumes of distribution to their ability to enter intermolecular water spaces from which inulin is excluded.

Using heavy water (D_2O) or tritiated water (THO) in a similar fashion, total water has been estimated at 50–60% of body weight, depending on the leanness and state of hydration of the individual (8, 6).

The tracer dilution transient: measurement of cardiac output by the sudden injection technique. A typical intravascular tracer dilution curve recorded from the arterial circulation following injection into the pulmonary artery is seen in Fig. 4.12. The tracer is diluted and dispersed during the traversal of the central circulation and some recirculated tracer masks the tail of the primary curve. The technique of semilog replot and monoexponential extrapolation introduced by Hamilton et al. (11) is still considered adequate

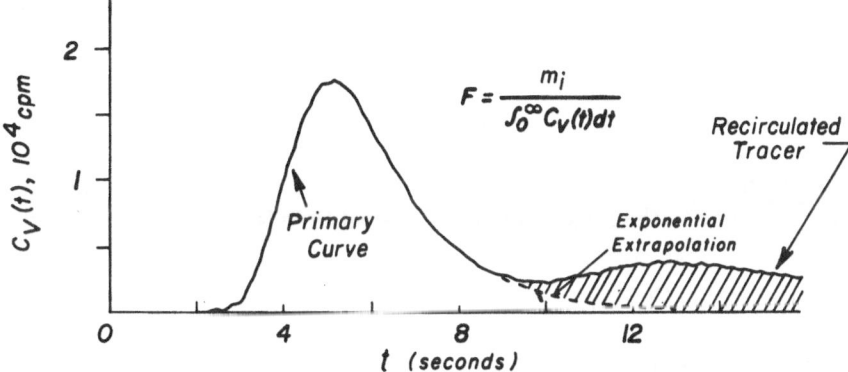

$$F = \frac{m_i}{\int_0^\infty C_V(t)\,dt}$$

Fig. 4.12 A tracer dilution curve with a monoexponential extrapolation to eliminate recirculation.

to account reasonably well for the tail of a curve. Flow is calculated from the area under the first pass or primary curve:

$$F = \frac{m_i}{\int_0^\infty C_v(t)dt} \qquad (4.12)$$

The flow, F (ml/sec), equals the amount of tracer injected (cpm), m_i, divided by the area underneath the concentration time curve, sec. x cpm/ml. The standard assumptions made in using this equation include (1) stationarity (constant flow and constant distribution of flow) and (2) linearity (responses of the system must summate linearly); these have been outlined more fully by Zierler (19, 20). There are a few indicators, like temperature, hydrogen, and xenon, whose recirculation is minimal, and for whose curves the extrapolation is not required.

The conditions of stationarity and linearity are not restrictive because the method is dependent simply on conservation of mass: what was injected, m_i, comes out as the integral of the product of flow times concentration in the outflow at the sampling point.

Radioactive microspheres as well as the complete gamut of diffusible and nondiffusible tracers can be used for measuring cardiac output when the injection is made into left atrium or left ventricle and the sampling is from a peripheral artery. All that is required is that there be good mixing of the tracer in the blood before any branching occurs.

The volume of a flowing system: the mean transit time volume. In any constant volume and constant flow system, the use of tracers to measure the volume is a theoretically simple calculation based on conservation of mass. The volume of the system is the flow, F (ml/sec), times the mean transit time, $\bar{t}$ (sec):

$$V = F \cdot \bar{t} \qquad (4.13)$$

With an impulse injection, or an extremely short sudden injection at the entrance to the system, the mean transit time is:

$$\bar{t} = \frac{\int_0^\infty t \cdot C(t)dt}{\int_0^\infty C(t)dt} \qquad (4.14)$$

For practical purposes, this equation may be expressed as:

$$\bar{t} = \frac{\sum\limits_{i=1}^{i=N} (t_i \cdot C_i \cdot \Delta t_i)}{\sum\limits_{i=1}^{i=N} (C_i \cdot \Delta t_i)} \tag{4.15}$$

In this expression, the continuous curve C(t) is treated as a sequence of points along the curve, each indexed by the subscript i. The N points in time, t_i, are taken with a time interval, Δt_i, between each one. The summation in the denominator is simply the area underneath the primary curve and the summation in the numerator provides the weighting or relative amount of the tracer existing over each particular time interval, Δt_i, so that the result is the weighted average or mean transit time. As in measuring flow (Fig. 4.12), it is the primary curve that is used as C(t) or as the sequence of concentration values, C_i.

This dynamic technique is useful for finding volumes of components of the circulation and is most readily applied to discrete organs with a single inflow and a single outflow. The volume, V, is different for different tracers. It is the volume of distribution of the component of the system which is being "traced", i.e., the "mother substance". The volume of distribution is that volume accessible to the tracer during its passage from the injection point in the inflow to the organ to the sampling point in the venous outflow. When the tracer, such as labeled erythrocytes or albumin, is limited to the vascular system, the mean transit time volume will simply be the blood volume of the organ. When the tracer escapes from the vascular system, the calculation of V is equally valid, so long as the primary curve can still be identified before recirculation masks it too severely. Thus, if one injects ^{24}Na as a slug injection into the arterial inflow and samples venous outflow, the mean transit time volume for sodium would be the total extracellular fluid; that is, the blood volume between the injection and sampling points plus the extracellular, extravascular, or interstitial fluid space. The problem of excluding recirculation is severe because recirculation is ordinarily fairly rapid compared to the time for washout of the sodium from the extravascular space. If one simultaneously samples at the inflow and outflow from the organ, recirculation can be accounted for and the mean transit time obtained accurately using deconvolution techniques (1, 16).

When the tracer penetrates the cells as well as the interstitial space the volume of distribution includes both these spaces. Again the theory is applicable but the practice becomes more difficult, as recirculation will tend to mask the tail of the curve. ^{131}I-iodoantipyrine, and tritium-labeled water are appropriate tracers for measuring this volume.

External detection (residue function) technique for determining flow per unit volume. The detection of gamma-emitting tracers within the organ or within the segment of the circulation using an external sodium iodide detector can provide an estimate of the mean transit time of the organ or of the segment of the circulation. This holds as long as recirculation either does not occur or can be excluded by some empirical artifice such as the monoexponential extrapolation which was applied to intravascular dilution curves in Fig. 4.12. The mean transit time is given by the area underneath any of the three curves of the residue function, R(t), in Fig. 4.13.

$$\bar{t} = \int_0^\infty R(t)dt \tag{4.16}$$

This is the basis of the "height/area" technique introduced by Zierler (1965) for estimating flow per unit volume, where R(o) is unity and K is the efficiency factor:

$$\frac{F}{V} = \frac{1}{\bar{t}} = \frac{1}{\int_0^\infty R(t)dt} = \frac{height}{area} = \frac{K \cdot C(0)}{\int_0^\infty K \cdot C(t)dt} \tag{4.17}$$

In this expression, V is the volume of distribution of the tracer. If one wishes to know the flow per unit volume of the organ itself, one has to know something more about the tracer; that is, one should have an estimate of the steady state volume of distribution of the tracer relative to the total organ volume. This can be described as an apparent partition coefficient, λ, which is the ratio of the volume of distribution for the tracer within the organ to the total organ volume, V_{org}:

$$\lambda = \frac{V}{V_{org}} \tag{4.18}$$

For sodium in heart or skeletal muscle, λ is only about 0.30. However, for antipyrine, which is more lipid soluble and penetrates both capillary and cell walls, λ is close to 1.0 for most organs. The substitution of λV_{org} for V in equation 4.17 leads to the expression given in Fig. 4.13, which is the one commonly used.

This general expression has been used for estimating flow per unit volume from washout curves from various organs. A modification to re-

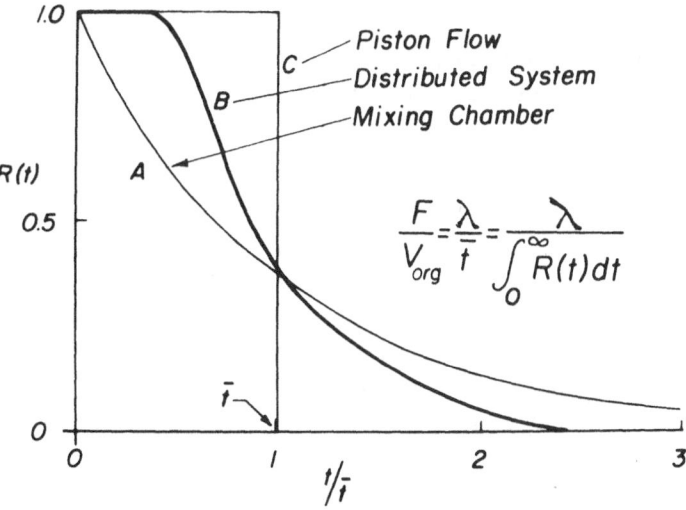

$$\frac{F}{V_{org}} = \frac{\lambda}{\bar{t}} = \frac{\lambda}{\int_{0}^{\infty} R(t)dt}$$

Fig. 4.13 The area under the residue function, $R(t)$, is the mean transit time, $\bar{t}$. Curves A, B, and C have the same area and thus the same $\bar{t}$. Curve C represents a system having all mean transit times equal. Curve A represents a perfect mixing chamber having a true exponential washout. Curve B represents an intermediary system having both delay and dispersion.

duce the length of time required for recording the curves was used by Høedt-Rasmussen and Lassen (12) in the brain and by Bassingthwaighte, Strandell, and Donald (5) in the heart. In the latter study it was shown to give estimates comparable to those obtained by measuring the flow and the organ weight.

Exponential washout: a specific form of the residue function.   When diffusion or mixing is so rapid that concentrations are uniform throughout a system, $R(t)$ has the form of a single exponential:

$$R(t) = e^{-t/\tau} \tag{4.19}$$

or

$$C(t) = C_0 e^{-t/\tau} \tag{4.20}$$

where $C(t)$ is the count rate at time t after the initial value C_0. These are the same form of equations as for radioactive decay, equation 4.1. The amount of tracer washed out from a perfect mixing chamber per unit time is a constant fraction of the whole. The mean transit time for a first order system (monoexponential) is τ, its time constant, which is the time for the concentration to fall to 37% of the original concentration. From this, the flow calculation from an exponential washout curve such as curve A in Fig. 4.8 is:

$$F = \frac{\lambda V_{org}}{\tau} \qquad\qquad\qquad (4.21)$$

Two or three exponentials are often required to fit R(t), as is the case for curves recorded from the brain (12) and from isolated heart preparations (5). One of the main reasons why these curves differ significantly from monoexponential ones is that there is substantial heterogeneity in the regional perfusion within an organ. Achieving a good fit of data with two or three exponential curves does not mean that there are only two or three discrete levels of flow within an organ. However, fitting the curve more accurately by a weighted sum of exponentials rather than with a single exponential does give a more accurate estimate of the overall average organ blood flow. Kelly et al. (14) provide greater detail in explaining the application and interpretation of two-exponential analysis. If this type of analysis is applied to situations where the washout is from monoexponential regions in parallel (but not in series), the equation can be interpreted in terms of physical quantities.

The semilog plot of a two-exponential washout is shown in Fig. 4.14. Of the tracer in the organ at time zero, 60% washed out rapidly, with a mean transit time τ_1 of 3 minutes, while 40% of the dose washed out slowly, with a mean time τ_2 of 15 minutes. For any aggregate of parallel exponential processes, their summation, R(t), is concave upward on a semilogarithmic plot.

4.5.2 Applications of tracer exchange techniques to the estimation of transmembrane fluxes

The tracer principle. The main advantages in using tracers to measure transmembrane fluxes are (a) to make measurements without changing the concentration of mother substance in steady state situations, (b) to measure concentrations too low or otherwise inconvenient for chemical measurement, and (c) to measure unidirectional fluxes under conditions where two-way fluxes are occurring.

The use of a tracer reduces the apparent complexity of a system. For example, the flux (moles/sec) of most ions or sugars traversing a cell mem-

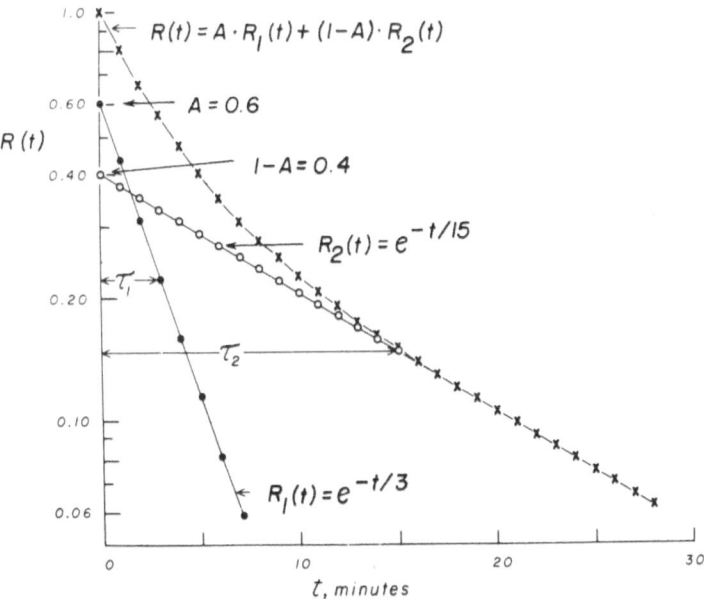

$$R(t) = A \cdot R_1(t) + (1-A) \cdot R_2(t)$$

$$A = 0.6$$

$$1 - A = 0.4$$

$$R_2(t) = e^{-t/15}$$

$$R_1(t) = e^{-t/3}$$

Fig. 4.14 The washout from two parallel regions which each have a monoexponential washout.

brane is a complex process involving other substances and often involving energy exchange. Since such fluxes are influenced by the concentration of the solute of interest, the mother substance for the tracer, one cannot obtain an accurate measure of the flux by adding or subtracting mother substance and measuring the change in concentration on the other side of the membrane. In contrast, addition of tracer to one side of a membrane and measurement of concentration on the other allows precise estimation of flux without changing the total concentration of mother substance by more than one part in a billion or a million. This rationale can be expressed more formally. Consider the system in Fig. 4.15 where the flux of mother substance J across a barrier has a rate constant, K, for exchange which is dependent on the concentrations of mother substance on each side of the barrier C_1 and C_2 and on other influences:

$$J_{1\rightarrow2} = K_{12}(C_1, C_2, \text{etc.}) \cdot C_1 \tag{4.22}$$

$$J_{2\rightarrow1} = K_{21}(C_1, C_2, \text{etc.}) \cdot C_2 \tag{4.23}$$

and

$$J_{net} = J_{1\rightarrow2} - J_{2\rightarrow1} \tag{4.24}$$

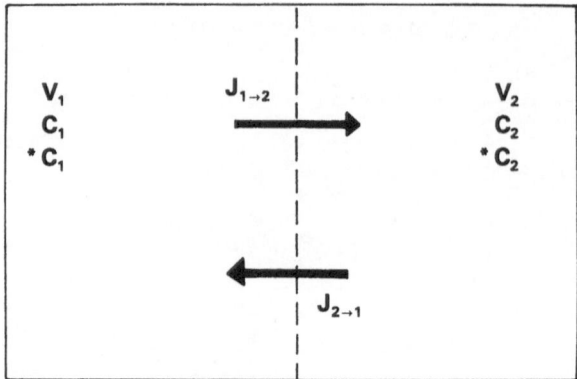

Fig. 4.15 A system of two mixing chambers of volumes V_1 and V_2 separated by a transport barrier across which solute flux can occur. When $^*C_2$ is small compared to $^*C_1$, then the changes in $^*C_2$ can be used to estimate $J_{1 \to 2}$.

where $J_{1 \to 2}$ indicates unidirectional flux from side 1 to side 2, and J_{net} is the net flux from side 1 to side 2 in the presence of opposing flux $J_{2 \to 1}$. The terminology $K(X)$ indicates that the exchange rate K is dependent upon X, that is, upon C_1, C_2, and other possible influences.

An appropriate tracer (e.g. ^{24}Na for mother substance Na) will be transported in exactly the same fashion as the mother substance. Thus, using $^*C_1$ to denote the tracer concentration on side 1, the total flux $J_{1 \to 2}$ is:

$$J_{1 \to 2} = K_{12}(C_1 + {}^*C_1, C_2, \text{etc.}) \cdot [C_1 + {}^*C_1] \qquad (4.25)$$

since $^*C_1$ is negligible compared to C_1 the flux of mother substance is identical to equation 4.22. The tracer flux is:

$${}^*J_{1 \to 2} = K_{12}(C_1 + {}^*C_1, C_2, \text{etc.}) \cdot {}^*C_1 \qquad (4.26)$$

in which $K_{12}(C_1 + {}^*C_1, C_2, \text{etc.})$ is identical to $K_{12}(C_1, C_2, \text{etc.})$ as for the mother substance. Thus the K_{12} for this particular steady state is independent of the tracer concentration and flux, but is measurable from the tracer information alone:

$$K_{12}(C_1, C_2, \text{etc.}) = \frac{{}^*J_{1 \to 2}}{{}^*C_1} \qquad (4.27)$$

Then the flux of mother solute is calculated using C_1 in equation 4.22, or if K_{12} is not needed, then directly:

$$J_{1\to2} = \frac{*J_{1\to2}}{*C_1} \cdot C_1 \qquad (4.28)$$

This development was simply to say that tracers behave as tracers, not perturbing the system, and yet serving as tools for measurement.

The measurement of unidirectional flux. The next question is how and under what conditions one can measure $*J_{1\to2}$ or any other unidirectional flux. Consider the simple yet very common situation illustrated in Fig. 4.15. In certain experiments a biological membrane will be mounted between two chambers and tracer added to one. In others, V_1 might be the volume of cells dispersed in a suspending medium of volume V_2. The flux $*J_{1\to2}$ can be measured precisely when $*C_2$ is small:

$$*J_{1\to2} = *J_{net} - *J_{2\to}. \qquad (4.29)$$

In the early moments of an experiment, $*C_1$ is much larger than $*C_2$, perhaps by a factor of a thousand or a million, so that $*J_{2\to1}$ is negligible in comparison and $*J_{1\to2}$ is essentially equal to the net flux alone. The net flux can be obtained from the change in $*C_2$ over a short time interval $t_2 - t_1$, given that V_2 is known:

$$*J_{1\to2} = V_2 \frac{*C_2(t_2) - *C_2(t_1)}{t_2 - t_1} \qquad (4.30)$$

This is the classical "initial velocity" approach to the flux measurement and is most useful, given that $*C_2$ is much less than $*C_1$.

The use of initial velocity to measure capillary permeability. The experiment shown in Fig. 4.10 was designed for estimating the transport rate of sucrose from blood across the capillary endothelial barrier. In this situation the exchange constant K is the product, PS, of the surface area times the permeability. Instead of measuring the rise in $*C_2$ over a short period of time, one measures the diminution in $*C_1$; this is done by comparison of the concentration of ^{14}C-sucrose with ^{131}I-albumin, which does not penetrate the membrane, to estimate an instantaneous extraction, $1 - *C_{suc}/*C_{alb}$, over the first few seconds of the experiment. This is based on the same idea as the initial velocity experiment, namely that the extravascular concentration of tracer is low compared to that in the blood. Because the time available for extraction diminishes at high flows, the flow F (ml/min) also comes into the calculation of PS:

$$PS = -F \log_e \left(\frac{*C_{suc}}{*C_{alb}} \right) \tag{4.31}$$

Further details and an appraisal of this approach are provided by Guller et al. (10).

The kinetics of two-compartmental exchange.   When tracer is added to one compartment of a two-compartment system with exchange between them, the concentrations $*C_1$ and $*C_2$ both change. Initially, the changes are linear, as presumed in the section on initial velocity experiments. However, as time progresses $*C_1$ falls from its initial value C_0, and $*C_2$ rises to become significant compared to $*C_1$. The back flux from V_2 to V_1 is now no longer negligible for it reduces the rate of rise of $*C_2$. This is shown in Fig. 4.16 for the special case of purely passive exchange. When the whole curve is provided by the experimental data, the relative volumes, V_2 and V_1,

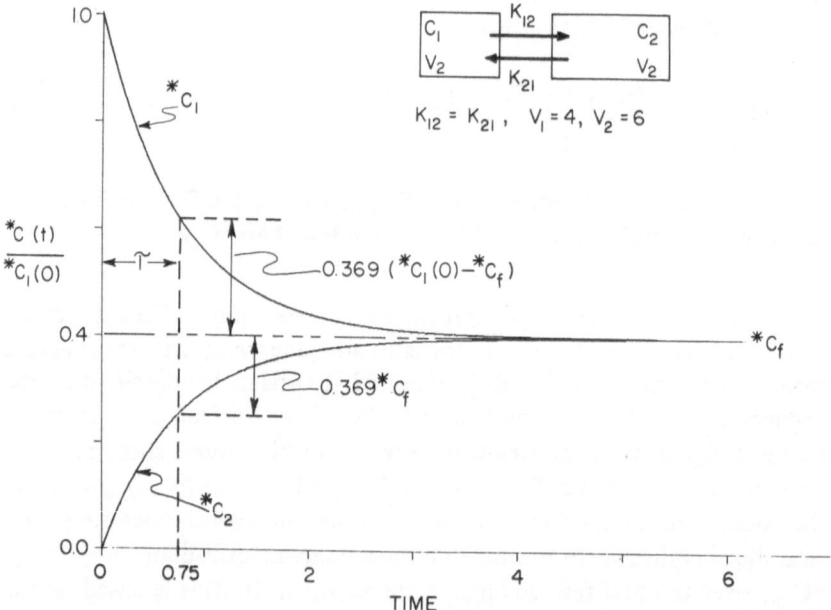

Fig. 4.16   The time course of tracer concentration after addition of tracer to chamber 1. The final concentrations, $*C_f$, in each chamber will become equal only when the rate constants K_{12} and K_{21} are equal.

and the rate constants can be estimated graphically from the data:

$$\frac{V_1}{V_1 + V_2} = \frac{C_f}{C_0} \qquad (4.32)$$

and

$$\tau = \frac{V_1 V_2}{K_{12}(V_1 + V_2)} \qquad (4.33)$$

When V_1 is measured then the two unknowns V_2 and K_{12} can be obtained:

$$K_{12} = \frac{V_1(1 - C_f/C_0)}{\tau} \qquad (4.34)$$

and

$$V_2 = V_1(C_0/C_f - 1) \qquad (4.35)$$

There are a number of useful references for more complex analyses of this sort. D.S. Riggs (18) presents diffusional and compartmental exchange in a very useful way that carries the reader beyond the introductory phase. J.R. Jacquez (13) provides a more theoretical and general treatment of compartmental analysis. Some theoretical approaches and many practical examples are provided in a symposium edited by Kniseley et al. (15).

4.5.3 Neutron activation analysis

A technique which is useful, but not strictly speaking a tracer technique, is neutron activation analysis (NAA). This is useful for determining the concentration of various elements in a sample. If a sample is irradiated by a beam of neutrons in a nuclear reactor, some atoms in the sample will absorb neutrons; these atoms usually become unstable and will then decay. Not all elements are susceptible to neutron activation, but many of clinical interest such as sodium, arsenic, potassium, cobalt, and iron can be analyzed by using NAA. To quantitate the concentrations, standards containing precisely known concentrations of the elements of interest must be activated under the same conditions as the samples, preferably simultaneously. The samples can then be compared against the activities of the standards to give absolute concentrations. Because so many elements are susceptible to activation by neutrons, the number of radionuclides produced in a given sample is large, and the resulting spectrum is complex. For this reason, the use of a GeLi detector coupled to an MCA is required. When carefully used, NAA is an extremely sensitive measuring technique which can detect concentrations down to a single part per billion.

Acknowledgements

The authors are indebted to Sylvia Danielson who has edited and prepared the manuscript and to Hedi Nurk who prepared the illustrations. Dr. G. Allen Holloway, Jr., has assisted by criticizing the manuscript and participating in discussions.

References

1. Bassingthwaighte, J.B.: (1966): Plasma indicator dispersion in arteries of the human leg. Circ. Res., *19*, 332–346.
2. Bassingthwaighte, J.B. (1970): Blood flow and diffusion through mammalian organs. Science, *167*, 1347–1353.
3. Bassingthwaighte, James B. (1977): Physiology and theory of tracer washout techniques for the estimation of myocardial blood flow. Progress in Cardiovascular Diseases *20*, 165–189.
4. Bassingthwaighte, J.B. and Holloway, G.A., Jr. (1976): Estimation of blood flow with radioactive tracers. Sem. Nuc. Med., *6*, 141–161.
5. Bassingthwaighte, J.B., Strandell, T. and Donald, D.E. (1968): Estimation of coronary blood flow by washout of diffusible indicators. Circ. Res., *23*, 259–278.
6. Body Composition in Animals and Man (1968), National Academy of Science, p. 1–521, Washington, D.C.
7. Davson, H. (1970): A Textbook of General Physiology, The Williams and Wilkins Co., vol. 1, p. 1–1030, vol. 2, p. 1031–1694, Baltimore.
8. Edelman, I.S. and Leibman, J. (1959): Anatomy of body water and electrolytes. Am. J. Med., *27*, 256–277.
9. Gaudino, M. and Levitt, M.F. (1949): Inulin space as a measure of extracellular fluid. Am. J. Physiol., *157*, 387–393.
10. Guller, B., Yipintsoi, T., Orvis, A.L., and Bassingthwaighte, J.B. (1975): Myocardial sodium extraction at varied coronary flows in the dog: Estimation of capillary permeability by residue and outflow detection. Circ. Res., *37*, 359–378.
11. Hamilton, W.F., Moore, J.W., Kinsman, J.M., and Spurling, R.G. (1931): Studies on the circulation. IV. Further analysis of the injection method, and of changes in hemodynamics under physiological and pathological conditions. Am. J. Physiol., *99*, 534–551.
12. Høedt-Rasmussen, K., Sveinsdottir, E., and Lassen, N.A. (1966): Regional cerebral blood flow in man determined by intra-arterial injection of radioactive inert gas. Circ. Res., *18*, 237–247.

13. Jacquez, John A. (1972): Compartmental Analysis in Biology and Medicine: Kinetics of Distribution of Tracer-Labeled Materials. Elsevier Publishing Co., p. 1–237, Amsterdam.
14. Kelly, P.J., Yipintsoi, T., and Bassingthwaighte, J.B. (1971): Blood flow in canine tibial diaphysis estimated by iodoantipyrine-^{125}I washout. J. Appl. Physiol., 31, 38–47.
15. Kniseley, Ralph M., Tauxe, W. Newlon, and Anderson, Elizabeth B. (1964): Dynamic Clinical Studies with Radioisotopes. Proceedings of a symposium held at the Oak Ridge Institute of Nuclear Studies, October 21–25, 1963. Atomic Energy Commission Division of Technical Information, p. 1–634, Oak Ridge, U.S.
16. Knopp, T.J., Dobbs, W.A., Greenleaf, J.F. and Bassingthwaighte, J.B. (1976): Transcoronary intravascular transport functions obtained via a stable deconvolution technique. Ann. Biomed. Eng., 4, 49–59.
17. Lawson, H.C. (1962): The volume of the blood – A critical examination of methods for its measurement. In: Handbook of Physiology, Sec. 2, Vol. 1, Circulation, Hamilton, W.F. and Dow, P., eds., American Physiological Society, p. 23–49, Washington, D.C.
18. Riggs, Douglas Shepard (1972): The Mathematical Approach to Physiological Problems: A Critical Primer, p. 1–445. The M.I.T. Press, Cambridge, MA.
19. Zierler, K.L. (1958): A simplified explanation of the theory of indicator-dilution for measurements of fluid flow and volume and other distributive phenomena. Johns Hopkins Hosp. Bull., 103, 199–217.
20. Zierler, K.L. (1965): Equations for measuring blood flow by external monitoring of radioisotopes. Circ. Res., 16, 309–321.

5
IMAGE FORMATION

Harold Wayland

5.1 Introduction

There has been a rapidly increasing sophistication of automated image interpretation, computer image enhancement, and the conversion of "invisible" patterns formed by acoustic waves, x-rays, ultraviolet and infra-red radiation into visible images. In spite of such developments, most medical personnel, including those involved in research, will eventually be dependent on an image which can be seen by the human eye and interpreted by the human brain. The eye itself is a remarkably sensitive optical instrument as long as the light is of such a nature as to be directly detectable. Only recently have electronic sensing systems equalled, and now surpassed, the human eye for sensitivity at low light levels. But photographic film cannot equal the eye for basic sensitivity, although it can record pictorial information at extremely low light levels if the object remains stationary long enough to permit long exposure times.

For a large number of medical applications, pictorial presentation can be made at sufficiently high levels of illumination that the statistical fluctuations associated with the quantum nature of light – its being made up of a finite number of photons – will be of no concern. Initially we shall explore such basic concepts as image formation by lens systems using the ray concept of light, rays of which travel in straight lines. We will then have to introduce the wave nature of light in order to understand the limits of optical resolution – the ability to see and distinguish small objects. In order to explore the mechanisms of the photographic process and of the workings of television cameras, we will then have to introduce the concept of light being made up of discrete photons or packets of energy.

No attempt will be made to discuss the subtleties of human vision. This discussion will be confined to two primary modes of pictorial recording and presentation: photography and closed circuit television. In either case, however, an optical image must be formed, and the imaging system will introduce limitations in the fineness of detail which can be observed or recorded. In a well designed system, the capabilities of the recording system should be matched to the information content of the image to be recorded.

Additional information on the material discussed in this chapter can be obtained from such handbooks as Cornsweet (1), Parrent and Thompson (2), Rose (3), and Smith (4).

5.2 Image formation on the ray theory

A basic understanding of image formation by a lens system can be obtained by considering light as if it travels in straight lines emanating from the object to be imaged. These rays of light will travel at a constant velocity in any homogeneous medium. In a vacuum its velocity is approximately 300,000 km/sec, and in air its velocity is only slightly slower than in a vacuum. In water, however, its velocity will be reduced to about 225,000 km/sec while in various glasses, depending on their composition, it will range from about 200,000 km/sec to about 167,000 km/sec. In a vacuum the velocity of light is completely independent of its wavelength or color, but in all other media it will depend on the actual wavelength. This will be considered in more detail when we discuss errors or aberrations in image formation.

Ordinarily we do not concern ourselves directly with the velocity of light in the glass, quartz or fluorite from which lenses are made, but with what is called the index of refraction of the material, which is given by the ratio of the velocity of light in a vacuum to that in the medium being considered. Since the velocity of light can never exceed that of its velocity in a vacuum—a basic tenet of Einstein's theory of relativity which has withstood all attempts to demonstrate otherwise – the index of refraction of a transparent medium will always be unity or greater.

The concept of index of refraction actually developed before its relationship to the velocity of light was understood. We will use this empirical concept in our initial approach to image formation by a lens. Suppose a region of index of refraction n_1 is separated from a region of index of refracton n_2 by a plane surface S (Fig. 5.1). A ray of light making an angle i to a line perpendicular to the boundary (called a "normal" to the surface) will be bent on passing the boundary so that it travels in the second medium in a different direction to the normal (at an angle r) such that

$$\frac{\sin i}{\sin r} = \frac{n_2}{n_1}$$

(5.1a)

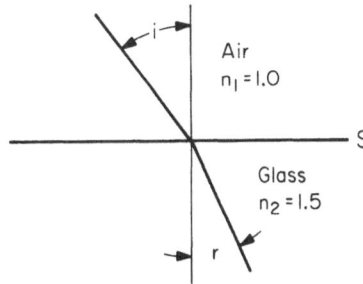

Fig. 5.1   Refraction of light.

or

$$\sin r = \frac{n_1}{n_2} \sin i \tag{5.1b}$$

If, now, $n_1 = 1.00$ (air), $n_2 = 1.50$ (glass) and the angle of incidence i is 30°
so that $\sin i = 0.50$, we have

$$\sin r = \left(\frac{1}{1.5}\right)(0.5) = 0.333$$

or

$$r = 19°27'$$

A general rule to keep in mind is that a ray of light passing from a
medium of lower index of refraction to one of higher index of refraction
will be bent toward the normal. If the ray of light were reversed in Fig. 5.1
the paths would remain the same, and the ray would be bent away from the
normal when it emerges from the glass into the air.

Equation 5.1a is known as Snell's law and is the basis for elementary
calculations of the properties of lenses. A piece of glass with appropriately
curved surfaces – i.e. a lens – can be used to form an optical image, as
shown schematically in Fig. 5.2. The ray OBB'I will be undeviated since it
strikes both the front and back surfaces of the lens along the normal,
while the ray OA will be deviated along AA' by the first surface, and then
along A'I as it reemerges into air on the back side of the lens. If the two
surfaces are appropriately curved, other rays will also be bent in such a

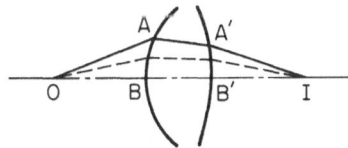

Fig. 5.2   Refraction by a spherical lens.

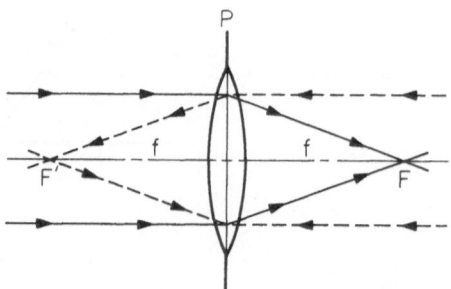

Fig. 5.3 Focal points of a "perfect" thin lens.

way as to intersect at I, as indicated by the dotted line, forming a point image at I of a point source at O.

Fortunately, thin lenses with both surfaces parts of spheres are capable of forming tolerably good images if the spherical curvatures are not too great and the angles which either the incident or emerging rays make with the axis of the lens are both small. No attempt will be made in this chapter to get involved in the intricacies of lens design: manufacturers of both microscopic and photographic lenses can furnish the relevant data for various applications. It is the purpose of this chapter to discuss basic principles in order to lay a foundation for permitting a reasonable choice of the equipment needed for various applications, as well as to assist the user in effecting the necessary compromises among conflicting requirements.

If a parallel bundle of light rays strikes a "perfect" lens parallel to the axis of the lens, these rays will be brought to focus in a point F (Fig. 5.3) at a distance f from a plane, called the principal plane of the lens. In this simplified approach, a parallel bundle of rays in the opposite direction will be focused at a point F' the same distance f from the principal plane, but on the opposite side, as indicated by the dotted lines, provided the medium on both sides of the lens is the same. Because of the principle of optical reversibility, a ray passing through F' from left to right and striking the lens will emerge from the lens as a ray parallel to the lens axis.

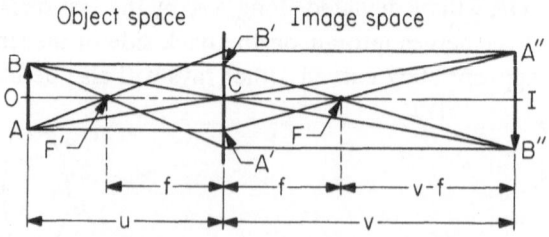

Fig. 5.4 Ray construction of an image for a thin lens.

These concepts can be used to construct the position and size of an image formed by such a perfect lens. Referring to Fig. 5.4, a ray from point B in the object space which is parallel to the lens axis and strikes the principal plane of the lens at B′ will be refracted through the focal point F in the image space. A ray from B which passes through the center of the lens C, will be undeflected. These rays will intersect at B″, where an image of B will be formed. Another ray path which can be easily constructed is one passing through the focal point F′ in the object space. The ray BF′ will emerge from the lens as a ray parallel to the lens axis in the image space. A similar construction can be applied to any points on the object AB. In Fig. 5.4 the image of the tail of the arrow, A, is formed at A″. If we compare similar triangles ABC and A″B″C we see that the image in this case is magnified such that

$$M = \frac{A''B''}{AB} = \frac{v}{u} \tag{5.2}$$

Since A′B′ has the same size as AB in Fig. 5.4, comparing the similar triangles A′B′F and A″B″F we obtain

$$\frac{A''B''}{A'B'} = \frac{v - f}{f} = \frac{v}{u} = \frac{A''B''}{AB} \tag{5.3}$$

from which, by algebraic manipulation, we obtain the basic lens formula

$$\frac{1}{u} + \frac{1}{v} = \frac{1}{f} \tag{5.4}$$

These same basic formulas can be adapted to thick lens systems if the measurements for the image and object distances are made from appropriate principal planes, which will be spatially separated, whereas in the simple lens approximation the principal planes are coincident. For the "perfect" complex lens, with the object and image both in the same medium the construction shown in Fig. 5.5 can be used. The ray BB′, parallel to the optic axis, can be considered as emerging from the principal plane P′ in such a direction that it passes through the emergent focal point F. The ray BF′ passing through the conjugate focal point F′ strikes the principal plane P at D and emerges from P′ at the point D′ parallel to the optic axis. The intersection of this ray with B″F locates the image of B at B‴. The ray BC

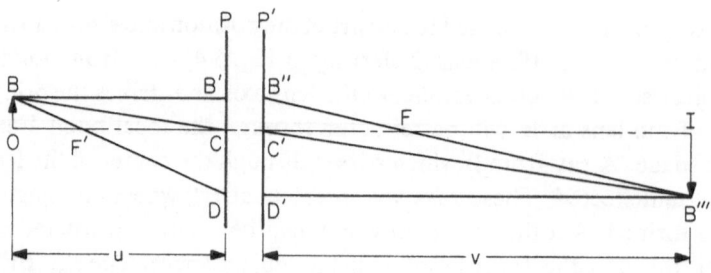

Fig. 5.5   Ray construction of an image for a thick lens.

emerges from the principal plane P′ at its center C′, so that C′B‴ is parallel to BC.

Such perfect lenses as we have considered would image a point precisely into a point. A finite object will be imaged into an object such that the linear dimensions are in the ratio of the magnification v/u. (Note that this could be a diminution if u > v.) We see from these constructions, that as the object approaches F′ from the left (so that u > f) the image recedes further and further to the right, and becomes increasingly larger. Superficially it appears that, since we can get an indefinite amount of magnification from even a simple lens, it would hardly seem worth while to construct complicated lenses.

Unfortunately, magnification alone can be extremely deceptive. There is an unfortunate tendency in the medical and biological literature to refer to the magnification of an optical system with apparent disregard for the effectiveness of this magnification – that is for the ability of the system to resolve and record the details. It is important to distinguish between "useful magnification" and "empty magnification". A simple way to become aware of this distinction is to examine a picture which has been reproduced through a half-tone screen. In "high fidelity" printing the pattern of dots is sufficiently fine that it cannot be distinguished by the unaided eye (Fig. 5.6a). There appears to be detail in this picture which would be brought out by further magnification. An additional two-fold magnification (Fig 5.6b) brings out some additional detail, although the structure introduced by the half-tone screen can now be distinguished with the unaided eye. Magnification by an additional factor of two (Fig. 5.7a) gives a less pleasing picture, and does not increase our ability to extract information from the picture. What appeared to be sharp edges in the previous pictures now take on a fuzziness which we cannot certainly identify as being associated with the actual structure of walls of the cells or nucleus. A still further two-fold magnification (Fig. 5.7b) confirms that much of the observed indistinctness of the cell boundary is associated with the nature of the reproduction process. It would be impossible to

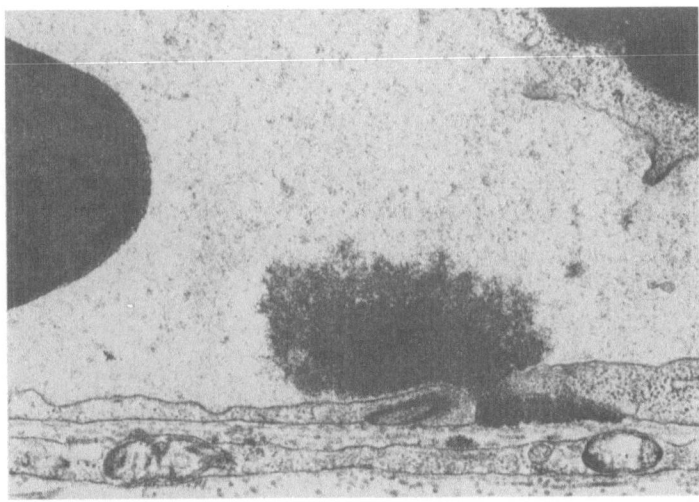

a

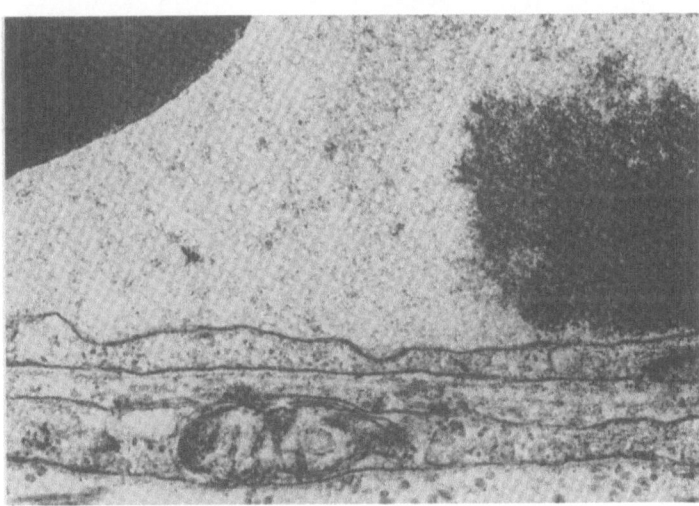

b

Fig. 5.6 (a) Electron micrograph of capillary wall from screened nagative.
(b) Same, enlarged 2 ×.

determine the width of the cell wall to a greater accuracy than about half
the width of one of the dots of which the pictorial representation is com-
posed. In fact, Fig. 5.7b has been magnified beyond the point of diminishing
returns –it is actually less useful than a lower magnification.

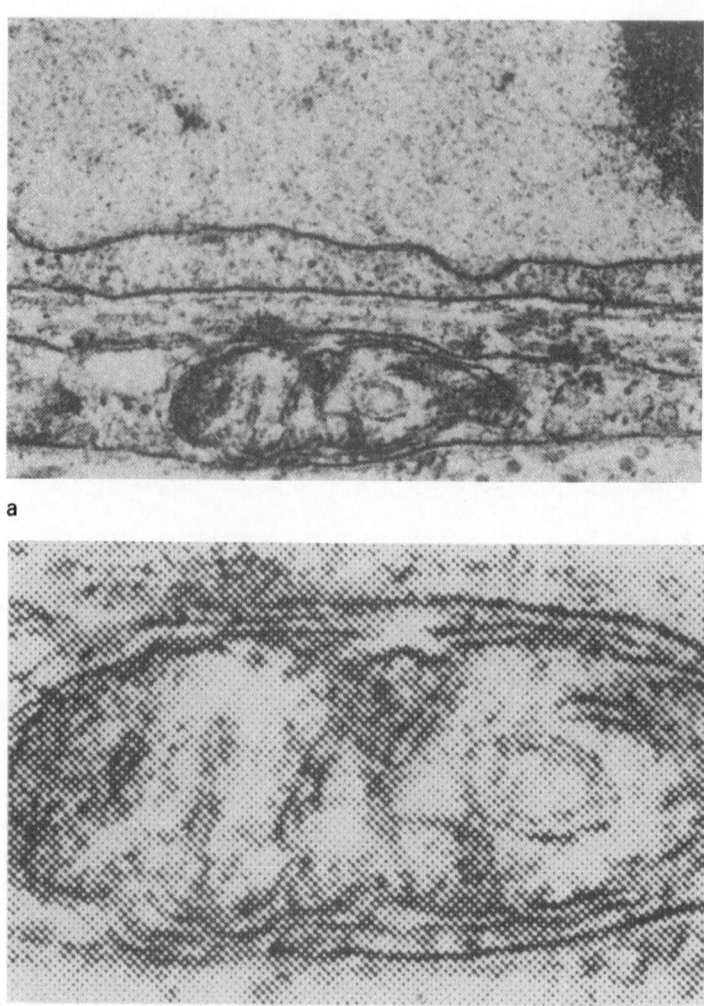

a

b

Fig. 5.7   (*a*) Same as 5.6(*a*), enlarged 4×. (*b*) Same as 5.6(*a*) enlarged 8×.

5.2.1 *Light gathering power of a lens*

With the same lens system extra magnification introduces another potentially serious complication: the intensity of the image diminishes as the
square of the increased linear magnification, so that Fig. 5.7b, with an
8-fold linear magnification over Fig. 5.6a required approximately 64 times
the exposure to make a satisfactory photographic print from the same
negative.

For an object at a fixed distance from a lens system, the amount of light collected by the lens from any point on the object will depend on the effective opening or aperture of the lens. Actually it depends on the solid angle contained in a cone with its apex at the point on the object and its base being the effective opening of the lens. Since the object distance will vary from application to application, a simple convention has been adopted to designate the "speed" of camera lenses – the f-number of the lens. When we say we have an f/2.0 camera lens this means that the ratio of the focal length of the lens to the maximum usable aperture of the lens is 2.0; hence if the focal length of the lens is 50 mm, a common value for 35 mm cameras, the maximum opening of the lens will have a diameter of 50/2.0 = 25 mm. In the case of camera lenses, this aperture is usually variable, so that the light gathering capability of the lens can be modified by varying the aperture. Other properties of the optical system will also change with varying the f-stop. The way in which this affects resolution and depth of field will be discussed in other sections of this chapter.

For microscope objectives a different relationship between focal length and aperture is usually used, a quantity called the numerical aperture, and usually abbreviated as NA. Since microscope objectives are often used with the objects in media other than air – oil immersion and water immersion objectives being readily available – the index of refraction of the medium in which the object and the front element of the objective are immersed comes in the definition. Referring to Fig. 5.8a, if the effective lens opening has a diameter $D = 2a$ and the lens has a focal length f in a medium of index of refraction n, the numerical aperture is defined as

$$NA = n \sin \theta = n \frac{a}{(f^2 + a^2)^{1/2}} = \frac{na}{f} \frac{1}{\left[1 + \left(\frac{a}{f}\right)^2\right]^{1/2}} \qquad (5.5)$$

Using this formula for an f/2.0 camera lens operating in air (n = 1) we find f/2a = 2.0 or a/f = 1/4, hence the NA = 0.243. As a first approximation the numerical aperture of a camera lens operating in air is given by the reciprocal of twice the f-number. To this approximation, the f/2.0 lens would have an NA = 0.25.

5.2.2 Depth of focus and depth of field

The concepts of the depth of field and depth of focus of a lens system are an outgrowth of an acceptance of a finite size of blur in an image formed by an optical system. Such a blurred circle is sometimes called the circle of

124

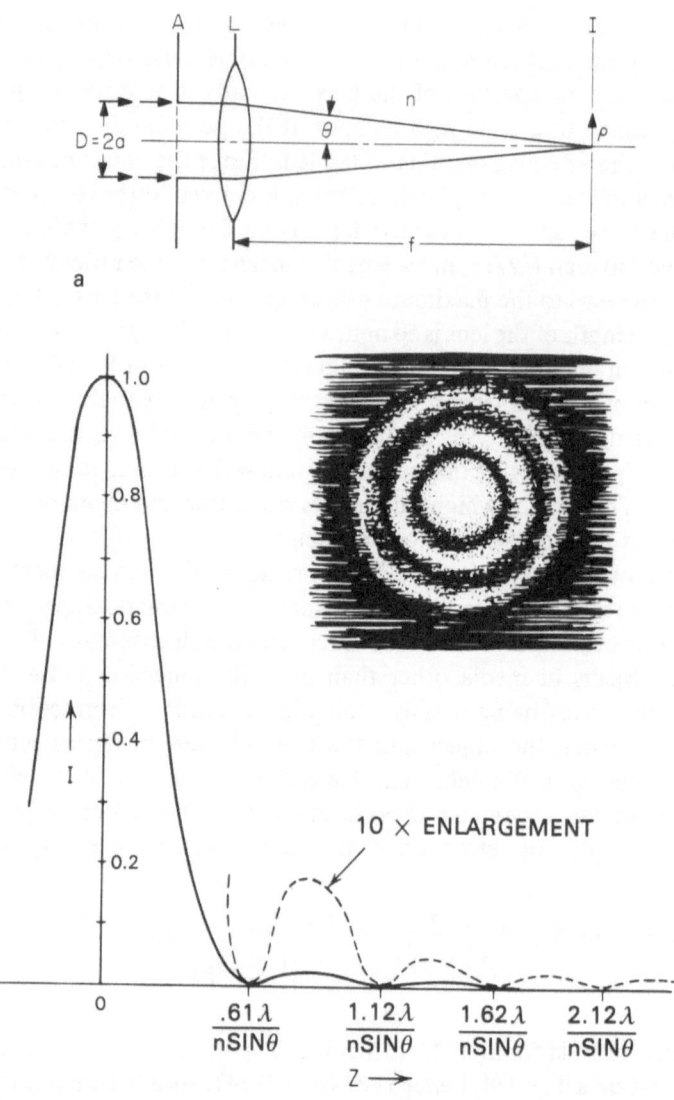

Fig. 5.8 (*a*) Parallel light striking lens of aperture A. (*b*) Diffraction pattern and Airy Disc. I = intensity; Z = the distance from the centre of the Airy Disc. From Modern Optical Engineering by Warren J. Smith. Copywright 1966 by McGraw-Hill, Inc. Used with permission of McGraw-Hill Book Company.

confusion. The two terms "depth of focus" and "depth of field" are often confused. The accepted definitions refer to the acceptable shift of the image plane (the film plane in a camera) as the depth of focus, since this is in the region in which the image is focused. The acceptable shift in longitudinal position in the object plane – in the field under observation – is called the depth of field.

Consider a perfect lens, which would focus a point source at O (Fig. 5.9a) into a point image at I, both points being on the axis of the lens. If we are willing to accept a certain amount of blur in the image – a practical necessity in real life, since perfect lenses and perfect recording systems do not exist – the longitudinal distance through which we can move the plane of focus on either side of the point of sharp focus and not exceed the acceptable blur pattern is called the depth of focus δ of the lens system. The acceptable diameter of the blur pattern or circle of confusion is, of course, arbitrary, and will depend on the properties of the eye for visual observation and on the character of the photographic emulsion for photographic recording. We see from Fig. 5.9a that the permissible shift is symmetric about the plane of sharpest focus. If the aperture of the lens is reduced, as shown by the dotted lines in Fig. 5.9a, the depth of focus is increased inversely with the decrease in diameter of the aperture. This is by no means all gain, however, since reducing the diameter of the aperture by a factor of 2 decreases the light gathering power by a factor of 4 and, as we shall see when we apply the wave theory of light to the problem of resolving power, we also introduce a significant loss in resolution.

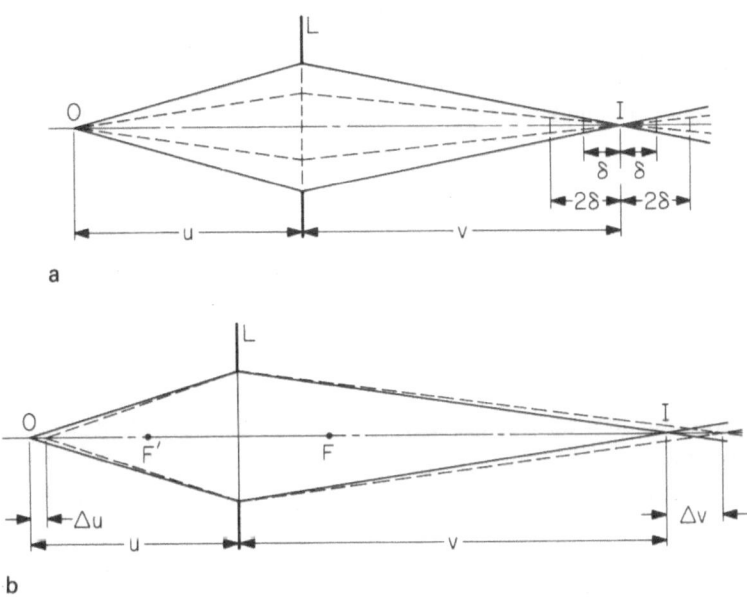

Fig. 5.9 (a) Depth of focus. (b) Depth of field.

To explore the concept of depth of field, let us examine Fig. 5.9 b. Assume that we keep the plane of observation fixed at a distance v from the principal plane of the lens which is the conjugate focal point to an object at a distance u from the lens in the object space. If the object is moved a distance Δ u towards the lens, the image will move away from the image plane by an amount Δ v which is approximately

$$\Delta v = - \left(\frac{v}{u}\right)^2 \Delta u = - M^2 \Delta u \qquad (5.6)$$

where M is the lateral magnification of the system (equation 5.2). (The negative sign means that if the object approaches the lens its image recedes from the lens.) We thus see that the "longitudinal magnification" is approximately the square of the lateral magnification. For a fixed lens aperture and a positive magnification we see that the acceptable depth of field will be less than the acceptable depth of focus – in fact, for high magnification microscope systems the depth of field is extremely small if we use a system with a large NA to obtain sufficient resolution to justify the large magnification.

The effect of shifting the object away from the lens is not completely symmetric with shifting it towards the lens. The interested reader can readily show by simple geometric construction that, for a given size of circle of confusion in the image plane, the object can be moved a greater distance away from the lens than towards the lens. This lack of symmetry is seldom of practical importance, particularly since there are many other aberrations in image formation which must be considered in the real situation.

5.3 Optical resolution

The limitations in ability to resolve detail in the half-tone reproduction illustrated in Figs. 5.6 and 5.7 were inherent in the particular method of reproduction. Similar limitations will arise from the grain structure of photographic film and in the structure of the light-sensitive surfaces of television cameras as well as in the manner of forming the image on a television monitor by sweeping a phosphorescent screen with an electron beam. In addition to these technological limitations there are even more fundamental limitations to the ability to resolve fine detail with an optical system: the finite wavelength of light and the finite size of the lenses used to form the image. These limitations would exist even if we could design and construct a theoretically perfect lens. The nature of these limitations will be discussed before the other aberrations associated with practical lens design and construction are summarized.

The English astronomer G. B. Airy first calculated the type of image of a luminous point source which would be formed by a perfect lens with a circular aperture. If monochromatic light is used, it consists of a central bright region of finite radius (often called the Airy disc) surrounded by dim circular rings separated by dark circular bands. If a plane wave of light of wavelength λ strikes the aperture A of diameter D placed in front of a lens L of focal length f (Fig. 5.8a), and the space between the lens and the image plane I is filled with a substance of index of refraction n, the intensity of the image will vary with radial distance from the focal point by a complicated mathematical expression, which is represented graphically in Fig. 5.8b. The actual values of the radii of the bright and dark rings and the peak intensity in each ring are given in Table 5.1. The column labeled ρ_E represents the data for the Airy rings for the human eye assuming a pupil diameter of 4mm, a 20 mm distance from the lens aperture to the retina, the index of refraction of the vitreous humor n = 1.33 and green light of wavelength λ = 550 nm. For this case the Airy disc which would be formed on the retina of the eye would have a radius of 2.54 μm or a diameter of 5.08 μm. Since the peak intensity of the second bright ring is less than 2%

Table 5.1   Airy rings for a circular aperture.

Ring	ρ	Peak intensity (relative)	Energy in ring	ρ_E
Central maximum	0	1.0	83.9%	0
1st dark ring	$\dfrac{0.61\lambda}{n\sin\theta}$	0.0		2.54 μm
1st bright ring	$\dfrac{0.81\lambda}{n\sin\theta}$	0.0171	7.1%	3.41 μm
2nd dark ring	$\dfrac{1.16\lambda}{n\sin\theta}$	0.0		4.82 μm
2nd bright ring	$\dfrac{1.33\lambda}{n\sin\theta}$	0.0041	2.8%	5.54 μm
3rd dark ring	$\dfrac{1.62\lambda}{n\sin\theta}$	0.0		6.73 μm
3rd bright ring	$\dfrac{1.85\lambda}{n\sin\theta}$	0.0016	1.5%	7.69 μm
4th dark ring	$\dfrac{2.12\lambda}{n\sin\theta}$	0.0		8.81 μm

that of the central disc, it will seldom be observed without taking special precautions.

If two points in the object being observed are sufficiently close together that their Airy discs overlap we are faced with the problem of establishing a useful criterion as to when they can be recognized as two distinct points. If the separation of the centers of images of two adjacent points is

$$\delta = \frac{0.61\lambda}{n\sin\theta} = \frac{0.61\lambda}{\text{N.A.}} \qquad (5.7)$$

(since, by equation 5.5, NA = n sin θ) then the peak intensity of the image of one point falls on an intensity minimum for the other point, and it is possible to recognize that two points are present (Fig. 5.10). This criterion is known as Rayleigh's criterion for the limiting resolution of an optical system. In a case in which we have assurance from other information that we are dealing with two adjacent point sources

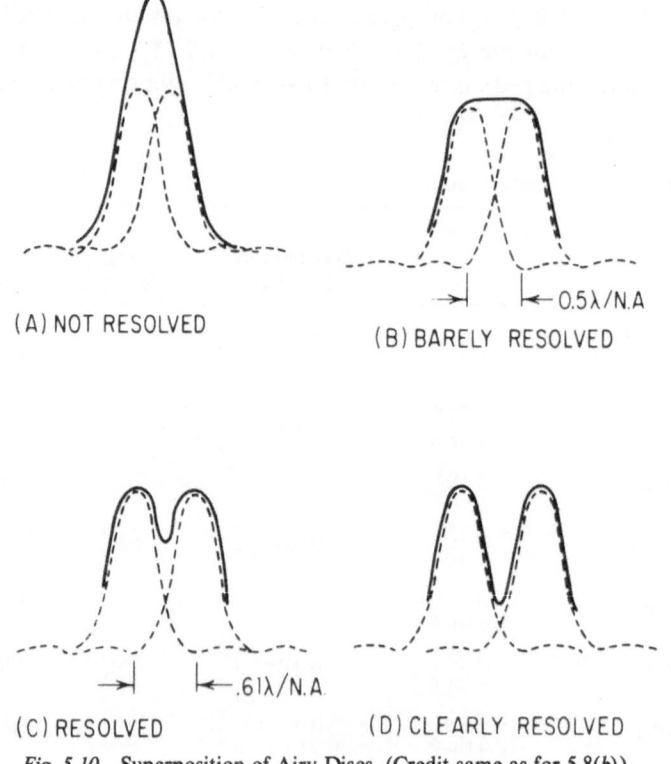

(A) NOT RESOLVED   (B) BARELY RESOLVED

0.5λ/N.A

(C) RESOLVED   (D) CLEARLY RESOLVED

.61λ/N.A.

Fig. 5.10 Superposition of Airy Discs. (Credit same as for 5.8(*b*)).

of light, as in the case of two stars, the actual shape of the diffraction pattern permits the recognition of the existence of two distinct sources even if they are too close together to meet the Rayleigh criterion. In most biological systems we have no such assurance – in fact, except in the case of phosphorescent or fluorescent bodies we seldom deal with self-luminous objects in biological applications. A careful mapping of the diffraction pattern due to fluorescent antibodies on neighboring antigenic sites could be used to ascertain separations less than required by the usual resolution criteria, but only in the case that other fluorescent loci are far enough away not to interfere. Whenever it is possible to examine the structure of the diffraction patterns some additional information can often be extracted from the image. Suppose, for example, an isolated point source were to move a small distance in the object plane so that the image has moved less than the distance normally required to resolve two adjacent point sources. With adequate recording equipment we could measure the displacement of the Airy disc, and deduce a movement less than the "resolving power" of the system. This concept has been used to observe small motions of cell walls, such as in the study of contraction of smooth muscle cells. The change in width of the cell may be correctly measured to a fraction of the nominal resolving power of the optical system. Unfortunately, the absolute width of the cell cannot be measured with an accuracy better than approximately the effective resolution of the system, which may be several times the incremental accuracy in measuring displacements.

5.4 Aberrations of spherical lenses and mirrors

Since spherical surfaces are relatively easy to produce, most lenses are made up of coaxial spherical surfaces. Such a lens which is thicker at the center than at the sides is called a positive lens, since it will bring a parallel beam entering the lens from one side to a focus on the opposite side. It may be biconvex, with the same or different curvatures on the two sides (Fig. 5.11a); plano convex (a plane can be considered as a part of a sphere of infinite radius) (Fig. 5.11b); or a concavo-convex – also called a meniscus lens (Fig. 5.11c). If the lens is thinner at the center than at the edges it will cause a parallel bundle of rays to diverge as if coming from a focal point on the side of the lens on which the beam strikes. Such a lens is called a negative lens, and can be either plano-concave (Fig. 5.11d), biconcave (Fig. 5.11e), or even concavo-convex (Fig. 5.11f).

Spherical lenses and mirrors introduce a variety of aberrations or faults

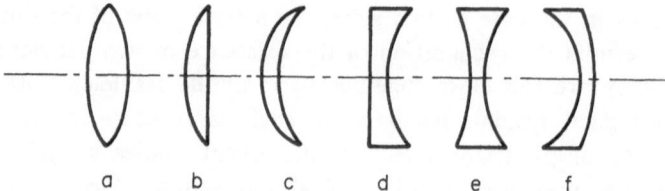

Fig. 5.11 Types of simple lenses.

due to the shapes of the surfaces employed, the relative positions and sizes of stops or apertures, and the position of the object. Both spherical lens and mirror systems have five such aberrations: spherical aberrations; astigmatism; coma; curvature of the image field; and distortion. Since the index of refraction of glass and quartz varies with wavelength, lens systems are also subject to chromatic aberration when used in any but monochromatic light (Table 5.2).

5.4.1 *Spherical aberration*

Spherical aberration is related to the fact that rays striking a lens close to the optical axis are brought to focus at a different location than those striking it at a considerable distance from the axis, hence it is particularly noticeable with lenses of large aperture compared to focal length. Referring to Fig. 5.12, a narrow beam of central rays parallel to the axis will be brought to focus at what we have hitherto called the focal point F of a "perfect" lens. This point is called the "Gaussian focus" of the lens. An annular bundle of rays A, also parallel to the optic axis, but striking the lens near its rim will be brought to focus at the point L, the limiting focus for this lens, which is closer to the lens than the point F. Other annular rings of light will be brought to focus between the focus of the limiting rays

Table 5.2 Variation of index of refraction with wavelength for various glasses and fused quartz.

Color	Wavelength in nanometers	Index of Refraction			
		Light crown	Light flint	Heavy flint	Quartz
	400	1.5238	1.5932	1.8059	1.4699
Blue	450	1.5180	1.5853	1.7843	1.4635
	500	1.5139	1.5796	1.7706	1.4624
Green	550	1.5108	1.5757	1.7611	1.4599
	600	1.5085	1.5728	1.7539	1.4581
Red	650	1.5067	1.5703	1.7485	1.4566
	700	1.5051	1.5684	1.7435	1.4553

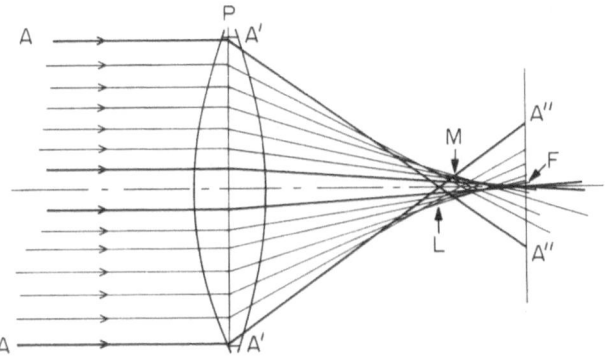

Fig. 5.12 Spherical aberration.

and the Gaussian focus. The envelope of all the rays as they approach the axis comes to a point at the Gaussian image point. The diameter of the cross section of the entire bundle of rays will be a minimum in the plane determined by the intersection of the limiting rays with this envelope. This area is the least circle of aberration. Its position and size will depend on the aperture of the lens. It can be reduced by stopping down the lens, but at the cost of loss of illumination. Combinations of lenses can be used to reduce spherical aberration, although this may introduce other aberrations in the image.

5.4.2 *Astigmatism*

Astigmatism is generally associated with an asymmetry in the curvature of the lens of the eye, so that lines parallel to one principal direction of curvature will be brought to focus in a different plane from lines parallel to the other principal direction of curvature. A spherical lens also displays a similar phenomenon for objects which are not on the central axis. A point object lying in a vertical plane of symmetry of the lens, but below the optical axis (Fig. 5.13) will not be imaged into a point by a spherical lens. The tangential fan of rays in Fig. 5.13, which lie in the plane containing the object point and the optic axis, will be brought to focus in a line perpendicular to this plane, indicated as the tangential image. The location of this image is variously known as the tangential, meridional or primary focus. The sagittal fan, which consists of rays in a plane at right angles to the tangential fan, will be brought to focus in a vertical line further from the lens for a biconvex lens. This position is known as the secondary or sagittal focus. The envelope of rays – the astigmatic bundle – will vary in shape from an ellipse with its major axis horizontal on either side of the

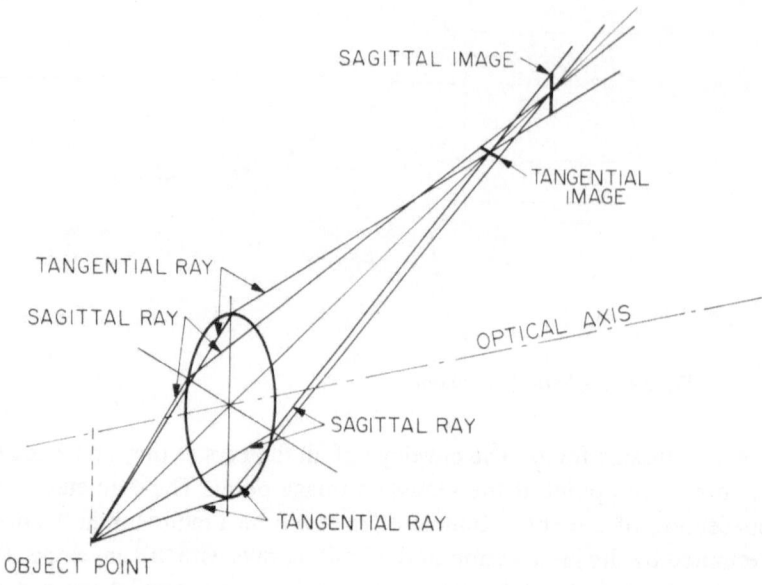

Fig. 5.13 Astigmatism.

tangential image to an ellipse with its major axis vertical on either side of the sagittal image. The closest approach to a point image will lie between the tangential and sagittal images where the envelope of the astigmatic bundle is circular.

5.4.3 *Coma*

For many optical systems a point object off the axis will be imaged into a comet shaped patch (Fig. 5.14). Such an aberration is called "coma" and it is more serious for precise location of position than a circular patch since it is difficult to estimate the position of the center of gravity of the image along its long axis. It arises from the fact that for an off-axis object the effective magnification differs with different annular zones of the lens. If the tail is toward the optic axis we speak of positive coma, and if the point is away from the axis the coma is said to be negative. Coma is often quite noticeable at the edges of the field in astronomical photographs taken with large reflecting telescopes.

5.4.4 *Curvature of field*

For a point object on or near the axis, there will be a point representation in an image plane passing through the conjugate focal point. The image of

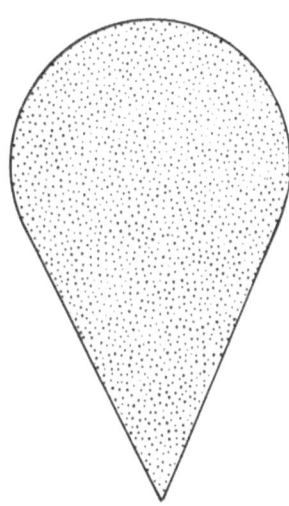

Fig. 5.14   Coma.

points lying in a plane which are further off the axis will, however, be blurred. The surface of minimum blur will not, in general, be a plane but a surface of revolution centered on the optic axis (Fig. 5.15). This phenomenon is known as curvature of field. For photography, since we normally wish to use a photographic film which is flat, it is important to eliminate field curvature, and lens designers make considerable effort to achieve a flat field for photographic systems.

When the image is formed on a fiber optic surface, this surface can be curved to correspond to the curvature of field. This has been effectively used in image intensifiers in which a curved fiber optic surface covered with a phosphor is bombarded by electrons which have been focused by an electrostatic lens, the design of which can be greatly simplified if a curved focal surface can be accepted.

5.4.5 *Distortion*

For object points off the axis, the image may be formed either further from the axis or closer to it than given by the elementary formula for magnification (equation 5.2). If the object consists of a rectangular grid (Fig. 5.16a) a perfect lens would form an image which was also a rectangular grid. If the magnification increases as we go off the axis, the resulting image shows what is called "pin-cushion" distortion (Fig. 5.16b). A decrease in magnification with distance from the axis results in "barrel" distortion (Fig. 5.16c).

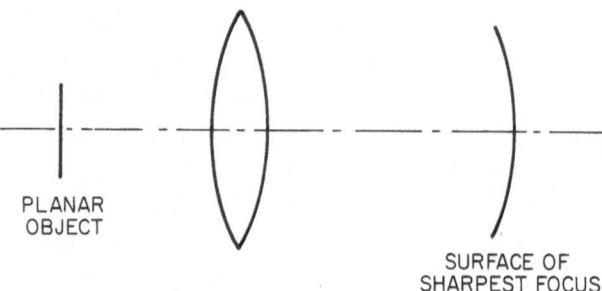

Fig. 5.15   Curvature of field.

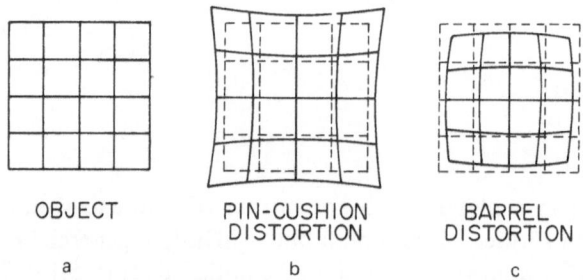

| OBJECT | PIN-CUSHION DISTORTION | BARREL DISTORTION |
| a | b | c |

Fig. 5.16   Distortion.

5.4.6 *Chromatic aberration*

Except for a vacuum (and, to a good approximation, air), the velocity of
light varies with its wavelength, and hence the index of refraction also
varies with the wavelength, as can be seen by an examination of Table 5.2.
Since the index of refraction is greater for the shorter wavelengths, they
will be more strongly refracted than the longer wavelengths, so that a
simple positive lens will bring blue light to focus closer to the lens than the
focal point for red light (Fig. 5.17a) – a phenomenon called longitudinal
chromatic aberration. Even if this aberration has been corrected (which
will require a compound lens made up of elements of different indices of
refraction and different dispersions, or relative change of index of refraction with wavelength), the magnification may vary with color, leading to
what is called lateral chromatic aberration (Fig. 5.17b). This is seen as a
vari-colored fringe at the edge of objects viewed in white light.

Chromatic aberrations do not occur in optical systems made up entirely
of mirrors operating in air.

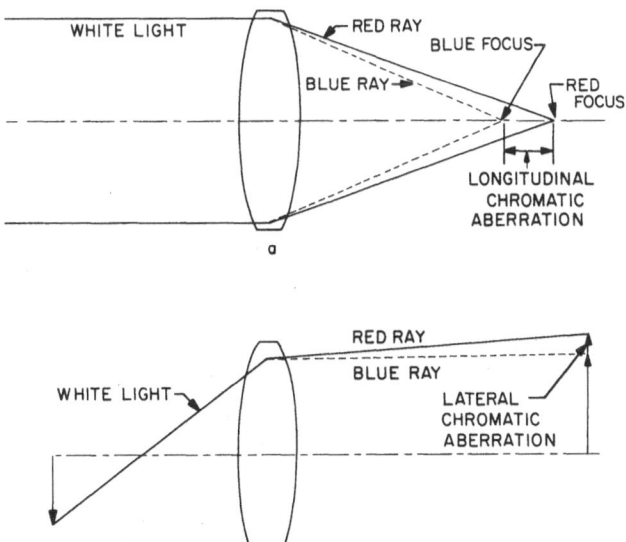

Fig. 5.17 Chromatic aberration.

5.4.7 *Correction of aberrations*

Good optical design seeks the best compromise between resolving power, light gathering power, and magnification commensurate with each particular application – including attention to cost. By carefully chosen combinations of lens elements made from glasses of various indices of refraction, appropriate spacing between elements, and positioning of limiting apertures, compound lenses can be designed which minimize the aberrations for a given application or range of applications. The use of high speed computers has greatly increased the ease and flexibility with which complex lenses can be designed, even allowing modifications to be made to account for slight changes in the optical properties from one batch of optical glass to another. Although the quality of lenses for many applications has improved significantly in the past few years, the demands of mass production and mass marketing have, in many cases, severely limited the diversity of lenses available – for some specialized applications, microscope objectives made many years ago will prove to be more suitable than those currently available.

Much excellent information is available from lens manufacturers and any attempt to evaluate those currently available would soon be out of date. An extremely useful characterization of a lens system is given by the modulation transfer function (MTF) which will be discussed in the next section.

5.5 Resolution of lenses: the modulation transfer function

Practically we need both a means of testing the overall performance of a lens or system of lenses and some convenient form in which to express the quality of the lens. A common form of target to test the performance of a photographic objective consists of an array of alternating light and dark bands of equal width, sets of patterns with different spacings being provided to permit determination of the closeness of spacing which can be detected. Such a test chart is reproduced in Fig. 5.18. For each spacing a set of horizontal and vertical bars is provided to permit checking the lens for astigmatism. The number associated with each group is the reciprocal of the distance in millimeters encompassed by one dark and one bright strip, i.e. the number of cycles per millimeter. In the finest pattern in Fig. 5.18 there are 5 such cycles per millimeter. In this case we speak of there being 5 line pairs or five "optical" lines per millimeter. In television parlance it

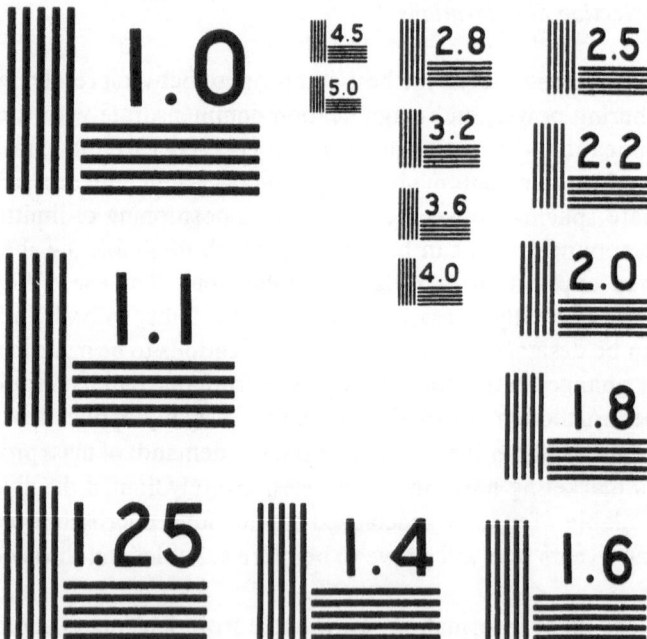

Fig. 5.18 Optical Test Chart. The numbers refer to the spacing of line pairs, in millimeters, on the original chart.

is common to speak of each light or dark line separately, so that five line pairs or optical lines would correspond to ten television lines. We will normally speak of cycles, since this will be more convenient when we deal with other than simple bar patterns. For the bar patterns of Fig. 5.18, a cycle consists of one dark and one light bar.

Since even the black bars reflect some light and the white bars are not perfect reflectors, the light intensity variation across such a bar pattern will ideally be represented by the square wave pattern of Fig. 5.19. The modulation or contrast of this object is defined as the ratio of the change above and below the average intensity to the average intensity. Referring to Fig. 5.19 we see that

$$\text{Modulation (Contrast)} = \frac{\dfrac{I_{max} - I_{min}}{2}}{\dfrac{I_{max} + I_{min}}{2}} = \frac{I_{max} - I_{min}}{I_{max} + I_{min}} \tag{5.8}$$

When such a bar pattern is imaged by a lens, the boundaries between the light and dark regions are blurred. Even for a perfect lens they would be blurred by diffraction, and for real lenses they will be further blurred due to lens aberrations. As the bars get closer and closer together the blur due to the bright areas encroaches further and further into the dark areas, so that the contrast of the pattern becomes less and less as the lines get

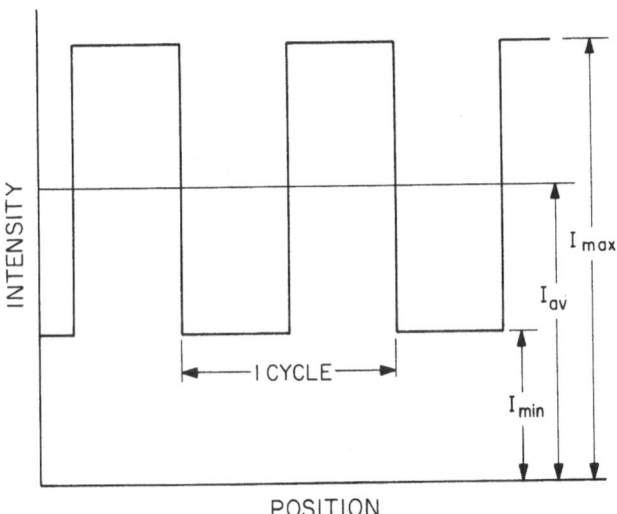

Fig. 5.19 Modulation of a square wave.

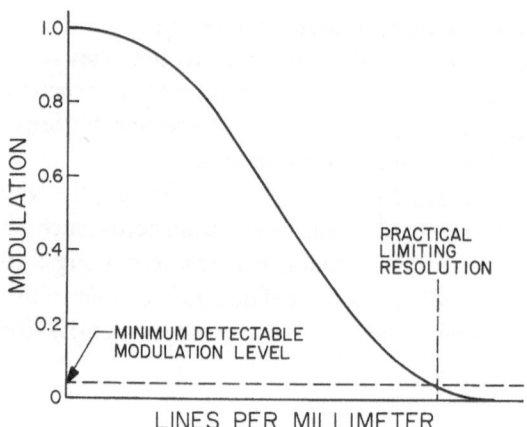

Fig. 5.20   Modulation vs resolution.

closer and closer together until the modulation or contrast actually goes to zero. Obviously, this will depend on the initial modulation of the image. A hypothetical curve, assuming unity modulation for low frequencies (i.e. assuming I_{min} for the target is zero) is shown in Fig. 5.20. The effective resolving power will be less than the number of lines per millimeter at which the modulation precisely vanishes, depending on the means of detection. For the human eye, a practical limit is reached when the modulation drops below 0.04, although this will vary somewhat with the individual observer.

In the case of the square wave pattern just discussed, the form of the illumination distribution pattern loses its sharp edges by being imaged by a lens, so that the pattern of distribution in the image, although periodic, is no longer a square wave. If the object pattern has a distribution in the form of a sine wave, the image formed by a lens system will also be in the form of a sine wave of the same frequency, but a different modulation, regardless of the shape of the "spread function" or energy distribution in the image of a point or line source formed by the lens. This has led to the introduction of an extremely useful tool for analyzing the behavior of optical systems, the (Optical) Modulation Transfer Function. We will find this concept of the Modulation Transfer Function particularly useful when we come to consider photographic and television recording, but we will introduce it in connection with lens behavior.

The Modulation Transfer Function (MTF) is defined as the ratio of modulation in the image (m_i) to that in an object (m_0) with a sine wave modulation as a function of spatial frequency (cycles per unit of length) of the sine wave pattern (Fig. 5.21). If v is the frequency in cycles per millimeter

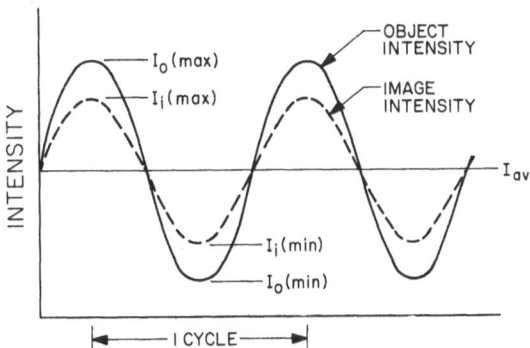

Fig. 5.21 Degradation of sine wave modulation by a lens.

$$\mathrm{MTF}(v) = \frac{m_i}{m_0} \qquad\qquad\qquad (5.9)$$

Such a curve can be obtained empirically by measuring I_0(max) and I_0(min) for a sinusoidal grating with a photometer and the corresponding quantities in the image.

The MTF can be calculated theoretically for an aberration free system in which only diffraction effects limit the resolution. Rayleigh's criterion for resolving power of a lens was that the objects had to be separated by a distance $\delta = 0.61\lambda/\mathrm{NA}$, for which $v_0 = 1/\delta = \mathrm{NA}/0.61\lambda$. Another criterion which is frequently used makes δ slightly smaller, so that

$$v_0 = \frac{2\mathrm{NA}}{\lambda} = \frac{1}{\lambda(\mathrm{f\ number})} \qquad\qquad (5.10)$$

which makes it easier to compare the resolution of microscope objectives, for which the NA is usually given, with that of photographic lenses, for which the f-number is commonly available. Since the numerical aperture and f-number are both dimensionless, v_0 will have its dimensional characteristics determined by the units in which the wavelength, λ, is expressed. For MTF graphs, we usually want v_0 in cycles per millimeter, so that λ must be expressed in millimeters (Green light, which has a wavelength of about 550 nm will have a wavelength of 5.50×10^{-4} mm). If we accept the limiting resolution of a lens which is limited by diffraction, the theoretical curve is given in Fig. 5.22. From this the diffraction limited MTF curve can be constructed for any prescribed NA or f-number.

As an example of the use of such a curve, let us consider the image formed of a square wave grating by an f/5 (NA = 0.1) lens which is only diffraction limited, having no other aberrations. Any function which re-

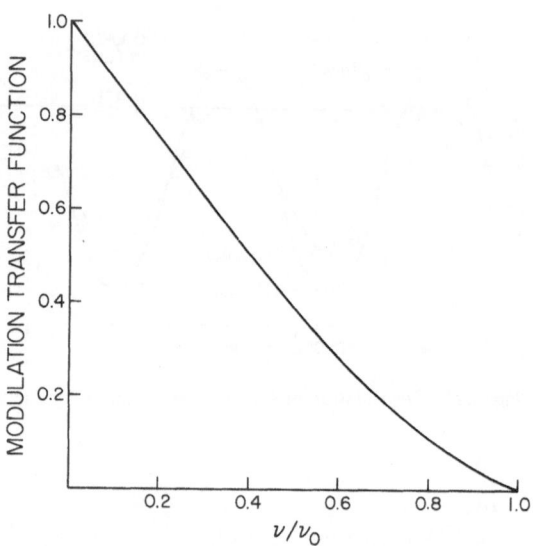

Fig. 5.22   Theoretical modulation transfer function for diffraction limited lens.

peats itself regularly – called a periodic function – can be expressed as a series of sine and cosine terms, the so-called Fourier series for the function. Fig. 5.23 shows the intensity variation of part of a square wave grating with a periodicity of 60 cycles per mm, a maximum intensity of 3 units, a minimum intensity of 1 unit, and an average intensity of 2 units, giving a modulation of 0.5. The Fourier series for this square wave is given by the expression

$$I = I_{av} + \frac{4}{\pi}\Big[\sin (120\pi x) + \frac{1}{3} \sin 3(120\pi x)$$
$$+ \frac{1}{5} \sin 5(120\pi x) + \ldots\Big] \tag{5.11}$$

where I_{av} is the mean value of the intensity, as in Fig. 5.19, and x is the distance along the grating in mm and the argument of the sine function is expressed in radians. The first sine function has the same periodicity as the square wave, the second represents what is called the third harmonic, the next term the fifth harmonic and so on. For green light of wavelength 550 nm and a diffraction limited lens with an f-number of 5, equation 5.10 gives $\nu_0 = 364$ cycles/mm: i.e. the lens cannot even theoretically resolve a sinusoidal modulation for which the amplitude peaks are closer than 1/364 mm. The fundamental spatial frequency of the square wave grating of Fig. 5.23 being $\nu = 60$ cycles/mm, we have for this lens

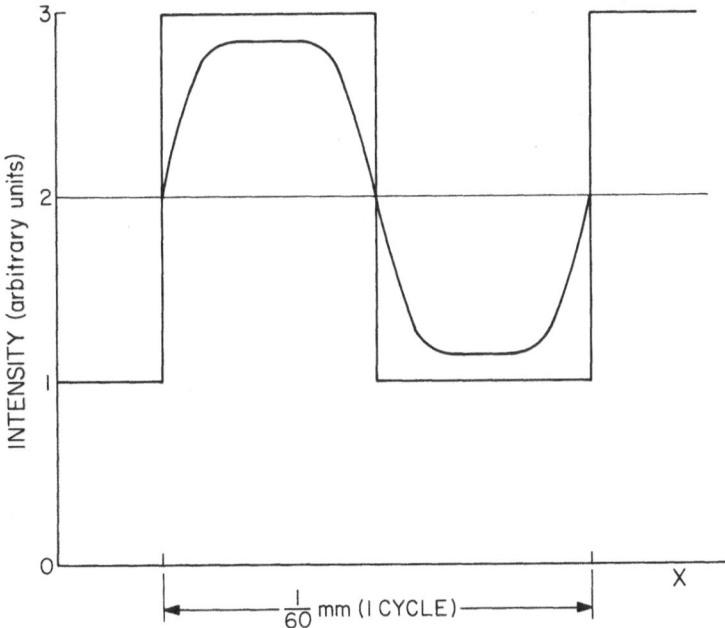

Fig. 5.23 Degradation of square-wave image by diffraction-limited f/5 lens.

$$\frac{v}{v_0} = \frac{60}{364} = 0.165 \qquad\qquad (5.12)$$

For the curve in Fig. 5.22 we see that for $v/v_0 = 0.165$ the modulation of
the image will be 0.79 of that of a sinusoidal object of that relative frequency. Making similar calculations for the other harmonics, and noting
that the constant term is not attenuated, while the seventh harmonic (the
fourth sine term in the series of equation 5.11 has a spatial frequency of
420 cycles per mm which is too high to be passed by the lens, the actual
shape of the intensity modulation in the image can be calculated. The
result is shown by the curve in Fig. 5.23. We see that, whereas the modulation of the object is 0.5, that of the image has been reduced to 0.43. Even
more important, however, the sharp edges of the patterns of the square
wave have been rounded off considerably.

Modulation transfer functions can be constructed for systems other than
lenses. We will explore this concept for photographic film and television
systems in the next chapter. The great power of the MTF arises from the
fact that the MTF for a given spatial frequency in a combined system,
such as a camera lens plus a photographic film, is given by the product of

the MTF for each system at that spatial frequency. One must bear in mind that the MTF for a lens is based on the modulation frequency of the object, while that for film will be based on the frequencies in the image it is recording. This will be further illustrated in the chapter on photography.

It would be inconvenient to have to read off the value of the MTF for each frequency of interest from a series of curves and then multiply them together to obtain each point for the response of the composite system. Fortunately, the logarithm of the product of two numbers is the sum of the logarithms of each of the numbers. By plotting the MTF's on log-log paper the resultant curve can be obtained by a simple subtraction of one curve from another – subtraction being the appropriate operation since MTF's are usually numbers less than 1, and the logarithm of a number less than 1 is negative.

Such a logarithmic plot for the f/5 diffraction limited lens used as an example in Fig. 5.23 is shown in Fig. 5.24. The basic curve is for an object modulation of unity. If the object modulation at each frequency is only

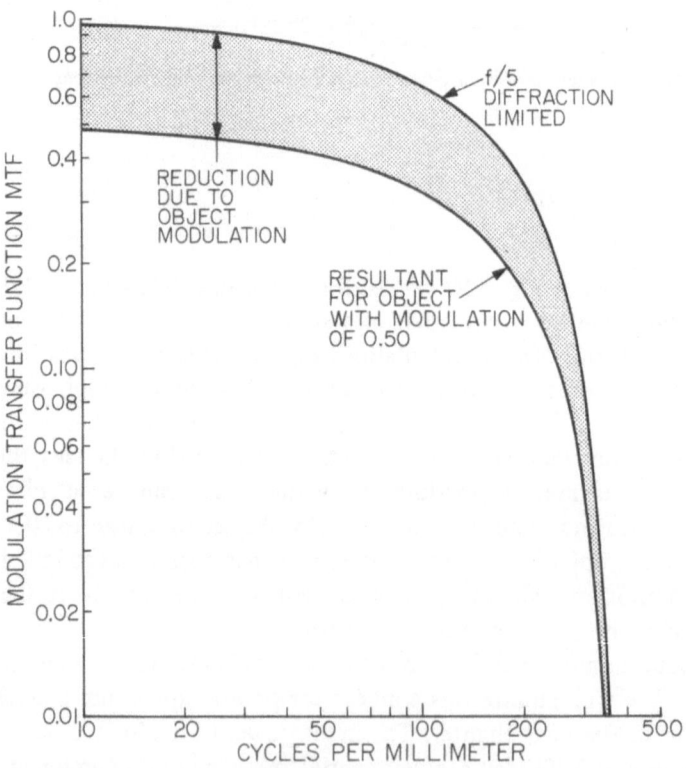

Fig. 5.24   Use of logarithmic plot of modulation transfer function.

0.5, we can obtain the appropriate response curve by adding the logarithm of 0.5, which is done graphically by moving each point on the original curve down by a distance corresponding to the logarithm of 0.5, obtaining the lower curve in Fig. 5.24.

5.6 Choosing lenses

From the discussions on the sources of imperfections in images formed by lenses, it should be apparent that even if we could build a lens which was limited only by the fundamental laws of physics, it would be impossible for any single lens design to meet all needs. The types of compromises made in the design of camera lenses will be generally quite different from those made for microscope objectives. The most obvious difference is that for cameras the object is usually relatively far from the lens, while the image is usually considerably smaller than the object and is formed relatively close to the back principal focal plane of the lens. For the microscope objective, the object is close to the front principal focal plane and the enlarged image is relatively far from the back principal focal plane.

For most applications, we want a high degree of variability in a camera lens: the ability to vary the light gathering capability by modifying its effective aperture; a good degree of chromatic correction, over the visible range, since they are usually expected to record scenes similar to what the eye would see; the capability of being used over a rather wide range of object-lens distances. Camera lenses are usually far from being designed to be diffraction limited, but this is seldom a serious compromise, since the film in general does not have high enough resolving power for this to be important.

There is one major exception to the general rule in designing camera objectives: for use in photo-etching techniques for micro-circuit application. Extremely carefully designed camera objectives are available, which often closely approach being diffraction limited. They are designed, however, to be used with a single wavelength of light (often the 408 nm line of the mercury-arc) and at a single magnification ratio. Microscope objectives are seldom required to have the versatility one expects of a camera objective, so that they can be designed with highly specialized uses in mind. Although some objectives are available with variable apertures, and some with adjustments for tube length (i.e. for optimising the correction for the distance of the image formation from the lens) or adjustments for coverslip thickness, most microscope objectives have a very limited range of applicability, and the particular objective should be chosen with the application in mind.

Too frequently, we tend to be misled by emphasis on the "magnifying power" of the objective. What is most important is its ability to resolve the detail in which we are interested. Since depth of field is lost approximately as the square of the increase in numerical aperture (which is a good measure of resolving power), the balance between depth of field and resolving power requires careful attention. For the pathologist working entirely with histological sections, carefully mounted on flat glass slides, flatness of field and high resolution may be extremely valuable attributes. For the ophthalmologist looking at the conjunctiva, or the clinician interested in the microcirculation in the nail fold, depth of field may be sufficiently important to justify considerable compromise in resolution, and, since the details of interest do not lie in a plane, flatness of field becomes an expensive and useless luxury, since a lens designed for a high degree of flatness of field requires more lens elements, and the more individual pieces of glass in a lens the greater the loss of light as well as the greater the cost.

For fluorescein angiography it is not even necessary that the lens be particularly well corrected for chromatic aberration, since the fluorescent emission is confined to a rather narrow band of wavelengths.

Unfortunately, it is not usually possible to buy a lens which is ideal for many purposes. Availability is heavily influenced by the market, and, although the quality of microscope objectives for the most common application (particularly for examination of histological specimens) has increased markedly in the last few years, those particularly well suited to other applications – particularly to intravital microscopy – are becoming more difficult to find.

Even having chosen the objective, whether for a camera or a microscope, which seems best designed to meet the needs at hand, one should insist, whenever possible, on comparing several objectives of the same design in the particular application of interest: quality control is far from perfect even by the manufacturers with the finest reputations. It is also easy to be misled by price. The most expensive lens is not necessarily the best –particularly for a highly specialized application. Caveat emptor.

References

1. Cornsweet, T.N. (1974): Visual Perception. Academic Press, New York and London.
2. Parrent, G.B. Jr. and Thompson, B.J. (1969): Physical Optics Notebook. Society of Photo-optical Instrumentation Engineers. Redondo Beach, California.
3. Rose, A. (1970): Vision: Human and Electronic. Plenum Press, New York and London.
4. Smith, W.J. (1966): Modern Optical Engineering. McGraw Hill, New York.

6
PHOTOGRAPHIC AND TELEVISION RECORDING OF IMAGES

Harold Wayland

6.1 Introduction

Whether an optical image is "recorded" by the human eye, on photographic film, or by a television system, that is biologically, chemically or electronically, the recording process ultimately depends on the ability to detect the photons of the light emitted or scattered by the object to be observed. A finite amount of light means a finite number of photons, and the information content available in the optical image depends on the total number of photons reaching the image. The usefulness of this information will also depend on the spatial and temporal distribution of the photons reaching the image sensor. Ideally, we should like to be able to count the number of photons reaching each part of the optical image, and relate the rate of their arrival in each part of the image to properties of the object of interest to us such as gross and detailed shape, brightness, color, etc.

The human eye is a remarkably sensitive photon counter as long as the radiation is in the color or wavelength range to which it is sensitive. This range is illustrated in Fig. 6.1 along with the sensitivity curve for a typical panchromatic film. The eye is, of course, by no means so uniformly sensitive over the range, and the peak sensitivity is different for the rods (which are "color blind" but represent the important sensors at very low light levels) and the cones, which are responsible for color perception. Interestingly enough, the peak sensitivity for the cones corresponds reasonably well to the peak of the solar radiation passed by the atmosphere in mid-day, while the peak for the rods is shifted toward the blue which better matches the light scattered from the sky (Fig. 6.2).

The sensitivity distribution of photographic films can be tailored to meet a variety of requirements as shown in Fig. 6.3, although within the visible range photographic films are far less sensitive than the eye. Photography has some important advantages, however: it has a "memory" which is far more consistent and reproducible than human memory, and photographic film can be exposed for long periods of time (if the image can be held sufficiently stationary) so that the effects of the arriving photons can be summed up or integrated to permit recording at extremely low light levels.

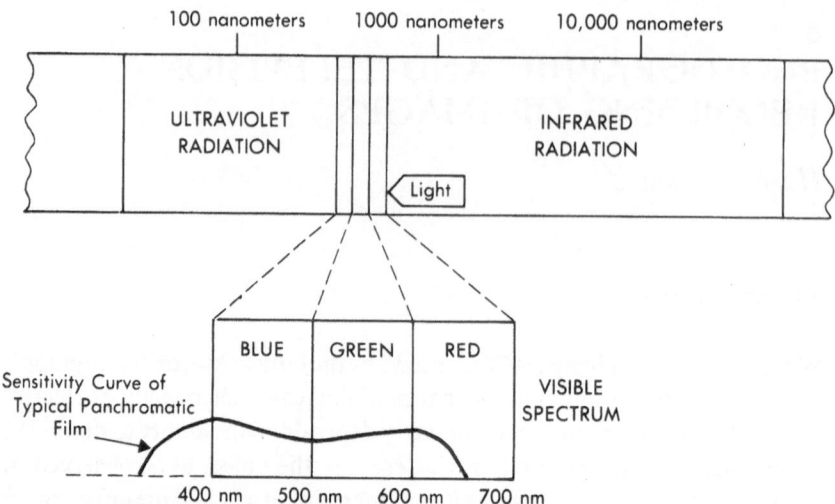

Fig. 6.1 Photographic and visible sections of electromagnetic spectrum. From "Kodak Professional Black and White Films" Publication F – 5 p. 34. A copyrighted publication of the Eastman Kodak Company. By permission.

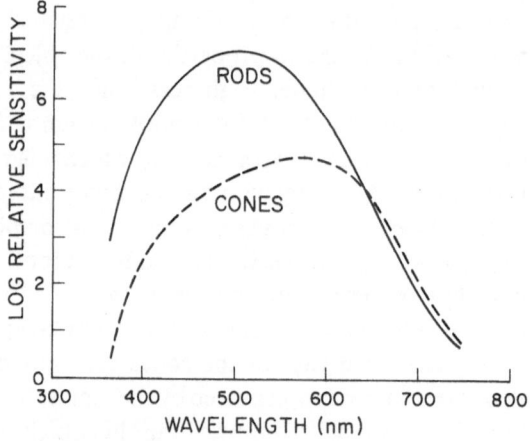

Fig. 6.2 Spectral sensitivity of the eye.

Photographic films can also be designed to record light of wavelengths to which the human eye is completely insensitive.

Recently, television cameras have been developed which equal or exceed the human eye even in the "visible" parts of the optical spectrum. There are significant problems in obtaining permanent records of pictorial data gathered by a television system. The television broadcasting industry has provided systems of high quality (and high cost), but the permanence of

BLUE-SENSITIVE FILMS

Spectrogram to
Tungsten Light

ORTHOCHROMATIC-SENSITIVE FILMS

Spectrogram to
Tungsten Light

PANCHROMATIC-SENSITIVE FILMS

Spectrogram to
Daylight

INFRARED-SENSITIVE FILM

Spectrogram to
Tungsten Light

Fig. 6.3 Spectral sensitivities of various film types. (Source same as Fig. 6.1 p. 26.)

such records, except when transferred to photographic film, has not been
satisfactorily established.

A fundamental question behind the problem of extracting useful information from an optical system is just how much information can be
contained in a finite number of photons, and how much of this can we

extract with the system we are using. For medical applications it may seem rather academic to be concerned with the ultimate in information retrieval from an optical system. For the ophthalmologist who wishes to study the interior of the eye, however, this is far from an academic problem. He might well wish to record fine detail of the retinal structure or the retinal microcirculation at illumination levels sufficiently low to cause no damage and little or no discomfort to the patient. It would be helpful if his observational system were at least as sensitive as the system he is observing, but even this would not be ideal. At low light levels the eye can utilize about 10% of the photons striking the retina. To image the retina, however, we must ordinarily use light scattered by its surface, and the percentage of the light scattered from a given small region will only be a fraction of the incident light, and only a small part of that will eventually be collected by an optical system used to form the image. If the image is magnified, which would be essential if we wish to record a blood capillary, further reduction in illumination per unit area will occur on the sensing surface.

When photons of light strike a detector, they produce what might be called "photo-events". In the eye, the light sensitive cells, the rods and cones, are capable of converting optical energy by means of a photochemically triggered chain of events into a pattern of nerve impulses. In the photographic process, the absorption of some minimum number of photons by a silver halide grain produces a "latent" image, which can be amplified chemically to produce an opaque grain of silver. In a television camera the incident photons induce local changes in the electrical properties of a light sensitive target which can be detected by electronic means. Each of these systems has a high gain or amplification, so that the energy available for the ultimate sensing or recording is very much greater than that falling on the detecting surface. The chain of events leading to the amplified image is merely triggered by the incident light, and the bulk of the energy required for the final presentation is supplied from other sources.

The ultimate effectiveness of the overall system will depend on the efficiency with which the sensing target can utilize the incident photons, which ultimately depends on the quantum efficiency of the target, and the subsequent amplification of its basic pattern of responses in space and in time. Ability to discriminate detail will also depend on the magnitude of the signal in relation to any background fluctuations inherent in the system: i.e. to the signal-noise ratio. Literally the term "noise" is associated with sounds which interfere with the sounds the listener wishes to hear. This term has been taken over by information theory to apply to any fluctuations in a signal which occur irregularly with respect to the signal of interest and which tend to obscure that signal. In our case we will be concerned with purely optical noise associated with random fluctuations in the number of photons arriving at a detector; photographic noise associated with the grain structure of a photographic film; or electrical noise associated with fluctuations in the electric current into which optical information is transformed by a television camera.

Because of the inherently small amount of energy in an optical signal, all of these systems require high levels of amplification or gain to bring the final patterns of information up to a usable level. The amplification systems themselves will introduce further noise into the response to the signal, which may have a strong influence on the minimum signal to noise ratio which can be detected.

Since the problems of signal detectability in a noisy background are basic to an understanding of the limitations of detection and recording of pictorial data, some of these basic concepts will be discussed in the next section. The material discussed in this chapter can be supplemented by referring to the literature presented at the end of this chapter.

6.2 Detectability of an optical signal

Whether or not we can detect the difference in light intensity between a small area (the test element) in an optical image and the surrounding region will depend not only on the difference in average illumination per unit area between the test element and the surrounding region, but also on the variability in illumination in each region due to optical "noise" or intensity variations due to the source of illumination. The number of photons of light per unit area emitted in any direction by a conventional incoherent light source, such as an incandescent filament or a mercury or xenon arc, fluctuates in a random fashion about a mean or average value due to the basic mechanisms of photon emission from excited atoms. The signal to noise ratio in any test element in an image is defined as the ratio of the average number of photons in the element to a prescribed measure of the deviations of the number of photons from the average, which can be calculated from the nature of the statistical fluctuations in the emission of the source. For an incoherent light source (coherent sources such as lasers will have a different statistical behavior), if the average number of photons emitted in a given time is N, the signal to noise ratio is $N^{1/2}$. It is clear from this expression that the signal to noise ratio improves with increased luminous flux or, for a fixed flux per unit area, with an increase in the area of the test element.

It is sometimes stated that a signal to noise ratio of unity is sufficient to permit detection of a signal. This is hardly adequate for an optical signal – a threshold signal to noise ratio, k, ranging from 3 to 5 is more realistic to avoid false alarms – i.e. concluding a signal is real when it is merely due to a statistical fluctuation. Surprisingly, higher intensities require a higher figure for k.

If we consider an ideal photon-noise-limited system in which the number of photons per square centimeter is n, d the diameter of the test element in cm, and $C (\leq 1)$ is the contrast of the test element with the surround (the ratio of its difference in average brightness to the average brightness of the surround), then the characteristic equation is

$$(nd^2)^{1/2} C = k \tag{6.1}$$

The limitations on resolving power due to contrast and the statistical nature of photon emission is clearly brought out by rearranging equation 6.1. The resolving power can be expressed as the reciprocal of the diameter of a barely detectable test element

$$\text{Resolving power} = \frac{1}{d} = \frac{Cn^{1/2}}{k} \approx \frac{Cn^{1/2}}{5} \tag{6.2}$$

This is a more stringent criterion than is usually applied in testing optical systems, since a series of alternating dark and light bars of high contrast such as found in the usual resolution chart (Fig. 5.18) will be more detectable than a small isolated spot. The most important consideration brought out by equation 6.2, however, is the great importance of contrast in determining the resolving power of a system. In the past there has been a tendency for lens and film manufacturers to try to put their products in as favorable a light as possible by testing them against high contrast bar charts. Fortunately, modulation transfer function curves are becoming more commonly available, and these permit evaluation of the effect of contrast on the system performance (e.g. Fig. 5.24). Some film manufacturers are also making available resolving power data for high and low contrast bar charts even when MTF's are not available. For example, two commonly used black and white films in U.S. laboratories are Kodak Panatomic-X, for which the high contrast resolving power is given as 200 lines/mm at a luminance ratio of 1000:1 and the low contrast resolving power is only 80 lines/mm at a luminance ratio of 1.6:1; the comparable numbers for Kodak Tri-X are for high contrast 100 lines/mm and for low contrast 50 lines/mm. The MTF curves for these films are discussed later and illustrated in Figs. 6.10 and 6.11.

There are, of course, other sources of noise than photon noise which tend
to degrade the quality of an optical image. Photon noise, however, is fundamental to the nature of light, so it has been singled out for special treatment. Degradation of an image due to lens aberration has already been
discussed. In the subsequent sections we will discuss problems specific to
photographic and television recording.

6.3 The photographic process

6.3.1 *Some fundamental concepts*

There is a considerable variety of photochemical processes in which the
action of light can be used to change the color of the material in which it
is absorbed (a popular, but inefficient, system being the human skin subjected to a sun tan), and hence can, in principle, be used to form a photographic image. In spite of many attempts to find a more suitable material,
micron-sized silver halide crystals suspended in a gelatin matrix, and supported on a film base, still form the basis of black and white photography
after more than a century since its empirical discovery. A remarkable property of such photographic film is that it can be stored in the dark for considerable periods of time with little loss in its effectiveness. When exposed
to a pattern of light, the absorption of a few photons by any grain prepares
that grain for subsequent action by a chemical developer so that the overall
process yields a catalytic gain of the order of 10^9 – a highly remarkable
level of amplification. If we did not possess this remarkable product and
were to draw up a set of specifications for its development it seems highly
unlikely that any scientist or group of scientists would have the temerity
to undertake the task.

The silver halide crystals which form the sensitive elements of a photographic emulsion are sensitized to various parts of the spectrum by the
absorption of various organic dyes prior to being suspended in the gelatin
carrier. The range of size of the crystals, although all in the micron range,
the nature of the sensitization, and the thickness of the emulsion are all
important factors in determining such film characteristics as speed, contrast, shape of the exposure-density curve, graininess, and sensitivity to
various parts of the spectrum.

It is not known precisely how many photons must be absorbed by a silver
halide grain to make it subject to chemical transformation to silver by the
developing process. It certainly requires more than one and may require
as many as ten. The development is an all-or-none process: a grain is either

sufficiently sensitized to be reduced to silver by the developer or it is un-affected by the developing process. Since not all photons falling on the film are absorbed by the silver halide crystals, and multiple absorptions are essential to produce a developable grain, the overall quantum efficiency of a photographic film does not exceed 1%, and it seems unlikely that this level of efficiency will be significantly increased.

For a given photon flux larger silver halide crystals will have a higher probability of absorbing enough photons to activate them than will smaller grains, the speed of a film can be increased by increasing the grain size. Since the grains are randomly distributed in the emulsion, the same statis-tics will apply to determining the signal to noise ratio due to the film struc-ture as were applicable to random photon emission. This means that in a spot containing N randomly distributed grains, the root mean square devia-tion will be $N^{1/2}$ and the signal-to-noise ratio will be $N^{1/2}$. Since the number of grains in a given surface area of film varies roughly as $1/d^2$, where d is the average diameter of the grains, the signal-to-noise ratio due to the grain structure of the film will vary inversely with the diameter of the grains, so we pay for greater sensitivity not only with poorer resolution, but with poorer contrast.

6.3.2 Structure of photographic films

The sensitive emulsion of a photographic film is made up of a suspension of silver halide crystals of various sizes which have been treated with suitable organic dyes to give them the desired sensitivity to various wavelengths of light. The untreated crystals would be sensitive primarily in the ultraviolet. The size distribution is adjusted to give a useful balance between sensitivity and resolution. The sensitive material is suspended in a gelatin base, which is then coated onto a film base, usually cellulose acetate or, where high dimensional stability is required, a polyester. Since by no means all of the light falling on the film is absorbed by the light sensitive particles in the emulsion, some of that which penetrates the emulsion may be reflected from the back of the base with sufficient intensity in the highlights to cause further exposure of the emulsion in the neighborhood of the image of the highlights. Since this often appears as a halo of light surrounding the image of a bright region, the phenomenon is called halation. This is convention-ally eliminated, or at least substantially reduced, by one of three methods: a dyed gelatin layer may be coated on the back of the film base; a dyed gelatin layer may be coated on the base just below the sensitive emulsion; or the film base itself may be dyed. The sensitive emulsion is usually pro-tected on the side away from the base by a transparent coating of hard gela-tin.

Film manufacturers are constantly improving the capabilities and uniformity of quality of various films, and the user who has particular requirements is urged to request the technical data of the type made available to professional photographers rather than relying on the relatively uninformative data sheets usually packaged with the film.

6.3.3 *Properties of photographic emulsions*

Up to this point we have been primarily concerned with the fundamentals of image formation and the basic mechanisms of photographic recording, including some of the fundamental limitations in the photographic process. In this section we will be more directly concerned with the practical behavior of photographic film, with the aim of acquainting the user with the meaning and use of the technical data available from the manufacturers of photographic products.

Some basic terms which are frequently met in photographic work are listed and defined in Table 6.1. Many of the important aspects of the overall behavior of a photographic film can be deduced from its characteristic curve, which is a log-log plot of density vs. exposure (see Table 6.1 for definitions of these terms). A typical characteristic curve is shown in Fig. 6.4. The abscissa is linear in the logarithm to the base 10 ($\log_{10}$) of the exposure in meter-candle-second, while the direct measure of the exposure is also indicated on a separate scale. The ordinate is linear in density of the processed film, which is the $\log_{10}$ of the opacity. Scales for both the opacity and transmission are also shown alongside the density scale in Fig. 6.4. Such

Table 6.1   Some Useful definitions and units.

candela (cd) or standard candle (c): the basic standard light source which has been arbitrarily defined.

1 meter candle = illumination falling on a surface 1 meter from a candela.

1 meter-candle $\equiv$ 1 lux $\equiv$ 1 lumen/m^2

1 meter-candle-second = a measure of optical exposure

1 lambert = surface brightness of a 100% reflecting and perfectly diffusing surface on which a luminous flux of 1 lumen/cm^2 is falling or 10^4 meter candles.

1 lumen of green light = 1.5×10^{-3} watt
$$= 4 \times 10^{15} \text{ photons/sec.}$$

1 lumen of white light $\approx 4 \times 10^{-3}$ watt
$$\approx 1 \times 10^{16} \text{ photons/sec.}$$

Transmittance $= \dfrac{\text{intensity of transmitted light}}{\text{intensity of incident light}}$ (a fraction)

Opacity $= \dfrac{1}{\text{Transmittance}}$ (≥ 1)

Density $= \log_{10}$ (opacity) (≥ 0)

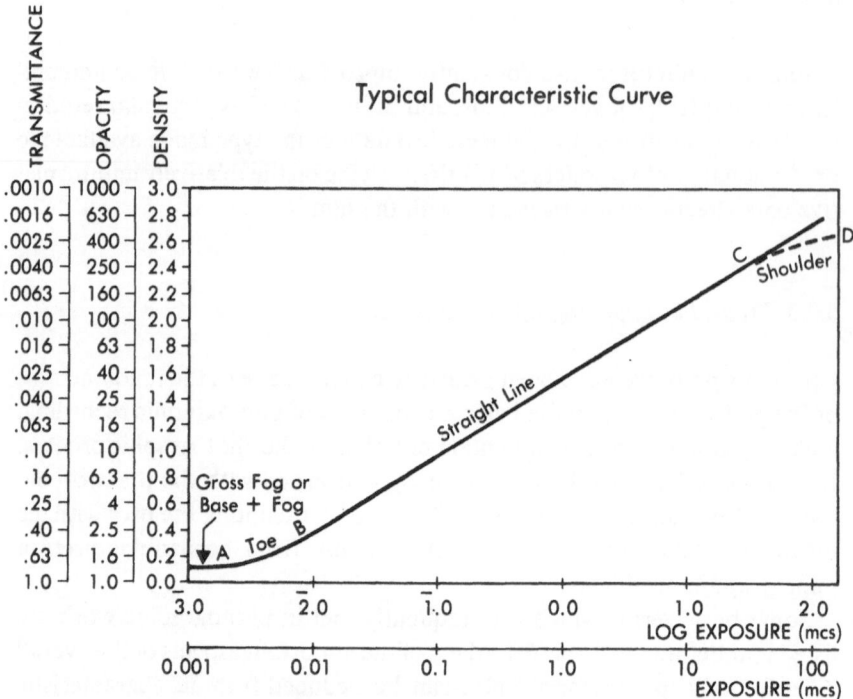

Fig. 6.4 Typical characteristic curve for photographic film. (Source as in Fig. 6.1 p.6.)

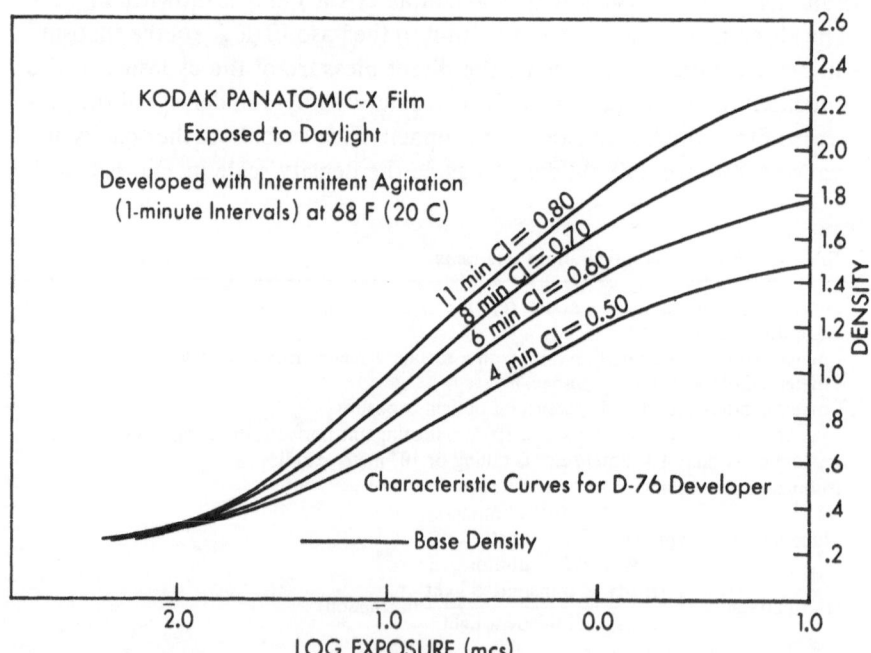

Fig. 6.5 Characteristic curves for Kodak Panatonic-X film. (Source same as Fig. 6.1 p.DS-10.)

a characteristic curve is not only a function of the photographic emulsion, but can be varied by modifying the developer and, with a given developer, by changing the developing time. The effect of changing the developing time is shown in the characteristic curves for Kodak Pan X and Kodak Tri X Pan films in Figs. 6.5 and 6.6.

The salient features of such a characteristic curve are illustrated in Fig. 6.4. Because of its shape, it is usually divided into three distinct regions, the toe (AB), the straight line (BC) and the shoulder (CD).

The toe.   In the toe region of Fig. 6.4 the part of the curve to the left of A is a horizontal straight line corresponding to a finite density or opacity. In section 6.3 it was pointed out that it takes several photons to activate each grain of silver halide, so it is not surprising that there is a minimum level of exposure before any grain will be activated. The finite background is due to the sum of the basic opacity of the film base plus background or chemical fog due to a small population of grains which are developable without exposure to light. Between A and B the curve begins to rise, with slope increasing as B is approached. In this region equal increments of log exposure will cause increasing increments in density as we approach B. This means that a given contrast in dark parts of the object will yield a smaller

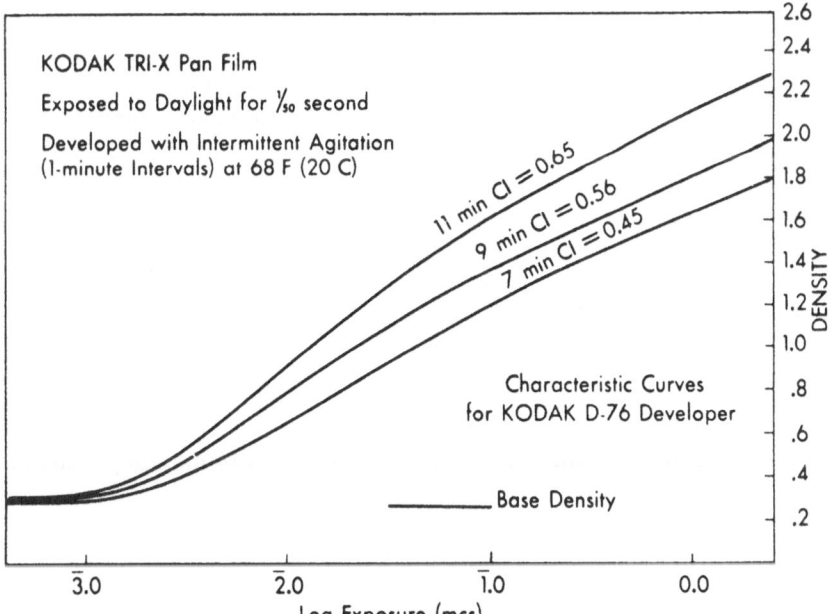

Fig. 6.6 Characteristic curves for Kodak Tri-X film. (Source same as Fig. 6.1 p. DS-18.)

contrast in the film than a similar difference in brighter parts of the object. This is consistent with the previous observation that contrast is reduced at low light levels.

The straight line.   The mid-portion of the characteristic curve in Fig. 6.4 is a straight line, which means that over this region equal increments in log exposure will result in equal increments in film density.

The slope of this linear portion, or the average slope if it is curved only slightly, is often called the "gamma" (γ) of the film, a high γ indicating that a given change in exposure will lead to a greater change in film density than a low γ. Because of the inherently arbitrary nature of the scales chosen for the coordinate axes, a fixed relation between the log exposure and density scales must be established for this slope to be comparable from one plot to another. The accepted standard requires that a gamma of 1 corresponds to a film for which a 10-fold increase in exposure corresponds to a 10-fold increase in opacity in the straight line portion of the curve. In order to read gamma of a film by a direct measure of the slope of the graph the distance along the abscissa corresponding to 1 log exposure unit must correspond to the distance along the abscissa corresponding to 1 density unit (1 log opacity unit). Fig. 6.4 is actually plotted to such a scale, so that this hypothetical film has a gamma of only 0.55.

Throughout this linear region the tonal scale will have a uniform compression for this film, since $\gamma < 1$. For photometric purposes a long linear portion of the characteristic curve greatly simplifies the interpretaion of the relationship between density changes in the image and brightness changes in the object, although the existence of a threshold below which changes in object brightness do not increase film density, followed by the compressed region in the toe of the curve, complicates the use of film for quantitative photometric measurements.

For pictorial presentation, a slight curvature in the middle portion of the characteristic curve, such as we see in Figs. 6.5 and 6.6, does not change the tone reproduction to a visible extent.

The shoulder.   The dotted line in Fig. 6.4 represents a region in which the slope of the characteristic curve decreases and, for sufficiently long or intense exposures, it will again become horizontal. For most applications, and particularly for pictorial use, the film should never be exposed sufficiently that even the brightest highlights fall in this region.

For pictorial work, best results will usually be obtained if the greatest density of the negative does not exceed 1.2. This is far below the useful range of most films for photometric work, although the greater linearity of certain types of television cameras make them highly competitive for imaging photometry.

Reciprocity effect. A perfect light sensing device would be one in which the total response is identical for a given amount of luminous energy whether it is due to a low energy flux for a long time or a high energy flux for a short time. The existence of a threshold of response of silver halide grains implies that, for extremely low levels of illumination there would be no response of a photographic emulsion no matter how long the exposure. If response is proportional to the product of illumination with the time of exposure we say that the reciprocity law is fulfilled. The failure of film to follow this law is sufficiently marked to be of great importance either in extremely high speed photography, where exposures are short, or in low light level photography, where exposures are long. For most black and white films the reciprocity law applies quite adequately for exposure times from about 1/5 second to 1/1000 second. Outside this range the failure of reciprocity to hold will show up as underexposure, change in contrast or both. When possible, it is usually preferable to choose a film which permits exposure within the range for which reciprocity holds. Most light metering systems, even those which measure the light intensity in the film plane, are calculated to give correct exposure settings only within the region in which reciprocity holds.

For extremely high light levels for which the indicated exposure is less that 1/1000 second a loss of density in the highlights and possibly even in the middle tones will result. If possible, the exposure should be increased by increasing the aperture stop, since short exposures are usually associated with high speed cinematography or stroboscopic illumination, where there may be no control over exposure time. An increase in development time will also increase the contrast in the highlights as seen from the curves in Fig. 6.5 and 6.6. For an exposure time of 10^{-5} sec the lens aperture should be opened up one stop beyond that indicated and the development time increased by 20%. For an exposure time of 10^{-4} sec the lens aperture should be increased 1/2 stop and the development time increased by 15%. Even at 10^{-3} sec it is usually helpful to increase the development time by 10%.

For extremely low light levels quite dramatic increases in exposure time over those calculated from light meter readings are required. Fig. 6.7 gives the average adjustment for low light levels. Although the graph shows corrections based on increasing the exposure time, increasing the lens aperture can also be used for part or all of the compensation. Decrease in development time will usually improve the quality for long exposures: 10% decrease in development time if the indicated exposure time is 1 second; 20% decrease in development time if the indicated exposure time is 10 seconds; and 30% decrease in development time if the indicated exposure is 100 seconds.

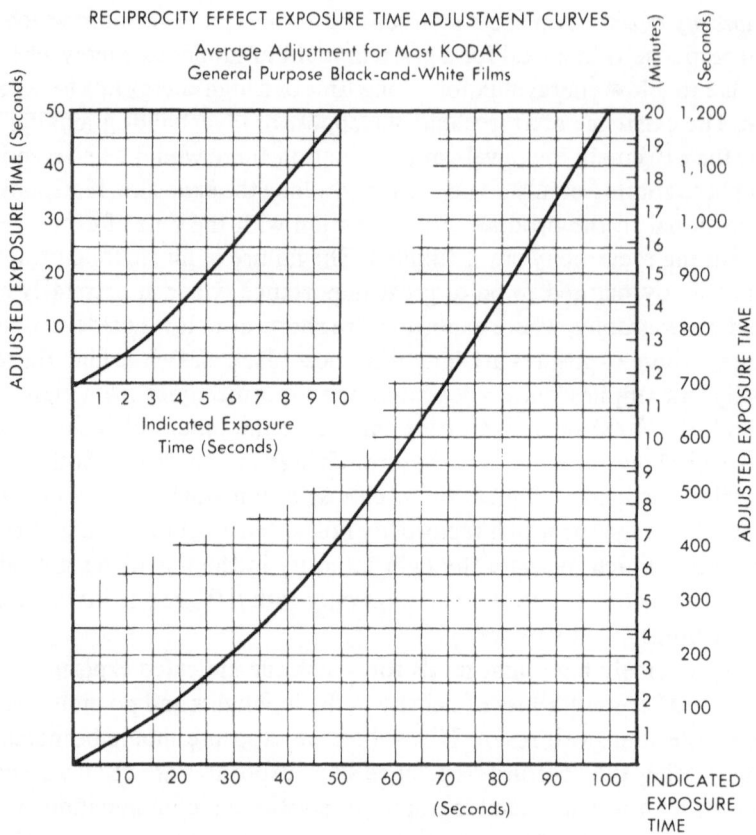

Fig. 6.7   Reciprocity effect. (Source same as Fig. 6.1 p. 28.)

6.3.4 *Resolution*

The resolving power of a film emulsion is measured by photographing resolution charts such as illustrated in Fig. 5.18. Such charts, with a high
degree of contrast, and being made up of linear elements which can be
much more readily recognized in a noisy background than patterns of an
irregular shape, or even circular spots, tend to indicate a resolving power
considerably better than is apt to be realized in practice. The effective resolving power for a given film depends on a variety of parameters such as
form and contrast of the test target, exposure, and, to some extent, the
development of the film. The resolving powers usually quoted are based on
high contrast bar targets with optimal exposure and development. Sometimes it is possible to obtain data on resolution for both high contrast and

low contrast targets: for example the Eastman Kodak Company in its publications for professional photographers gives resolution data for two luminance ratios, 1000:1 and 1.6:1. Using the definition of contrast or modulation given in equation 5.8 these correspond to modulations of 0.998 and 0.231. For Panatomic X film the quoted resolutions are 200 line pairs/mm for the high contrast target and 80 lines pairs/mm for the low contrast target.

The degradation in resolving power even for a bar target can be considerable when all factors are taken into consideration. Fig. 6.8 shows two photographs of the bar chart of Fig. 5.18 made with a 35 mm camera on Panatomic X film with the lens stopped down to f/16. In Fig. 6.8a the image on the film was reduced by a factor of 11.3 so that the barely resolved image of 4.5 line pairs/mm corresponds on the film to 51 line pairs/mm. In Fig. 6.8b the image on the film was reduced by a factor of 39, and the barely resolved chart of 1.25 line pairs/mm corresponds to 49 line pairs/mm on the film. Since the contrast in the target was high, and Panatomic X has a much higher potential resolving power than 50 line pairs/mm even for a low contrast target and these pictures were chosen as the best from several exposures, the degradation in resolving power must be largely attributed to the lens.

If the modulation transfer functions for the various components of a system are available, they can be combined graphically to give the overall response of a camera plus film. A hypothetical example is shown in Fig. 6.9. The solid line to the right shows the MTF for a diffraction-limited f/5.0 lens, while the dashed line shows the sort of degradation due to the usual compromises required in the design of a photographic lens. The effect of the film with properties similar to those of Kodak Panatomic-X is shown in the line with long and short dashes. If, now, the target has a contrast of 0.5 there is a further reduction, so that the response of the system is given by the lower solid line. Accepting a minimum detectable modulation of 0.04, this system is seen to have a resolution of 105 cycles/mm.

Modulation transfer functions for films can usually be obtained from the manufacturer. Two such curves are shown in Figs. 6.10 and 6.11. It is interesting to note that in the low spatial frequency range the peculiarities of the development process can increase the modulation of the image over that of the object, leading to an MTF greater than unity.

6.3.5 *Some comments on the use of photographic systems*

With the advent of built-in exposure meters, miniaturized electronic computers built into the newer cameras which adjust the exposure to corres-

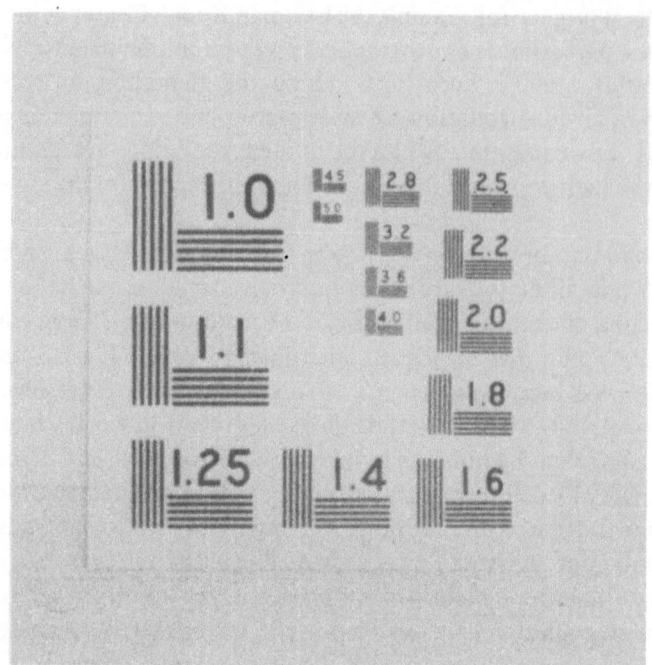

a

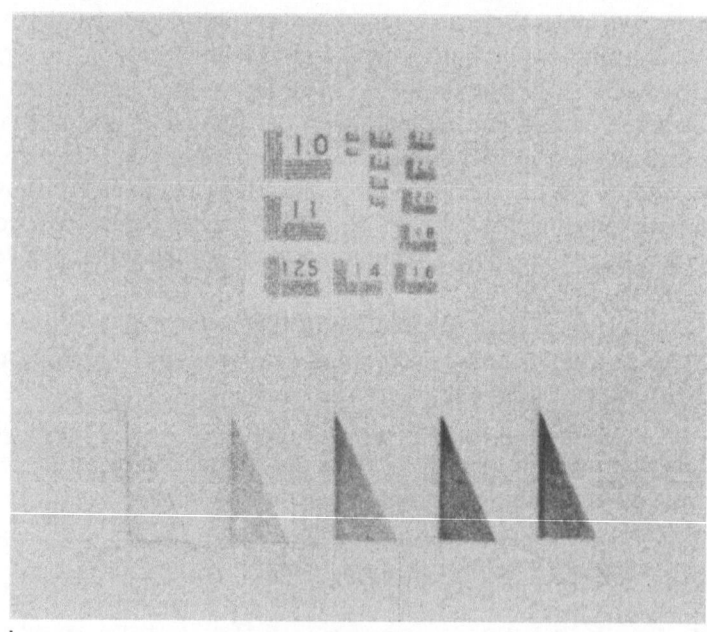

b

Fig. 6.8a) Resolution target photographed on Panatomic-X. Lens at f/16. 4.5 line target resolved. Equivalent to 50 line pairs/mm. *b)* Same as *a* photographed at a greater distance. 1.25 line target resolved. Equivalent to 49 line paris/mm.

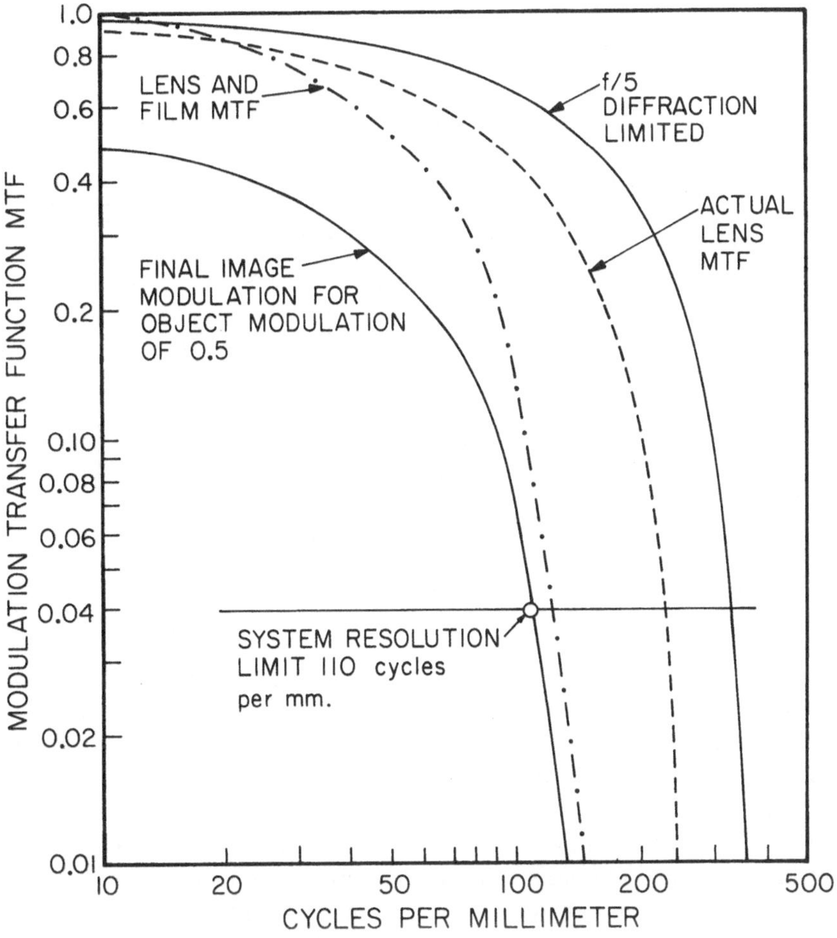

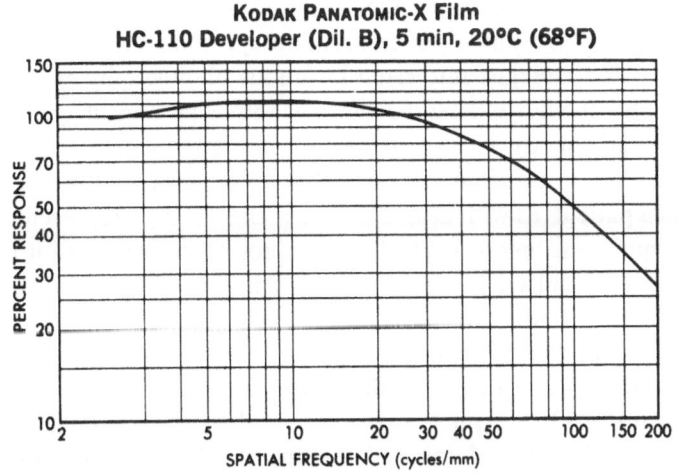

Fig. 6.9 System MTF and resolution.

Kodak Panatomic-X Film
HC-110 Developer (Dil. B), 5 min, 20°C (68°F)

Fig. 6.10 Contrast Transfer Function for Kodak Panatomic-X film. (Source same as *Fig. 6.1* p. DS-24.)

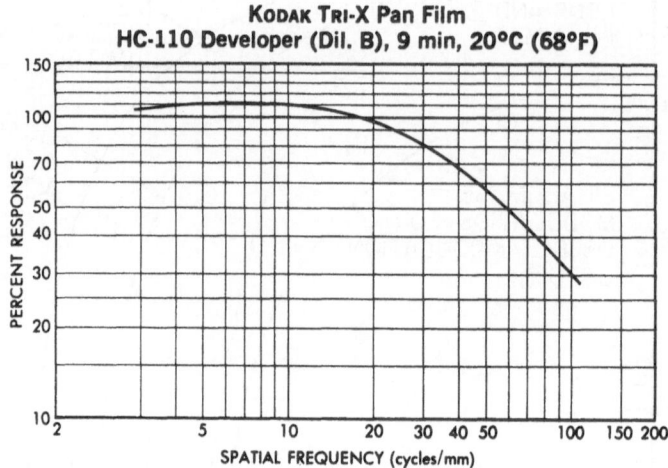

Fig. 6.11 Contrast Transfer Function for Kodak Tri-X Pan film. (Source same as Fig. 6.1 p. DS-24.)

pond to the amount of light available, and the highly reliable commercial processing of most film types, the medical practitioner should be able to carry on the bulk of his photographic work with a high assurance of obtaining a satisfactory end product. Even these automated devices produce unsatisfactory results at times, nor can everyone afford to purchase the latest in sophisticated equipment, so it is helpful to have a basic understanding of the power and limitations of the photographic process.

Such a basic comprehension is even more important in medical research, or clinical applications at the frontiers of established practice where existing technology must be adapted to the problem at hand, or where it is important to be able to make a sound technical judgement of new processes or devices. Photography of the retina of the eye is still far from satisfactory, and an ophthalmologist interested in fundus photography or fluoroscein angiography should understand the basic imaging and image recording problems sufficiently clearly to be able to judge if new products and methods have a sufficiently high likelihood of success to warrant trying them on a patient: as with drug detail men, the technical representatives of photographic and microscopic firms are not omniscient, and it is to the advantage of the physician to have a sound basis on which to judge the potential value of new products.

We have seen that the ability of a photographic film to utilize light is not highly efficient – its quantum efficiency is at best only about 1% as compared to up to 10% for the human eye, and there seem to be fundamental limitations to any considerable improvement. Electronic devices already exist with a quantum efficiency, for some useful wavelengths, of 70%, and there seem to be no fundamental limitations to approaching very close to 100% utilization of the photons striking a light-sensitive electronic target. The application of electronic aids to vision and image recording are proving more and more useful in a wide variety of applications. This will be discussed in the next section.

6.4 Electronic aids to image sensing and recording

6.4.1 *Some fundamental concepts*

Two types of electronic aids to optical imaging are proving useful in medical applications: the image intensifier and the television camera. In both types of devices the initial stage is to convert an optical signal into an electrical signal. This electrical signal can be processed in a variety of ways, but the ultimate objective is to produce a pictorial representation of the original object.

In the human visual process, all the stimulated photosensitive elements in the retina send their signal essentially simultaneously to the processing centers and on to the brain. The image intensifier is similar in its mode of action, in that the electrical signals from the entire sensitive screen are simultaneously transmitted to the phosphorescent screen, so that all parts of the image are processed simultaneously. In the television camera the sensor which converts the optical signal into an electrical signal is sequentially scanned electrically, so that the electrical signal due to the pictorial data is processed sequentially. When this electrical information is eventually converted into a picture this, too, must be built up by a sequential stimulation of the picture tube. For dynamic events the presentation rate must, of course, be rapid enough not to introduce bothersome flicker although there are applications in which isolated fields are of primary interest. In principle, even television systems could be devised to process the entire field simultaneously, but this would greatly increase the complexity and cost.

The current high level of electronic imaging technology has been made possible to a considerable degree by the popularity of television for use in the home, which has made it economically feasible for commercial firms to invest considerable sums in improvement of cameras, recorders and monitors. Even more important, however, for highly sophisticated instrumentation development has been the interest of the military in systems capable of forming useful images at low light levels, particularly in the infra-red. The space program has also sponsored and supported many useful developments in electronic imaging devices, but more particularly in improved methods of signal processing. These three sources of funds for research and development have permitted advances far beyond what any group, private or governmental, seems willing to finance for medical applications alone.

6.4.2 *Image intensifiers*

The basic principle of the image intensifier is remarkably simple: the optical image to be sensed is projected onto a photosensitive surface coated on a glass or quartz plate, with the sensitive coating in a vacuum chamber. On illumination, this surface emits electrons, the rate of emission from any small region being proportional to the intensity of illumination. These photoelectrons are accelerated in the evacuated space by an electric field, and by means of an electrostatic or magnetic lens are focused onto another transparent plate which is coated with a phosphor. The intensity of light emitted by the phosphor depends on the number of electrons striking it, so it will emit a pattern of light which is a more or less faithful image of that striking the photocathode. In practice, the two windows form the ends of a vacuum tube, with the photocathode and phosphor deposited on the vacuum side of these windows. One or both windows may consist of a coherent fiber optic plate, which has proved particularly advantageous for electrostatically focussed image intensifiers and for ease of manufacture of multistage systems. The technological developments which have permitted the production of highly sensitive, multistage image intensifiers have included improvements in the quantum efficiency of photocathodes, the reduction of thermally generated dark current, the improvement of the luminous efficiency of phosphors, and the development of compact high voltage sources.

The spectral response of an image intensifier is determined by the nature of the photocathode material. A wide range of photocathode materials is now availabe with peak responses in various parts of the spectrum showing a quantum efficiency of about 10%. Military demands have tended to emphasize response in the infra-red region for night vision applications.

The same problems of optical noise arising from the statistical distribution of photons from an optical source are fundamental to all methods of image detection at low light levels. Photoelectron emitters, when coated directly onto a glass or quarts substrate, have a finer structure than is found in photographic emulsions, so that this source of optical noise is not a limiting factor. Such photocathodes, however, spontaneously emit electrons, the "thermal noise" of the system. This can be markedly reduced by cooling the photocathode, although this would be necessary only for work at extremely low light levels, more apt to arise in astronomy than in medicine. If the photocathode is deposited on a fiber optic plate, the structure of the plate can give rise to a cell-like pattern, although the quality of fiber optics has improved to the point that this is not an inherent – albeit sometimes a practical – limitation.

Another potentially limiting factor, both in efficiency and resolution, is the light emitting phosphor. It not only should have a high efficiency of producing light when bombarded by electrons, but must be fine-grained so that its granularity does not degrade the image, and must have a long life under intense bombardment by high energy electrons – the accelerating voltage is usually of the order of 10,000 volts or more. Phosphors may also show a slow decay rate, limiting their usefulness in recording rapidly moving events.

The highest resolution is attainable with a single stage magnetically focused image intensifier with both the photocathode and phosphor coated on glass or quartz windows. Commercially available units are available with a resolution of some 90 line pairs per millimeter and an optical gain of a few hundred. Three stage magnetically focused image intensifiers rather typically have an overall optical gain of a million-fold, although the resolution will be considerably reduced – to about 40 line pairs per millimeter. The use of electromagnets for the electron lenses rather than permanent magnets will somewhat improve the quality of the final image. A major drawback to magnetically focused image intensifiers is their high cost.

Electrostatically focused image intensifiers are very much less expensive than the electromagnetically focused instruments. A typical design will have both the photocathode and phosphor coated on fiber optic plates. This permits the coated surfaces to be shaped to minimize the rather large spherical aberration associated with a simple electrostatic lens. If minimal optical distortion is important, it is essential to indicate this to the supplier. A good single stage electrostatically focused image intensifier will have a gain of 90–100 in white light and a maximum resolution of about 60 line pairs per millimeter. The use of fiber optic plates at both ends of an electostatically focused image intensifier greatly simplifies the construction of multistage systems, since it permits the output of one stage to be directly coupled with the input of the next stage through a layer of immersion oil or optical cement. This simplifies manufacture and quality control, since all units in a multistage system are the same.

Some of the properties of a typical line of electrostatically focused image intensifiers are shown in Fig. 6.12a, b, c and Fig. 6.13a, b.

Except for direct visual observation, a single stage image intesifier with a gain of less than 100 is of very little use. This can be emphasized by a relatively simple calculation. Suppose we were to image the output face of the tube at 1:1 with a f/1 lens. This means that the diameter of the lens is equal to its focal length f, but for a one to one magnification both the object and image have to be twice the focal length from the lens. The light from a point on the phosphor will be radiated equally in all directions, so to calculate the fraction of light gathered by the lens we compare the ratio of the area of the lens to that of a sphere of radius 2f. The area of the sphere will be $4\pi(2f)^2$ while that of the lens will be $\pi(f/2)^2$, so the lens can only pick up 1/64 of the light radiated from the phoshpor. Since compound lenses will lose a further fraction of the light – often as much as half of it –photographing a single stage image intensifier is a useless exercise unless it is used as an image converter – i.e. converting an image to which the film is not sensitive into a wavelength region where it is sensitive.

6.4.3 *Television*

As we pointed out in section 6.4.1, a television system presents an image in a sequential form since this has proven to be the simplest and most efficient way in which an optical image which has been converted into some form of electrical signal can be sampled and reconverted into an optical image at a location remote from the initial object. Basically, the optical image is formed on a photosensitive surface whose electrical properties are modified by exposure to light. In the currently common types of television camera tubes, this photosensitive surface is scanned by an electron beam. The electrical signal due to the charge distribution on the target is sensed by the electron beam and this modulation of charge is amplified and transmitted to the receiving system. The standard television monitor consists of a large evacuated tube, one surface of which is covered on the interior with a phosphor which emits light when bombarded by an electron beam. Such a beam is emitted by an electron gun in the same vacuum envelope as the phosphor. This beam is swept over the phosphor surface in synchronism with the sweep in the camera tube, and the intensity of the beam is modulated in a pattern mimicking the light and dark regions in the original image. No attempt will be made in this discussion to elaborate on the details of television transmitters and receivers.

In order to avoid noticeable flicker in the presentation of an image produced by such a sweep pattern, the sweep rate must be relatively high. In

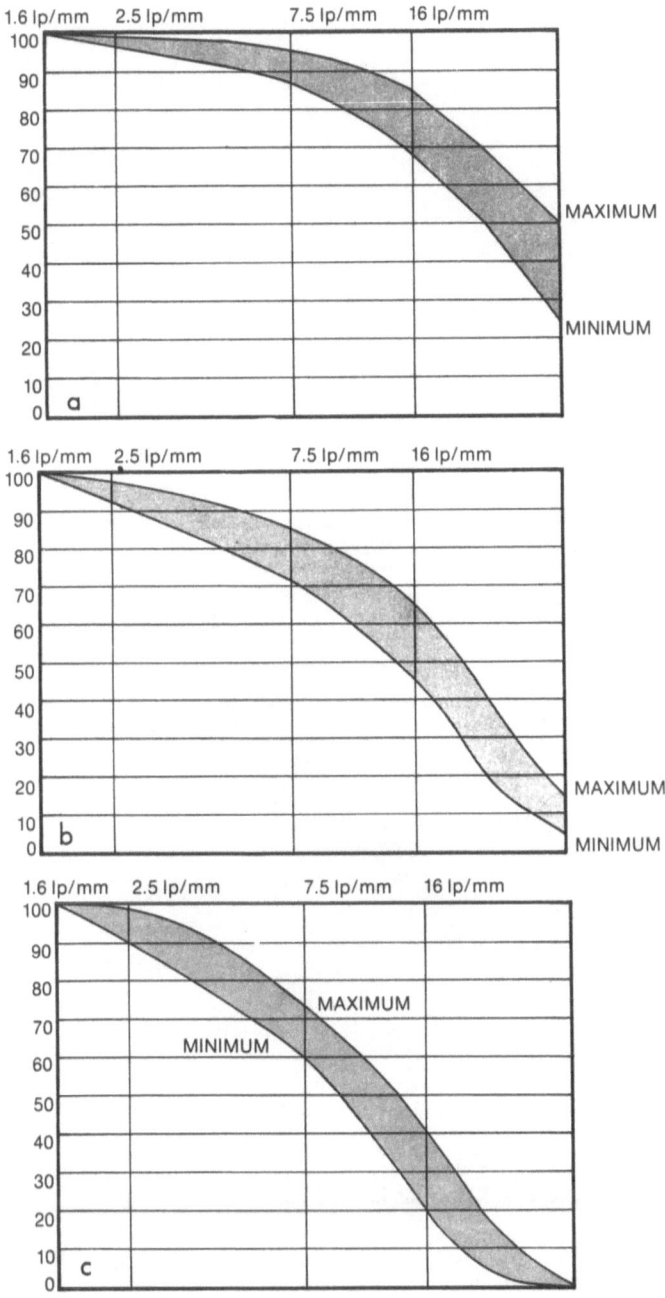

Fig. 6.12 Modulation Transfer Functions for 1 (*a*), 2 (*b*), and 3 (*c*) stage image intensifiers. 1p = line pairs. On the vertical axes the percent modulation is plotted. (Courtesy of
Electron Devices Division, Varo Corporation, Garland, Texas.)

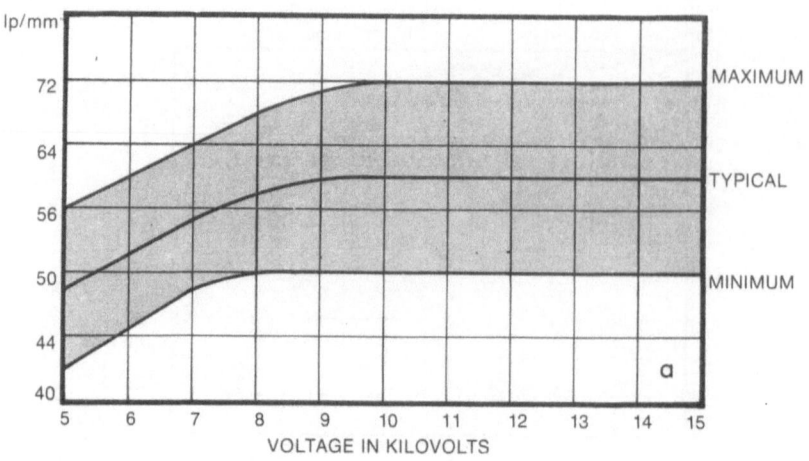

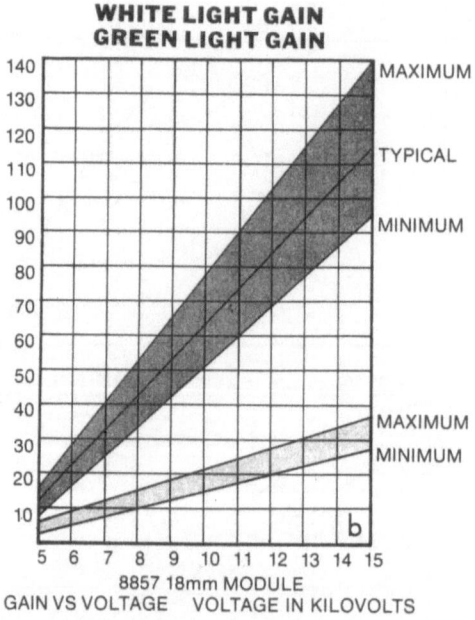

Fig. 6.13 Some properties of an electrostatically focussed image intensifier as a function of accelerating voltage. *a*) Resolution in line pairs/mm; *b*) optical gain.

standard practice the information is read off while the sweeping beam is moving in only one direction, and it is rapidly returned to the starting side, although shifted vertically downward, without transfer of information. In the standard television system in the U.S. the target (and hence the monitor) is scanned 525 times in 1/30 second to make up a picture frame. Such a frame is made up of two interlaced fields each nominally made up of 262½ lines, and each field being covered in 1/60 second to correspond to the 60 Hz line frequency common in the U.S. Since each field is swept from top to bottom, a few lines are lost between fields while the electron beam in the camera is being returned to the top of the picture. The transmission of electrical information is cut off during this "blanking" period. The two fields which make up a frame are geometrically interlaced, which gives a more pleasing picture.

Since the target is swept from top to bottom every 1/60 second, the entire image is not transmitted simultaneously; hence, for moving objects, a focal plane shutter effect may be observed. If the sensing beam were to remove all of the information from the target in the line of the sweep, there would still be an integration time of about 1/60 second during which optical information is integrated before the next sweep. This figure is given instead of the time between identical sweeps, since the electron beam sweeping the target usually covers a path which encroaches to some extent on the neighboring path for the interlaced field.

Except in the U.S., the basic frequency for alternating current power is 50 Hz, so that a different standard television sweep rate is used. European television has not been so highly standardized as it has in the U.S., but a fairly common system uses 625 lines per picture, made up of two interlaced fields of 312½ lines per field, with a total sweep time of 1/50 second per field.

Various "high resolution" television systems are offered with 800 lines per picture or even over 1000 lines per picture. Before paying the extra price for these systems a critical evaluation must be made as to whether or not the combination of optical system, camera and monitor or recording system are capable of producing and presenting or recording the information with sufficient resolution to justify the additional expense.

Over the past fifty years television camera tubes have achieved a gain in sensitivity of a millionfold. The quantum efficiency ranges from about 10% to close to 100%, and the low light level performance can now be made to exceed that of the human eye. The bulk of the camera tubes in common use today fall into two basic categories: those dependent on photoemission for the initial sensing of the image, and those depending on photoconductivity, although hybrid systems are also available. Newer developments in solid state devices are moving rapidly, so that extremely compact, highly sensitive cameras of this type will soon be commonly available.

This discussion must necessarily be limited to a few types of television cameras. Of the cameras using photoemission as the basic detection scheme, the image orthicon, which has a long been a standard in the television industry in the U.S., will first be discussed. Then photoconductive cameras (vidicon) with various sensing surfaces will be considered. One example of a hybrid system, the silicon intensifier tube (SIT) will be discussed in some detail because of the success the author of this chapter has had in applying it to biomedical research. Finally the rapidly developing field of solid state sensors will be discussed, and some likely future directions of development in this rapidly expanding field will be explored.

The image orthicon.   The image orthicon (Fig. 6.14a) was the first camera tube to have a sensitivity sufficient to approach photon-noise-limited performance. Since its initial transformation of the optical image to an electrical signal is by means of a photocathode, its intrinsic sensitivity (quantum efficiency of 10%) is reasonably close to that of the human eye over most of this camera tube's operating range. The operation of this tube is as follows: the scene to be converted into an electrical signal is focused on a semi-transparent conducting photocathode (at the left of Fig. 6.14a). The photoelectrons emitted by the photocathode are accelerated and focused onto a glass target only a few microns thick with sufficient ionic conductivity to allow charges on the two sides of the target to neutralise each other in less than one tenth second. The thinness of the target also reduces lateral spreading of the image to be insignificant during the time of sweep of a field. The beam of electrons used to scan the target is of low energy, and is decelerated by a decelerating ring (not shown in figure) which surrounds the target, so that the electrons approach the target at close to zero velocity, turn around and return toward the electron gun to an electron multiplier surrounding the gun. The beam current is of the order of the photocurrent from the photocathode and can be modulated by the charge on the target as much as 50%. The amplification of the electron multiplier, which is an integral part of the tube, adds additional gain to the signal. A major limitation in signal noise ratio arises from the fact that in the dark areas essentially the full beam current is returned to the photon multiplier, and the shot noise due to basic statistical fluctuation in the photon emission is a maximum for the dark parts of the picture. Although image orthicons have both good resolution and sensitivity characteristics, they are expensive compared to other tubes now available at considerably less cost, and, often, with better performance characteristics.

Vidicons.   The term "vidicon" will be used to cover a variety of camera types in which the sensing target is made up of a continuous photoconducting surface or, in the case of the silicon vidicon, of a silicon diode array.

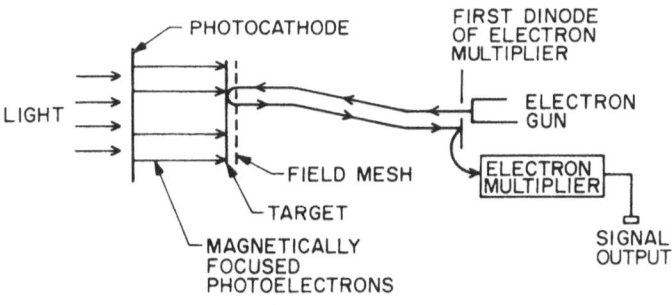

a. IMAGE ORTHICON

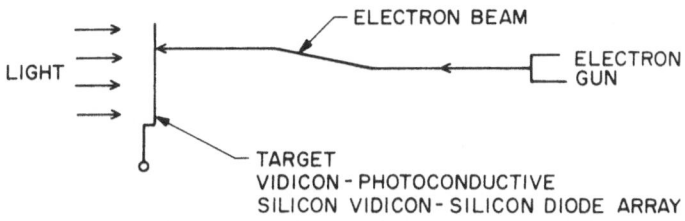

b. VIDICON

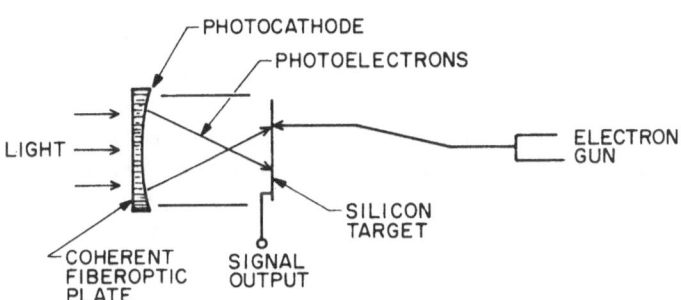

c. SIT

Fig. 6.14   Types of Television Tubes. (Courtesy of California Institute of Technology ©.)

Schematically the vidicon family is sketched in Fig. 6.14b. The light-sensitive target is made up of a material which changes its conductivity on exposure to light. Materials which have been used are amorphous selenium, cadmium selenide, gallium phosphide, cadmium sulfide, and lead oxide (which is used in the plumbicon). These substances all have a quantum efficiency of close to 100% for light which is strongly absorbed. The electron beam used to scan the target is similar to that used in the image orthicon, with a relatively low energy (of the order of 1000 volts), being focused and deflected by means of magnetic fields. Since an electron multiplier is not needed, the electron beam is not returned toward the gun, and the signal is taken off at the target. The fact that external amplification is needed introduces considerably more amplifier noise than with internal amplification, but in spite of this, these tubes show good sensitivity and high resolution with relatively low lag or image retention. Sensitivity as a function of wavelength, resolution characterics and persistence characteristics for a typical commercial vidicon are shown in Fig. 6.15.

One serious weakness of the vidicon is the fact that the target is rather easily damaged by overexposure. If not too badly overexposed it can be "cleaned up" by operating it for a few hours pointed at a uniformly illuminated wall or screen.

The silicon vidicon has approximately an order of magnitude greater sensitivity than the standard vidicon, and is not easily damaged by overexposure. It also has a much wider spectral response, and is particularly good toward the red end of the spectrum. It does not have as good resolution characteristics, and poorer image persistence characteristics than the standard vidicon. The relative performance can be studied by comparing Figs. 6.15 and 6.16.

The basic structure of the sensing system for the silicon vidicon is illustrated in Fig. 6.17. The optical image (or, in the case of the SIT tube to be discussed next, the electron image) is focused on the silicon target. This is made up of a closely spaced matrix of p-n junctions, about $14\,\mu$m center to center. The photons striking the target cause multiple dissociation of electron-hole pairs. The holes are collected at the p-side of the diode, charging the conducting top-hats positively. The target is scanned by an electron beam, as in the standard vidicon, neutralizing the positive charge on the top-hat as it passes. The neutralizing current furnishes the signal, which is processed in the same manner as that from the standard vidicon.

The silicon vidicon target has an excellent capacity to store information for long periods of time, although there will be a build-up of charge due to dark-current in a few seconds when operated at normal temperatures. When cooled to liquid nitrogen temperatures, the image can be accumulated for hours if it is not removed by scanning. This has been used effectively in astronomical applications, and might be considered as a possibility for autoradiography or other applications where long exposures are necessary. In this case the electron gun should not be activated until time to read out the signal. With a slow-scan readout, the data can be digitized and recorded on digital tape. In astronomical applications this system has proved to be linear over a range of integrated intensities of at least 1000, with no evidence of reciprocity failure such as is inherent in photographic film.

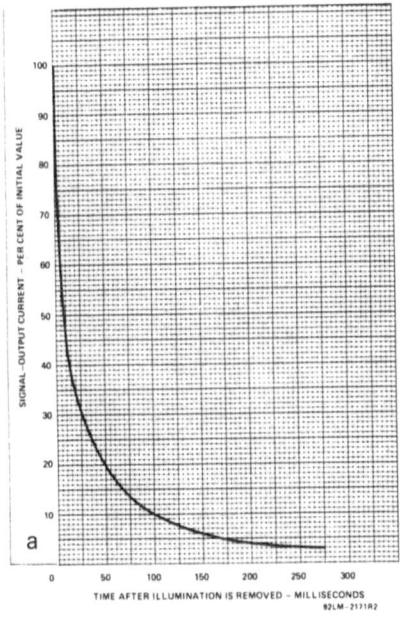

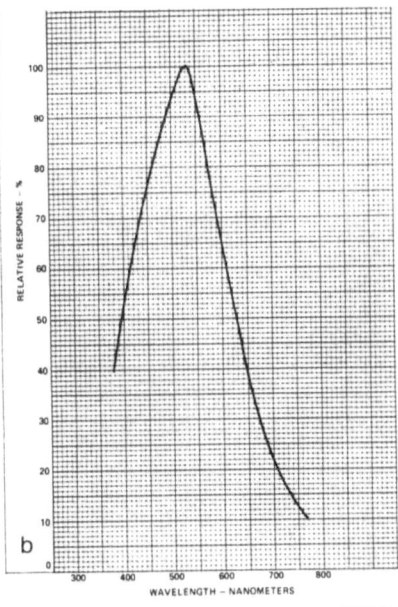

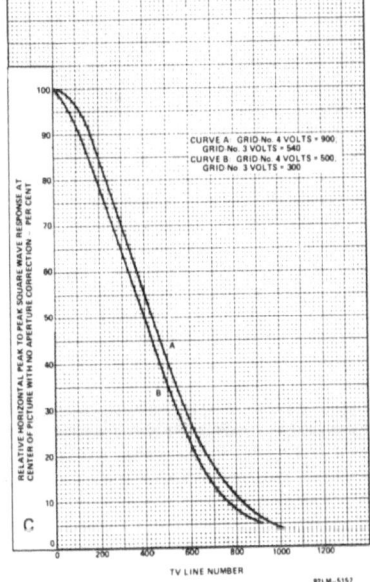

Fig 6.15 Characteristics of Vidicon Tube. (Courtesy of RCA.)

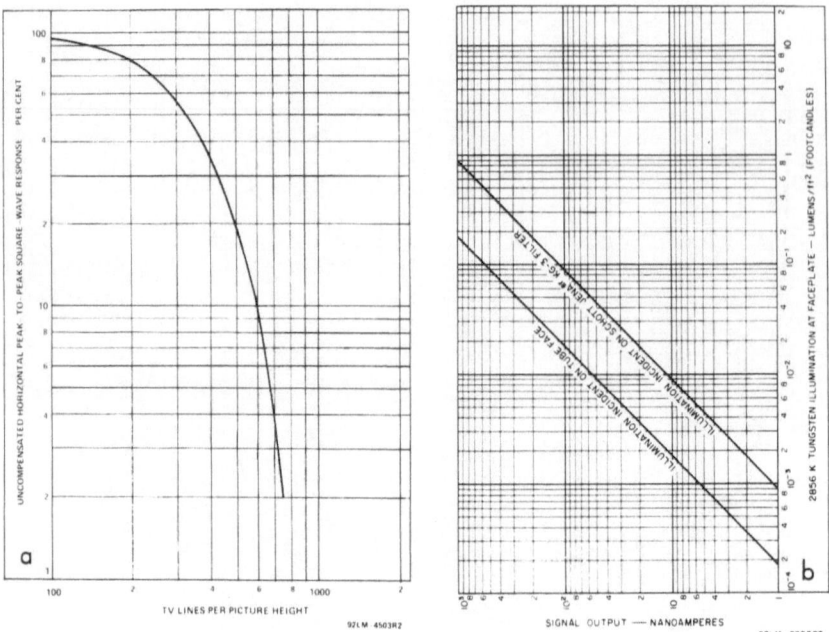

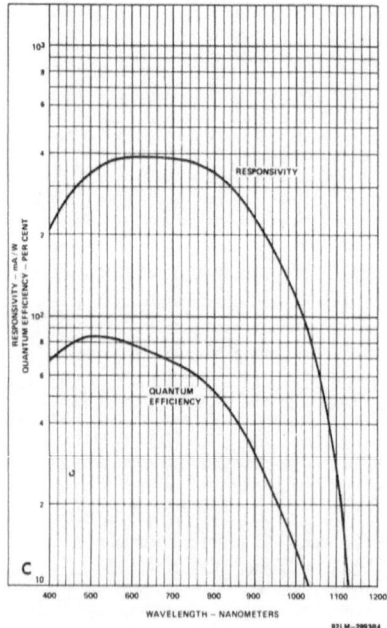

Fig. 6.16 Characteristics of
Silicon Vidicon. (Courtesy
of RCA.)

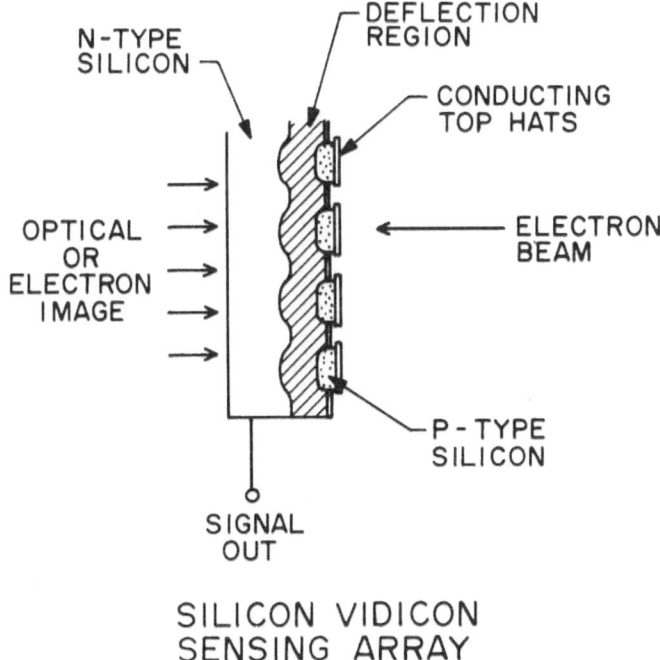

N-TYPE
SILICON

DEFLECTION
REGION

CONDUCTING
TOP HATS

OPTICAL
OR
ELECTRON
IMAGE

ELECTRON
BEAM

P-TYPE
SILICON

SIGNAL
OUT

SILICON VIDICON
SENSING ARRAY

Fig. 6.17 Sensing array of silicon vidicon. (Courtesy of California institute of Technology ©.)

The silicon intensifier tube (SIT) consists essentially of a single stage image intensifier (Fig. 6.14c) in which the electrons from the photocathode are electrostatically focussed directly onto a silicon vidicon target without the step of conversion into light by means of a phosphor. The accelerated electrons are far more efficient than photons in producing electron-hole pairs in the silicon diode array so that the overall sensitivity is extremely high in the region of efficient response of the photocathode. The contrast transfer functions, persistence characteristics and photocathode response are illustrated in Fig. 6.18.

SIT cameras are commercially available, but these have generally been specifically tailored for use in reconnaissance systems where the light level may vary over extremely wide ranges, so that a combination of electronic gain control and lens aperture control is built in. A camera based on a SIT has been in use in the author's laboratory for several years for quantitative imaging photometry of fluorescent tracers. We have built our own electronic circuits without any automatic gain control (although commercial camera firms should be able to modify their cameras to eliminate this feature) and find the system to be highly linear in response over a lumi-

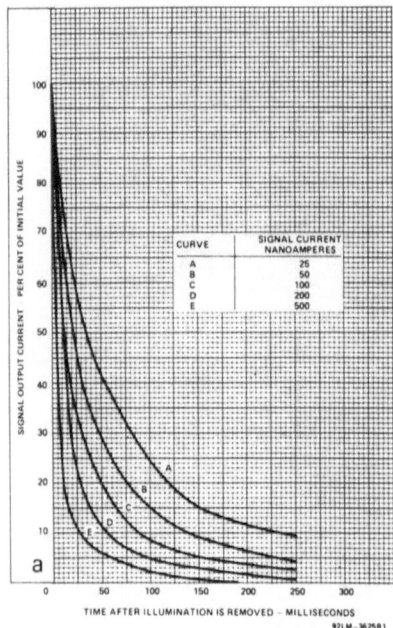

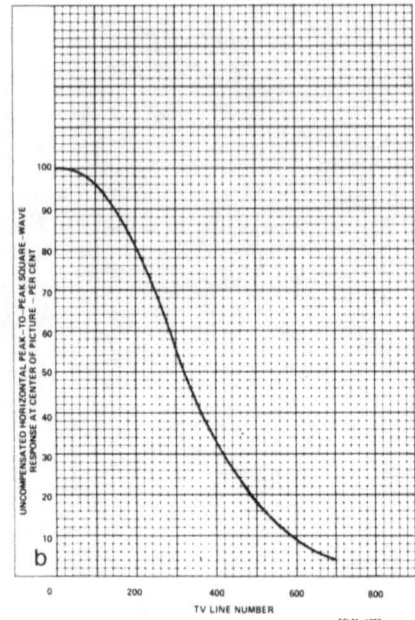

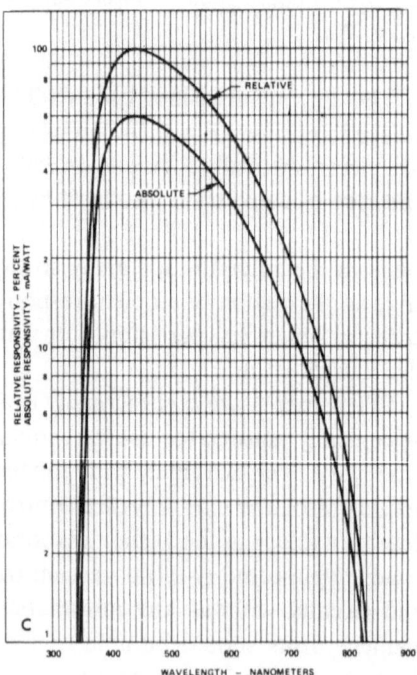

Fig. 6.18 Characteristics of SIT. (Courtesy of RCA.)

nance range of 1000 with no evidence of reciprocity failure. To give an indication of sensitivity, levels of fluorescence which we can record in real time (60 fields or 30 frames per second) with excellent contrast on the videomonitor or with videotape recording require exposures of about 15 seconds when the microscope image is projected directly onto Tri-X film (so there are no additional lens losses) instead of on the SIT camera target. We feel that this camera should possess immense potential for real time recording of fluorescence angiograms of the retina.

Solid state television cameras. The technological developments in solid state image sensors are being carried out so rapidly that any attempt to summarize the current state of the art would be out of date before this book is in print. There are a few such cameras on the market, but their primary virtue at the moment is their compactness, since the full capability of this technology is far from available commercially.

Such a sensor consists of a rectangular array of photosensitive elements which are mounted in a thin sandwich, one side of which is transparent so that the image can be formed on the array. These are essentially solid state detectors based on the properties of silicon, so the spectral response is similar to that of the silicon vidicon (Fig. 6.19). The resolution depends strongly on the closeness of spacing of the arrays, a technology which is rapidly improving.

An important feature of these devices is the way in which the signal is read out. The currently available RCA Silicon Imaging Device (SID) camera for which the spectral sensitivity and resolution curves are given in Fig. 6.19, has a 512×320 element array, of which 256×320 elements are used for image sensing and the other half for image processing. Two factors which have been important in making these devices practical have been improvement in the uniformity of behavior of the individual sensing elements and development of methods of readout of the signals. In currently available devices, the number of external connections is greatly reduced by using sequential readout across the horizontal sensing arrays of 320 elements each. Eventually it is expected that predetermined local regions can be read out without having to display the entire image. One particularly outstanding advantage of these sensors is that there is virtually no image persistence: each readout clears the sensor.

The sensitivity of such systems is now comparable with that of the silicon vidicon, although considerably more elements will be needed to make a marked improvement in the resolution.

Several groups are actively pursuing the development of an intensified solid state sensing system. Considerable progress has been made in de-

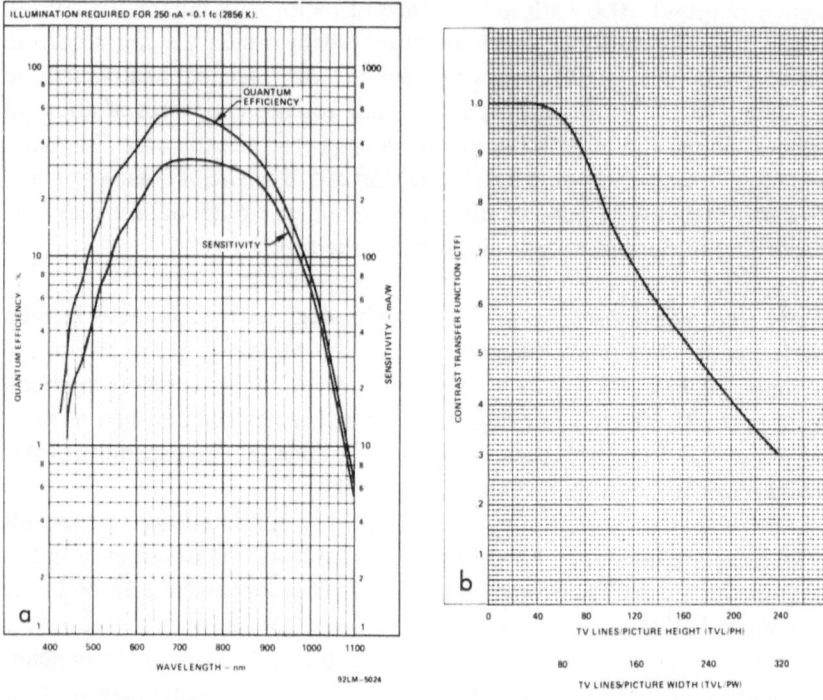

Fig. *6.19* Characteristics
of SID. (Courtesy of RCA.)

veloping arrays which can be bombarded with electrons to give a sensitivity
comparable to the SIT. Two configurations are being worked on. In one of
these, the solid state target will be placed in an evacuated chamber and the
electron image focused on it in the same manner as in the SIT (Fig. 6.14c).
The readout, however, will be done electronically rather than by means of
an electron beam scan, so the tube will be much shorter than the SIT.
Another approach is to use what is called proximity focus: the photocathode and the image sensor will be sandwiched in close proximity so that the
accelerated electrons will bombard the silicon array directly. This will
make for a very compact system, although the high voltage needed to acce-
lerate the electrons may pose some packaging problems, particularly if the
sensor is to become an integral part of an ophthalmoscope 'or endoscope.

Considerable effort is also being expended on digital storage and processing of the image data. It may be some years before these methods enter
clinical medicine, but they are already finding there way into biomedical
research.

6.5 Color recording

Because of the ability of all but the color blind to discriminate among extremely subtle differences in color, the use of color TV and color photography is often a much more effective way of conveying medically useful information than using a black and white system. This is particularly true of qualitative information, since color is usually a far better indicator of pathology than the shades of gray available on a black and white presentation. Closed circuit color television is certainly the most effective means of presenting live surgical procedures to a large audience.

No color recording system will precisely reproduce the colors of the object being recorded. A comparison of the color balance of a given scene taken with identical exposures on film from the same manufaturer's batch, but developed in different processing laboratories, will clearly illustrate this fact.

In medical applications color can be effectively used to assist in recognizing certain specific features rather than to record color differences with spectrophotometric accuracy. Although quantitative spectrophotometric measurements can be made from color film, calibration data should be recorded on the same piece of film on which the measurements are to be made. The technical aspects of such sophisticated use of color recording are beyond the scope of this introductory volume.

The use of color to discriminate among details is common in histology, particularly by the use of preferential stains. In living tissue such staining is not often feasible, but judicious use of color filters can often increase the ability to make useful discriminations. The observed colors will not be the "true" colors of the object recorded, but this is unimportant if the medically relevant information is made available.

With color television one can not only use color filters in front of the camera, but can also modify the color balance of the presentation on the monitor to accentuate features of interest. This latter operation is equivalent to modifying the processing procedure for color film, but this is a rather impractical procedure unless a large number of scenes requiring the same modification is required.

Neither color television nor color film is capable of as fine spatial resolution as the equivalent black and white system. As a rule of thumb, color television has only about one half the spatial resolution as a black and white system at high contrast, and even less at low contrast. On the other hand, contiguous features which have distinctly different colors, but would record at the same gray level in a black and white system, may be more readily discriminated with color than with a black and white system. Most color television systems in current use require more intense illumination than black and white systems, but it is possible to produce color cameras with extremely high sensitivity, although the cost will be several times as much as black and white systems of comparable sensitivity.

Although color films are available which can be "pushed" by special development procedures to have roughly the same maximum sensitivity as black and white films, such pushing results in a double loss: resolution and color balance. Within the normal ratings for exposure and development, with color films one is usually faced with the same type of trade-off between speed and resolution, although Kodak Ektachrome 200 Professional Film (which should be stored under refrigeration until it is to be used) is eight times faster than Kodachrome 25 Film, and has a slightly higher resolution at a contrast of 1000:1, but roughly the same resolution at a contrast of 1.6:1. The color balance is different between these films, and their relative ability to discriminate between features of importance must be tested for each particular application. Color films are constantly being improved, in speed, color balance, and resolution, so that each user should test out new products as they appear on the market to see if they lead to an improvement in his particular applications.

Both the effective speed and color balance of color film will depend critically on the nature of the source of illumination. Since the medical user is normally anxious to keep the total exposure of the subject to a minimum, flash illumination will usually prove to be more efficient than the use of steady light sources. Most flash lamps have a spectral output which is sufficiently similar to daylight that daylight films can be used without correction filters, although filters may be needed to accentuate the specific features of interest.

References

1. Cornsweet, T.N. (1970): Visual perception. Academic Press, New York-London.
2. Parrent, G.B. Jr., and Thompson, B.J. (1969): Physical Optics Notebook. Society of Photo-optical instrumentation engineers. Redondo Beach, California.
3. Rose, A. (1974): Vision: human and electronic. Plenum Press, New York-London.
4. Smith, W.J. (1966): Modern optical engineering. McGraw-Hill, New York.

7
STORAGE SYSTEMS

Henk G. Goovaerts, Henk H. Ros and Hans Schneider

7.1 Introduction

Recording of physiological signals has become a common tool in medicine during the past decades. An early example is the recording of the electrocardiogram (ECG) by Einthoven at the beginning of this century. He was one of the first physiologists to use a string galvanometer.

After World War II the recording of electrical signals has made a tremendous progress. Today, for instance, ECG registrations play an important role in the diagnosis of heart diseases. At present, the trend in RR intervals is an important parameter for the cardiologist. It will be clear that the construction of a so-called "tachogram" from a paper recording will be a time consuming method. In the afore-mentioned example the registrations can be made on magnetic tape recorders, so that the data obtained can afterwards be processed. Other examples are the recording of pressure and blood flow signals during surgery and the electromyogram. In a number of cases – where relations between several signal parameters must be assessed – storage in a memory system is essential. It is then possible to perform the calculations by electronic means. At present, most medical specialists and clinical investigators use some form of registration of the signals obtained from patients. Therefore, this chapter deals with the most important properties of recording devices. The intention is not to give a detailed description of storage systems, but a general review of available instruments. For detailed information the reader is referred to the literature listed in the references (1–7).

7.2 Graphic recorders

Most graphic recorders consist of the following basic components:
- an electromechanical device which converts an electrical input signal into mechanical movement.
- a stylus arm which transmits the movement from the electro-mechanical device to the stylus,

- a stylus that generates a written record on chart paper by moving across chart paper,
- a chart paper assembly consisting of a chart paper supply roll, a chart paper writing table and a chart paper take-up roll and
- a paper drive assembly to move the chart paper at constant speed across the writing table from the supply roll to the take-up roll.

The characteristics of graphic recorders can be broadly discussed in relation to the following aspects:
- *Recorder mechanism.* The basic mechanism underlying the conversion of input signal into a mechanical movement of the recording stylus uses either a galvanometric or a potentiometric principle.
- *Recording format.* The geometrical relationship between stylus movement and the chart paper will result either in a curved line or a straight line, transcribed on the chart paper when the stylus is moved to a new position (referred to as curvilinear or rectilinear recording, respecively).
- *Writing principle.* A written record is transcribed on the chart paper with either a inkpen stylus, a heated stylus or a rounded point stylus (referred to as ink writing, thermal writing and pressure writing, respectively).
- *General properties.* The performance of graphic recorders depends on general properties such as linearity, bandwidth, deflection range and recording speed.

7.2.1 *Recorder mechanism*

A commonly adopted conversion mechanism for recorders with a bandwidth greater than 5–10 Hz is found in the moving coil or galvanometric principle. However, in order to overcome the forces associated with stylus pressure and to provide good transient response, the conversion assembly used in a graphic recorder must develop considerably more torque than that used in a panel meter.

If one applies a step at the input of a system, the stylus will move abruptly to a new position. If this position is reached, overshoot occurs due to the inertia of the stylus. In order to prevent excessive overshoot, systems have to be damped. In Fig. 7.1 three different types of damping are shown.

A system is called critically damped if the stylus moves to its new position in the shortest time, without any overshoot. In this context it is important to ensure that the friction between stylus and paper contributes only a very small amount of damping, because otherwise the damping of the

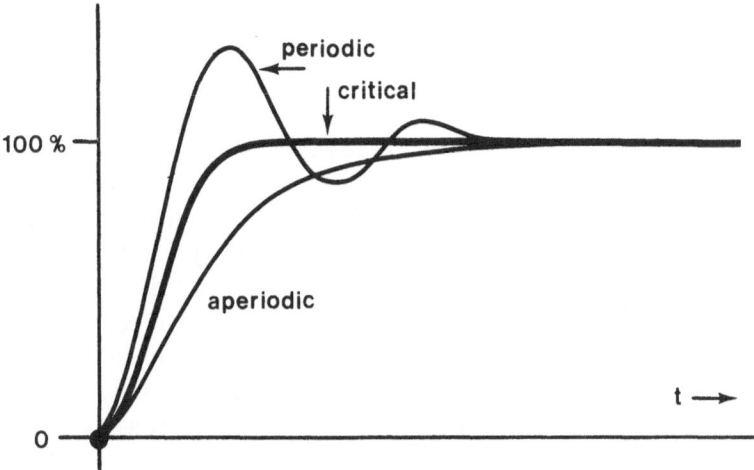

Fig. 7.1 Examples of three different types of damping in graphic recorders. The ordinate represents the relative deflection of the stylus in response to a stepwise change of the input signal.

system would depend on the type of chart paper being used. Most systems are or should be adjusted to a slight underdamping and thus show some overshoot, about 10%.

A second way of converting electrical signals into a proportional deflection of the stylus, is by making use of a servo system. Here the position of the stylus arm is detected through a contact mechanism, attached to the arm and in contact with a fixed slide wire or optically coupled to a light dependent resistor. In Fig. 7.2 the mechanism of a servo-recorder is shown.

The servo system will move the stylus arm and the stylus, until the po-

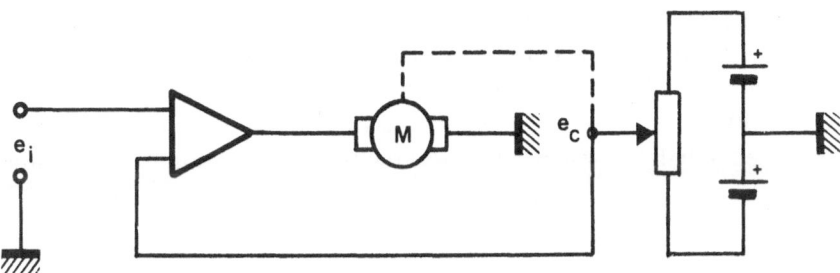

Fig. 7.2 Principle of a servo controlled system applied in potentiometric recorders. The resulting input signal $|e_i - e_c|$ will drive the motor M such that the wiper of the potentiometer is positioned on a step where $e_c = e_i$.

tential between stylus arm and input voltage is zero. The linearity and the accuracy of such a system depend on the characteristics of the slide wire or light dependent resistor and associated electronics. In general potentiometric recorders present a better linearity than galvanometric systems. Sometimes a variable capacitor is used as a position transducer of the servo recorder. This capacitor forms part of a bridge circuit supplied by an AC (Alternating Current) signal source. From this arrangement, the compensation voltage is derived through rectification. In general the frequency coverage of a servo system is limited to approximately 1 Hz. The amplitude associated wih this frequency is usually about 25 cm. One can increase the bandwith at the expense of a reduced amplitude. For a bandwith of 4 Hz, for instance, the amplitude achieved is 5 cm. In order to improve the performance several hybrid solutions are available. These use a position transducer attached to the galvanometric conversion system. The output of the position transducer is converted into a corrective signal and added to the galvanometer driving system.

7.2.2 *Recording format*

The geometrical relationship between stylus movement is either curvilinear or rectilinear. A pivoted stylus arm will transcribe an arc at its tip, when the arm is caused to rotate around its pivot point. A stylus at the tip of the arm will therefore transcribe a curved line on stationary paper. This recording format is referred to as curvilinear. In general, this format is easy and reliable but has two disadvantages:
- the configuration of the recorded signal is difficult to read and is less suitable for analysis,
- the chart paper used with curvilinear recorders requires a curvilinear grid.

Movement of the stylus in a straight line perpendicular to the direction of movement of the chart paper, would solve the problem – rectilinear recording. A simple solution is the passing of the chart paper over a fixed recording edge. This method, however, introduces a linearity error. This error will be less than 1%, if the total stylus deflection in both directions is limited to half the distance between the recording edge and the rotation point.

7.2.3 *Writing principles*

Ink writing. Ink-writing graphic recorders use an ink pen operating on a capillary system which is connected to the ink reservoirs. Other ink-writing styli are ballpoint or felt stylus.

Ink-writing recorders, having a capillary system, use ink requiring a substantial period of time to dry, thus preventing blockage of the ink pens. A different method of ink recording through a capillary system is achieved by using ink with a higher viscosity, brought on the chart paper under high pressure. This method requires specially treated chart paper. With the exception of the latter, the chart paper for ink recorders is relatively inexpensive. A disadvantage of ink-writing systems, is that, unless adequate care is taken, the ink capillaries will frequently block.

Thermal writing.   Thermal writing requires an electrically heated stylus and specially coated chart paper. For instance, one can use a black undersurface covered with a white, heat-sensitive coating. The heated stylus melts the coating locally as it travels across the chart paper, thus exposing the dark undersurface. The recorder requires little maintenance, but the chart paper is expensive. The frequency response of recorders with a thermal-writing stylus is usually lower than that obtained with ink-writing systems.

Pressure writing.   A pressure-writing system consists of a stylus which, while travelling across the reverse surface of the chart paper, presses the chart paper against a copying surface over a recording edge. Chart paper and copying surface are usually transported in opposite directions. A major advantage of pressure writing is that normal untreated paper can be used as chart paper. To obtain satisfactory writing at low chart speeds, an additional excitation of small amplitude is normally applied to the stylus. This garantees constant friction between stylus and chart paper, even at low recording speeds.

Ink-jet writing.   Ink-jet writing is accomplished through the use of a small ink jet attached to the recording mechanism, which provides a stream of small ink drops impacting on the chart paper. It is obvious that absence of friction and the small dimensions of the recording mechanism, provide this principle with a larger bandwidth. A disadvantage is again the blockage of the ink jet. Another disadvantage occurs at high frequencies where substantial phase errors arise. This will become clear if we compare the ink jet with a garden-hose that is rapidly moved from left to right and vice versa. Generally, the movement of the point where the jet of water touches the ground is delayed by one or more periods with respect to the movement of the garden hose. Finally, it is not advisable to apply ink jet recorders in conditions where longterm recordings are being made at very low chart paper speeds. The recording will become blurred due to surplus of ink.

A similar approach is made in the ink-jet printer/plotter where the ink-jet stream is acoustically broken into smaller drops. These are passed through a set of deflection plates, causing the drops to impact the paper in the pattern defined by the voltage applied to the deflection plates. However, the deflection range is small, being of the order of a fraction of a centimeter.

Photographic writing. A photographic-writing system consists of a small moving coil to which a mirror is attached. Rotation of the coil will deflect the beam from a light source, over an arc proportional to the amount of current that is fed through the coil. The reflected light beam is then projected on light-sensitive chart paper, moving at a constant speed. A disadvantage of photographic writing is that the recording will not be visible before the chart paper is developed. Using an ultraviolet light source instead of a visible light source, it is possible to display the recording on specially treated photographic chart paper, so that the recording is visible almost instantaneously. The bandwidth of such a writing system will be approximately 2 kHz for a deflection of about 25 cm. At an amplitude of approximately 6 cm, the bandwidth will be 10 kHz. The special treatment of the chart paper, required for photographic writing, makes this technique relatively expensive. Due to the low contrast of UV recordings, photographic reproduction of the recordings could be difficult. However, reproduction by means of hard copying is very satisfactory.

Electrostatic writing. Electrostatic writing is performed by means of a set of styli attached to the recording mechanism, which is fixed across the chart paper being transported across a back plate. In the writing mode, charge on the paper is produced by the potential difference between the writing stylus on the one hand and the paper and the back plate on the other. A driver circuit controls the stylus voltage. The fixed writing head, containing multiple conducting styli, is used to place electrostatic charges on the paper in the form of dots. The styli, fixed in a linear array, are individually activated by on/off signals, thus producing a permanent record. Resolutions of 30–40 dots per cm are common. The paper is usually moved over the writing head in precise increments; thus, the combination of scanning in the X-direction and stepping in the Y-direction digitally, covers every predetermined spot on the paper. The chart paper is usually treated with zinc-oxide. The charge-image on the paper is developed through a toning system. The toner particles adhere to the paper wherever a charge exists, resulting in a permanent, high contrast image on the paper.

7.2.4 *General properties*

In choosing a recorder one should take the following aspects into account:
- bandwidth,
- deflection range,
- linearity,
- recording speed.

Bandwidth.   In terms of the bandwidth, recording systems can be divided as follows:

potentiometric (servo) recorders	0–1 Hz
ink-writing recorders	0–100 Hz
pressure writing recorders	0–250 Hz
ink-jet recorders	0–1 kHz
photographic recorders	0–5 kHz
electrostatic recorders	0–5 kHz (depending on number of channels)

To assess the approximate bandwidth necessary to record a certain configuration without severe distortion, one can use the fastest rise time in the signal of interest, as a reference. This is shown in Fig. 7.3.
In the figure, t_1 corresponds to a half sine wave period. Therefore, the bandwidth of the recorder should be greater than:

$$f = \frac{1}{t_2} = \frac{1}{2t_1}$$

Usually the configuration to be recorded contains frequency components higher than $1/t_2$. Therefore, a more accurate recording of the configuration will be obtained if we raise the bandwidth requirements to approximately

$$f = \frac{5}{t_2}$$

Linearity.   A perfectly linear, writing system would produce an output (pen deflection) that is directly proportional to the input signal to the recording mechanism; i.e. equal increments of input should cause equal changes of trace amplitude in any region of the chart. The latter is especially important in situations where trend curves or other traces span

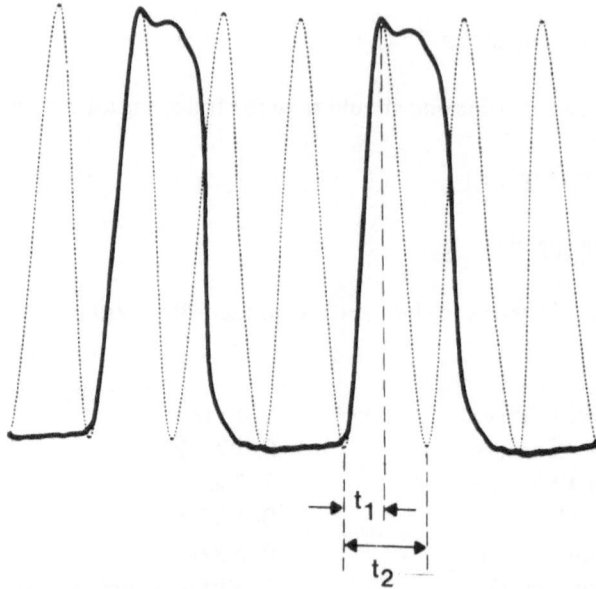

Fig. 7.3 Determination of the highest frequency components in a signal by means of an equivalent sine wave which is drawn by the dotted line. The solid line represents the recorded curve.

a major portion of the channel width. The concept of non-linearity is sometimes referred to as "the prevailing condition under which the input-output relation fails to be a straight line." This relationship can be expressed as an input-output curve. Non-linearity is expressed in several ways and it is necessary in any specification, to state direction and magnitude.

Some instruments behave like Fig. 7.4a; the error increasing with increasing input. On the other hand one finds situations like Fig. 7.4b. The error is largest for middle inputs but decreases towards the beginning and end of the scale. In both situations one is dealing with a maximum deviation in one direction. For simple quadratic (parabola) non-linearity, the best would be something like Fig. 7.4c. The error being positive in one part and negative in another part. This method gives an even better specification than the two methods mentioned before. It uses the best straight line for minimum error. Moving the calibration point inward from the channel edge, splits the error into a plus and minus deviation rather than all in one direction. This is the way in which non-linearity specifications are usually stated.

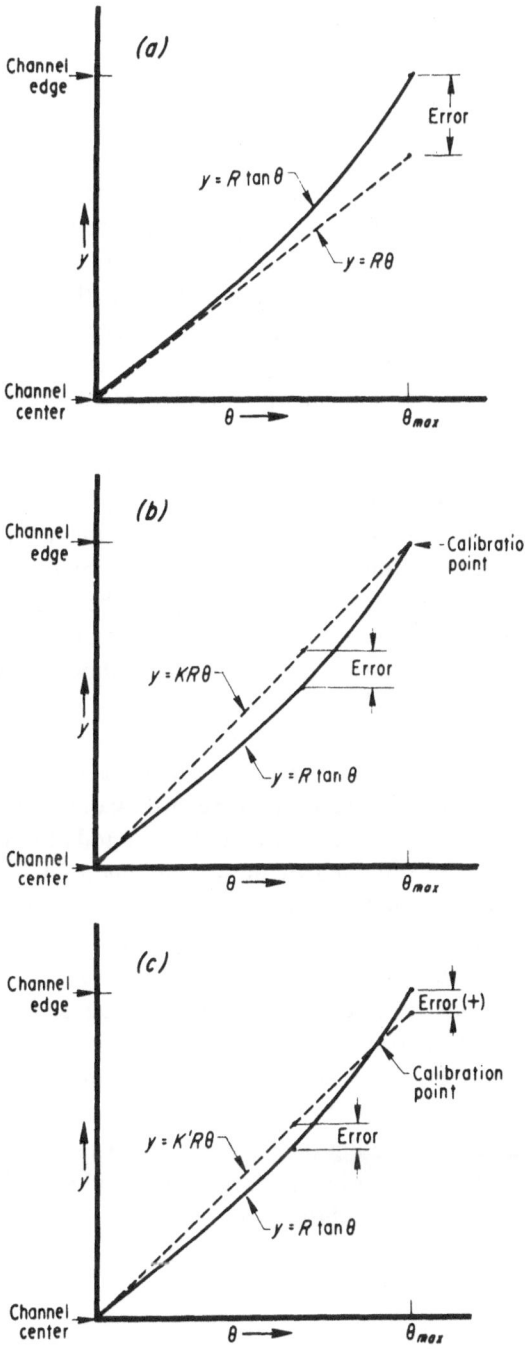

Fig. 7.4 Three common methods of specifying non-linearity (see text).

7.3 Storage from the oscilloscope screen

The oscilloscope is the most commonly used device for the display of analog data. The conventional display is a plot of current or voltage amplitude as a function of time. Generally, a time base is generated within the oscilloscope by the horizontal deflection system. The primary function of the horizontal amplifier is to convert the time-base ramp to an appropriate voltage that will drive the deflection plates of the cathode-ray tube (CRT). An oscilloscope CRT is a special purpose vacuum tube consisting of an evacuated glass or ceramic envelope, a heated cathode, various control electrodes and a phosphor coat which represents the screen. The signal to be displayed is amplified in the vertical amplifier system. To obtain various deflection sensitivities, the gain of the vertical amplifier must be variable both in calibrated steps and continuously. A great desadvantage of the oscilloscope is that it does not provide a permanent record of these data. Storage can be obtained if a storage oscilloscope is used. This type of oscilloscope stores a trace on the CRT phosphor; an erase control is activated manually or by remote control. Infromation displayed on a CRT can be photographed using either conventional photographic techniques or moving film photographic techniques. Conventional photography of information displayed on a CRT requires the use of a special oscilloscope camera. While this technique is entirely satisfactory for several applications, it does not provide a continuous recording of data as a function of time, but only a record of data over a discrete period of time. Continuous information might be obtained by photographing a CRT display with a movie camera. The movie film, however, can only be displayed by projection and does not provide a convenient record that can be studied with ease. A modification of movie photography is known as continuous-motion photography. In this technique continuous motion of film or sensitized paper past an oscilloscope CRT is achieved via a lens system using a special camera. Since horizontal information is provided by the moving film, the horizontal sweep of the oscilloscope display is turned off, leaving only vertical information deflecting the CRT beam. This recording system can be compared with a conventional chart recorder with vertical movement of the CRT beam acting similarly to the vertical movement of the stylus on the chart recorder. Film transport speeds are ranging from 0.05 mm sec^{-1} to 1000 mm sec^{-1}.

7.4 The magnetic tape recorder

7.4.1. *Introduction*

The magnetic tape recorder is an analog storage device. In these recorders electrical signals are stored in such a way that they can be reproduced in real time when required. It is advisable to be informed about the recording through simultaneous monitoring of the recorded analog signal and the actual input signal.

Most magnetic tape recorders have the capability of recording information at one speed and reproducing it at a different speed allowing time compression or time expansion of the stored information. Some basic analog recording techniques, used for storage of signals on magnetic tape, are direct recording (DR) and frequency modulation (FM). Converting analog data into digital data with a pulse-code modulation technique is another possibility. The proper transfer of signals onto the tape depends on the physical properties of recording head and magnetic tape. Accurate reproduction of the stored signal depends also on the equalization, if analog data are to be reproduced, and on the type of encoding used for the storage of digital data.

7.4.2 *Head properties*

In Fig. 7.5 a conventional record head is shown. The face of the head is contoured to allow a smooth transition of the tape across the head. If the current and therefore the flux is changed sinusoidally, the remanent flux on the tape varies in a similar manner. When the current reverses, the magnetic polarity also reverses. Increasing the frequency causes the spatial magnetization pattern wave to become shorter. As this spatial wave length becomes comparable to the gap length, the induced electromagnetic force, in the reproduce mode, decreases. For a gap length equal to one wave length, the resultant flux through the head is zero. The frequency at which this occurs is called the "extinction frequency" and depends on gap length and tape speed according to the following relation:

$$f_{ext} = \frac{s}{\lambda}$$

where s = tape speed in m/sec and λ = gap length in m. If the frequency is increased to reduce the wave length further, the output again slightly increases. Except for special applications, an output below f_{ext} is used.

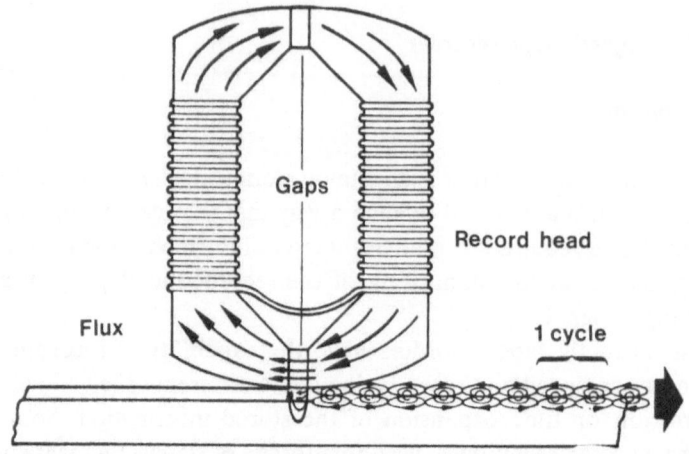

Fig. 7.5 Principle of tape magnetization by means of a recording head. The magnetic flux crossing the gap magnetizes the iron oxide on the tape, which is moved at constant speed.

Medium frequencies can be defined as those that produce wave lengths greater than the gap length but smaller than the head face. The head can be regarded as providing a low reluctance path for the flux between pole pieces and the head intercepts all the flux linkages between half wave length poles. If the wave length increases, that is, if lower frequencies are applied, the head does not provide a low-reluctance path for the total flux and an extra reluctance is introduced in the path of some of the flux.

7.4.3 *Tape characteristics*

Among a great variety of characteristics, only some are important in obtaining a good signal-to-noise ratio. Tape noise is mainly generated by surface imperfections causing amplitude modulation of the recorded signal. Long term variations are normally due to inconsistency of the oxide layer or changes in thickness. The so-called "drop-outs" can be caused by surface imperfections, particles of dust or complete loss of oxide on the tape.

7.4.4 *Direct recording*

In the direct recording mode, a bias signal with a frequency between 70–100 KHz is superimposed on the signal to be recorded. This is done to achieve better linearity between tape magnetization and magnetizing force, and hence recording current. The recording head converts the

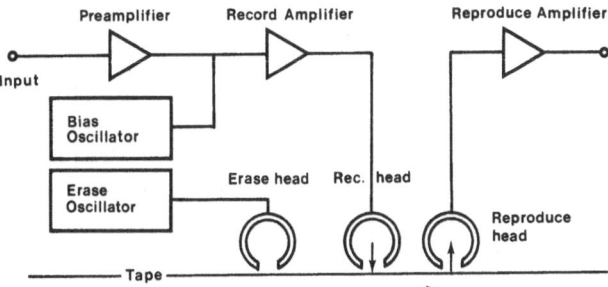

Fig. 7.6 Block diagram of a direct-recording (DR) system. The signal to be recorded is fed through a preamplifier. Then a bias signal is added. The resulting composite signal is fed into the recording head. Commonly, only one oscillator is used that provides both the erasure and the biasing signal.

current into a varying magnetic flux which changes the residual magnetism on the magnetic tape as it moves past the recording head. This magnetism on tape, in turn, produces an output voltage from the reproduction head. This voltage passes through an equalizing amplifier which corrects for transfer losses due to tape head properties.

Due to the head losses, the low frequency response is poor for frequencies below 100 Hz. Another disadvantage of direct recording is the variation in the output level in the order of about 10%, which is a severe limitation in many applications. In Fig. 7.6 the block diagram of the direct recording mode is given.

7.4.5 *Frequency modulation*

In the frequency modulation mode, which is shown in Fig. 7.7, the input signal is fed to a modulator where it is used to frequency modulate a carrier.

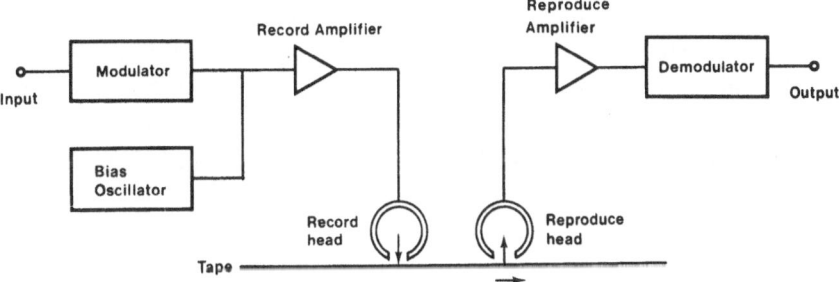

Fig. 7.7 Block diagram of frequency-modulation (FM) recording and playback. The signal to be recorded modulates a carrier that will be recorded on tape. Usually the recording head is driven into saturation. If so, the bias oscillator can be omitted. Reproduction is achieved by demodulation of the carrier recovered from tape.

The modulated carrier is then fed into the recording amplifier which provides the recording current necessary to drive the magnetic tape-head.
Saturation recording without bias is commonly used. The bandwidth of
such a system is proportional to the recording speed. A major advantage of
FM over DR recording is that FM recording permits frequencies down to
and including DC. In the reproduction mode, the signal derived from the
reproduced head is fed into a demodulator which converts the carrier
frequency into a DC voltage. The signal-to-noise ratio obtainable with
FM recording is between 40–50 dB. The deviation range of the carrier is
usually between 20–40%. Some carrier frequencies at different tape
speeds, are:

Table 7.1

$1\frac{7}{8}$ ips (4,76 cm/s) –	3375 Hz
$3\frac{3}{4}$ ips (9,52 cm/s) –	6750 Hz
$7\frac{1}{2}$ ips (19,05 cm/s) –	13500 Hz
15 ips (38,1 cm/s) –	27000 Hz
30 ips (76,2 cm/s) –	54000 Hz
60 ips (152,4 cm/s) –	108000 Hz

In the FM recording mode, amplitude stability between the signal applied
to the recording head and the signal received from the reproduction head,
need not be considered because all instability will either be removed in a
limiting stage in the reproduction amplifier or will be of no consequence
due to the insensitivity of the demodulator to amplitude modulated signals.

7.4.6 Pulse-code modulation

Recording of electrical phenomena by means of pulse-code modulation
(PCM) has the advantage that the signal-to-noise (S/N) ratio of the signal
to be stored, is established prior to the mechanical conversation. The
signal to be stored is converted into a digital signal representing the amplitude of the input signal. In Fig. 7.8 the simplified block diagram of the
pulse-code modulation recording part of a system is shown. The signal-to-
noise ratio is determined by the analog to digital (A/D) converter used
such that:

$$\frac{S}{N} = n.6 \text{ dB}$$

where n represents the number of bits into which the signal will be converted. For instance, if we apply a 12-bit A/D converter the S/N ratio of the

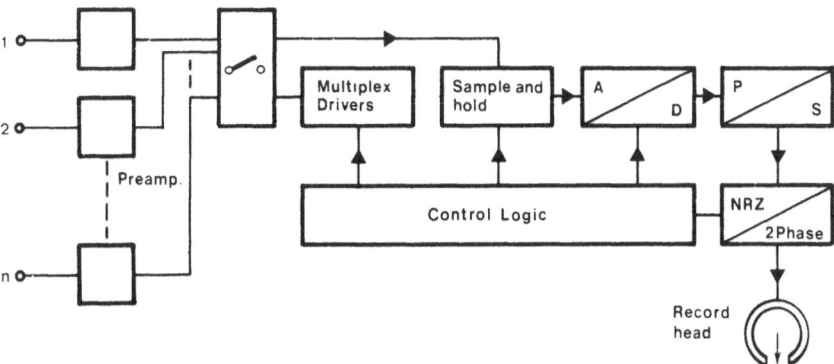

Fig. 7.8 Block diagram of a PCM recording system. From left to right the signals are multiplexed under control of the multiplex driver. Next, via a sample/hold circuit the signal is fed into an A/D converter, then the digital parallel information is converted into serial information and through a coding system (NRZ/2-phase) recorded on tape.

recording will be:

$$\frac{S}{N} = 12.6 = 72 \text{ dB}$$

This is much better than could be obtained with the direct recording mode, but also a significant improvement over the FM method, discussed previously. An additional advantage of the PCM system is that it is possible to multiply several channels on one recording track because of the sampling character of the data acquisition. In Fig. 7.9 the block diagram of a PCM reproducing part of a system is shown.

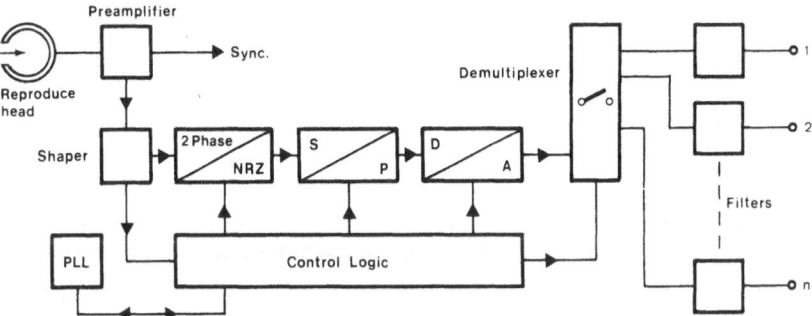

Fig. 7.9 Block diagram of a PCM reproduce system. The signal, recovered from tape is preamplified and a synchronization pulse is derived. After preprocessing in the shaper and the 2-phase/NRZ block the signal is D/A converted and fed into filters. The control logic is phase locked from the tape in a phase-locked loop (PLL).

7.4.7 Recording codes

Recording of PCM data requires a special method according to which the signal is stored on the tape. The amplitude of the signal to be recorded, is converted into a string of binary numerals consisting of n bits. In order to put these zero-one data on tape one could transform them into proportional pulse- amplitudes or -durations. A simple binary pulse duration code can be constructed using one pulse width for a zero and a second pulse width for a one.

An interesting variation in pulse code duration encoding is the use of two pulses to represent a signal instead of one. The information is transmitted in terms of the relative distance between a first (reference) pulse and a second (data) pulse. In this form of encoding most problems associated with pulse duration encoding are met, and it is not currently in widespread use.

A "return to zero code" is a code in which the information transferred is determined as a function of the position of a pulse in the transmission, rather than being a function of the shape or amplitude of the pulse itself. In such a code, the transmission time is divided into increments of time relative to a known starting time. To each increment is assigned an information value, and, if a pulse appears in that increment of time (time slot), the information value is affirmed. Thus, the time slots could be allocated for the conveying of information about ten switches. If a pulse appears in the slot associated with a particular switch, the switch is closed. If no pulse appears the switch is open. Such a code has the advantage that, as in pulse counting codes, it is sufficient to determine the presence of the pulse. Determination of the amplitude or width of the pulse is not necessary. Thus, an appreciably larger amount of noise is necessary to transform the signal enough to change its value. However, generating a serial binary code takes a certain amount of time dependent on the number of bits. It will be clear that such a time consuming procedure reduces the bandwidth of the signal to be stored because the number of samples fitting into a certain time interval diminishes with increasing length of the binary serial code. Increase of the recording speed enables storage on tape of signals at a larger bandwidth, but reduces the recording time.

The accuracy problem consists of determining correctly in which time slot the pulse occurs. If for instance, the receiver clock measures its time increments at a faster rate than is used in the transmitter, a pulse intended for a particular time slot is advanced in time by its own width. In this situation, it will be impossible for the receiver to determine whether the pulse is associated with the next of the previous time slot. In spite of this drawback some improvements have been made with this type of encoding as compared with the pulse codes in which one signal characteristic is used to represent a one, and another characteristic to represent a zero.

7.4.8 *Tape transport mechanism*

The basic components of the tape transport mechanism are an active capstan, an idler to move the tape at a constant speed past the magnetic heads, a tape supply reel and a take-up reel. The predecessor of such a tape-transport system is shown in Fig. 7.10a. With this configuration, however, interaction of the supply and the take-up systems (reels) with the tape in the head area causes too much speed variation, referred to as flutter, for instrumentation purposes. Isolating elements are therefore added to negate the influence of the two systems resulting in the open-loop tape drive as shown in Fig. 7.10b. Supply system flutter is controlled by a tape-driven inertia idler equiped with a flywheel to counteract erratic tape velocities. Because of its weight the inertia idler introduces another problem. An appreciable amount of time is required to bring the idler up to tape speed.

The closed loop tape drive in Fig. 7.10c. was introduced to avoid this slow starting and to provide shorter lengths of unsupported tape. The ultimate in isolation was the zero-loop tape drive which is shown in Fig. 7.10d. Here the tape is pressed against the heads by the capstan itself so that the idlers can be omitted.

Tape speed stability is an important consideration in instrumentation recorders. It will be clear that change in speed results in a change of frequency of the recorded signal. This is of particular importance when the FM method of recording is applied. The FM demodulator will convert tape speed variations into an additional signal superimposed on the recorded signal. A servo system within the recorder copes with this problem. The signal for the servo mechanism is derived from a tachometer coupled to the capstan or flywheel and delivers a frequency which is proportional to the rotation speed of the capstan. This frequency is compared with a frequency standard by means of a phase-sensitive detector. The resulting error signal is fed into an amplifier which controls the capstan motor.

Regardless of the accuracy of speed control in recording equipment, precise reproduction of recorded frequencies and time intervals can be achieved only by use of a speed control system in the playback mode, to eliminate the effect of geometrical changes in tape between recording and playback. Therefore, a reference signal has to be recorded on the tape. The servo loop controls the tape speed so that the reference signal re-

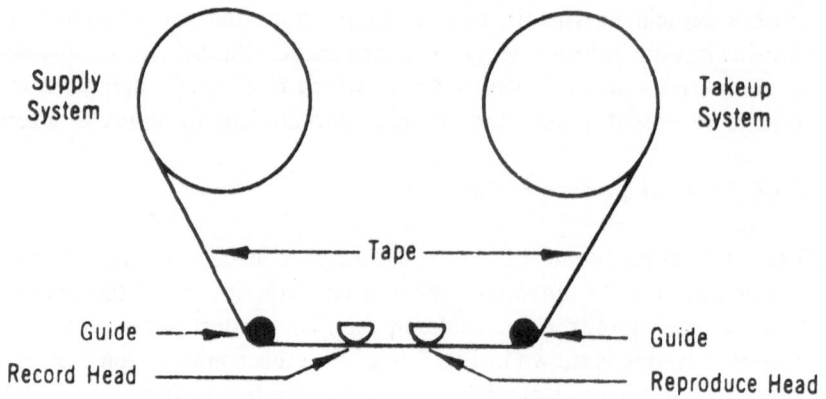

a Basic Magnetic Tape Transport

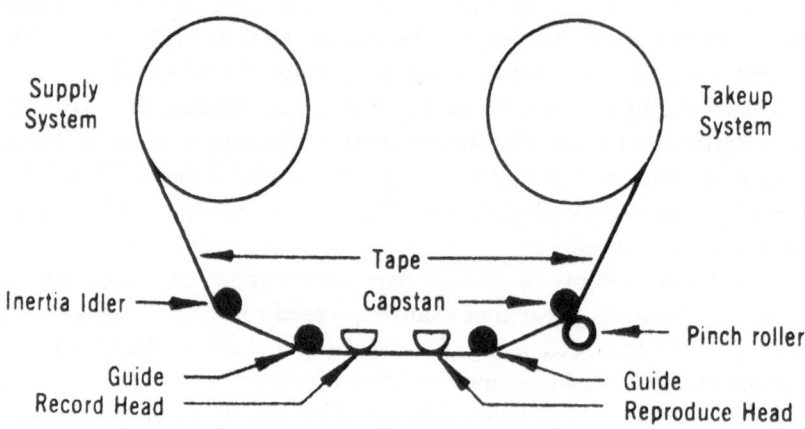

b Open-Loop Tape Drive

Fig. 7.10 Some examples of tape transport systems.

covered from the tape is locked in phase with the internal frequency standard of the instrumentation recorder.

7.4.9 *Time base error, flutter, and noise*

Faithful reproduction of a signal includes preservation of the timing of the original signal. The reproduction should be delayed "synchronously" with respect to the recording. Deviation from synchronism results in an ac-

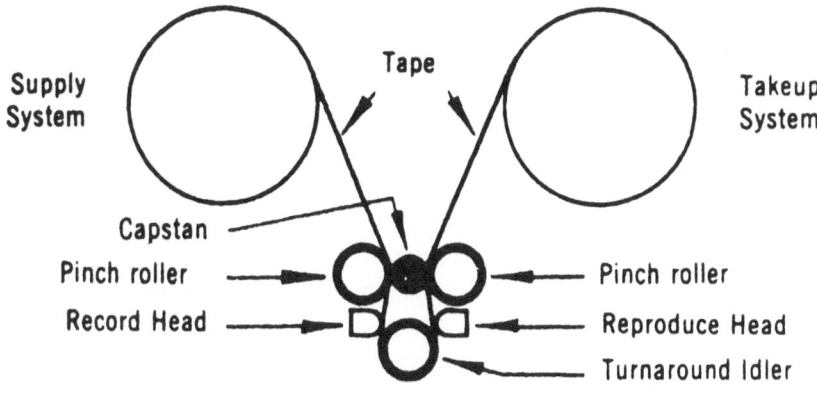

c Closed-Loop Tape Drive

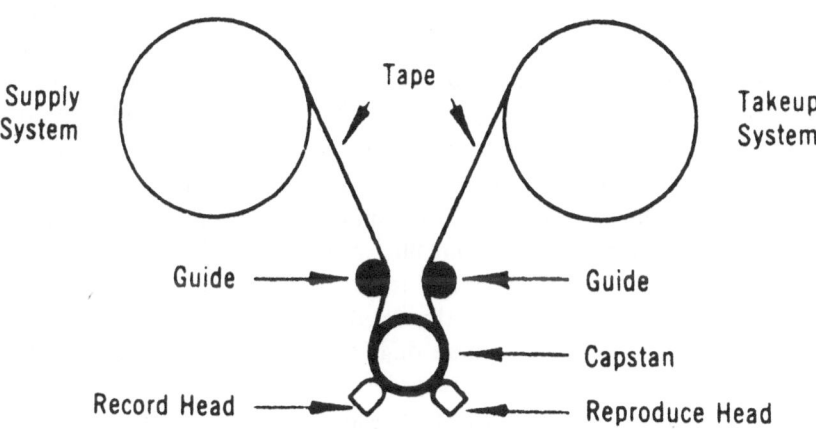

d Zero-Loop Tape Drive

Fig. 7.10   cont.

cumulation of time distortion called time base error, which is expressed
in percent variations from the absolute selected speed.

Relative tape speed instability is referred to as "wow" if occurring at
frequencies less than about 2 Hz and as "flutter" if occurring at frequencies
higher than about 2Hz; time base error is thus the time integral of "wow"
and/or "flutter." The value of these relative-speed errors is measured in
percent peak-to-peak.

Generally, flutter signals contain a variety of frequencies and amplitudes so that the combination is similar to broadband noise and often confused with it. The gadgets to keep the tape speed constant result in specific peaks in the flutter spectrum. To specify the effect of flutter with one figure can be extremely misleading, for it says nothing about the distribution of flutter in the spectrum of interest.

Noise in the magnetic recording system can be introduced by the tape as mentioned in 7.4.3 but can also be the result of cross talk in a multitrack recorder. Cross talk in a multitrack recorder is closely related to the physical spacing of the recorded tracks on the tape. It is caused by inductive coupling of the leakage flux between the gaps of adjacent heads and through coupling of leakage flux from the signals recorded on one track into the head of another track.

7.4.10 *Cassette recorders*

Due to poor tape stability, a common disadvantage of most cassette recorders, the signal-to-noise ratio of signals recorded on a cassette recorder by means of frequency modulation is significantly lower than in other systems incorporating supply and take-up reels. The velocity variations of the cassette-tape drive system result in a flutter component added to the signal which is derived from the tape. The flutter can be compensated for by recording a stable frequency reference singal on one track of the tape, recovering the flutter signal by FM demodulaton of the reference signal on playback and subtracting the flutter signal from all data channels. Cassette recorders usually provide 4 tracks on the tape, so, if one channel is used for flutter compensation this results in a reduction of 25% of data storage. In some situations where digitized data are acquired only over brief periods of time, the cassette recorder can be applied for storage of these data. In such situations the tape drive system consists of a stepping motor which moves the tape across the head during the recording of binary word.

7.4.11 *Digital recorders*

As mentioned in section 7.4.2 digital recording is used to avoid degradation of high-accuracy data. At least 10 bits are usually required for storage of such data. At high sampling rates, it is necessary to record all bits representing one data sample in a single line across the tape. Additional tracks are needed for a clock signal, parity check and channel identification. Therefore, storage of such data requires 12–16 tracks on a single-head stack. This type of recording is referred to as parallel recording. At lower sampling rates, it is possible to record one data sample on one track if the digital

Table 7.2 Comparison between methods of magnetic tape recording. Analog recording can be achieved in a direct way (DR) or by means of indirect recording. Indirect recording is possible through frequency modulation (FM) or by means of pulse-code modulation (PCM).

Type	Advantages	Disadvantages
Analog (Direct) 60 ips	– High-frequency limit, 100 kHz – Usually one signal per track – Simplest method	– Low frequency limit 50 KHz – 1% tape speed variation results in 1% error – Drop-out effect is serious – Stringent tape speed control required
Analog (FM) 60 ips	– Low-frequency limit, O Hz – Relatively insensitive to tape dropout – One signal for wide deviation or up to 18 inputs (narrow deviation) for one track	– High-frequency limit 10 kHz – 1% tape speed variation results in 2.5% error – Complex circuitry – High tape-transport quality required
Analog (PCM)	– Low frequency, O Hz – Multiplex up to 86 signals on one track – Less sensitive to speed vari- ation than FM (1% error for 1% speed change)	– Limited high-frequency response – Complex equipment required – Low tape utilization
Digital	– Low frequency, 0 Hz – High data accuracy – Insensitive to amplitude and speed variations – Compatible with computer operations – Start/stop capability – Extremely versatile	– Limited high frequency (150 kHz) – Data must be digitized – Sensitive to dropins and dropouts – Requires computer quality tape – Low tape utilization

information is fed in a serial mode to one channel of the recording head, which is referred to as serial recording. By combining the two methods, a continuous flow of recorded information is obtained. In Table 7.2 comparison is made between analog and digital magnetic-tape recording.

7.5 Transient recorders

In some situations, it is necessary to store an electrical signal of a specific event including its history and future. At a coronary care unit, for instance, information about ECG preceding a cardiac arrest is important. Measurements in a wide variety of fields such as stress analysis, nuclear research, and computer design make use of the storage facilities of transients, commonly referred to as transient recording. Magnetic tape recorders are not advisable because in general only a short portion of the signal has to be stored.

Storage of signals in a transient recorder is provided by digital memories. These memories are fed with a string of binary data that are generated by an A/D converter, which samples the input signal. Storing occurs when a trigger signal is applied to the system. In such systems, samples are pushed through a series of shift registers. The total length of the shift registers determines the amount of data to be stored. This length in combination with the sampling frequency of the A/D converter establishes the duration of the stored segment of data. Occurrence of an event that has to be recorded triggers the system which results in an inhibition of the input to the registers, hence no new data are entering. Reproduction of the stored data is achieved by recirculation of the shift register at the same clock frequency applied in the recording mode. Data compression or expansion can be realized by changing the clock frequency. An increase in clock frequency results in reproduction in a shorter time and vice versa.

7.6 Card and tape punchers

Like digital magnetic recorders, punchers can handle only digital data. The binary data, representing for instance a sample of a pressure signal, are punched on paper tape in such a manner that a hole is punched for a digital one and no punching occurs for digital zero's. The tape-feed drive is usually achieved by an incremental motor. The punch pins are operated by magnetic solenoid drives that will punch on 8-track tape. For economical use, the number of punch pins equals the number of bits of the digital word to be punched. If the digital input to the puncher is greater than 8 bits, usually two or more consecutive words are punched. Paper tape used in small computing machines never uses the 8-bit capacity.

Reading a punch tape can be accomplished in several ways. One can feed the tape through a reader consisting of several spring-loaded contacts which enable current to flow in a circuit if a hole in the tape allows electrical conduction between the undersurface and the spring loaded contact.

A common type of reader consists of an array of photodiodes that will be excited if light from a light source on the reverse surface falls on the array through a hole in the tape. One should therefore be careful to inspect a malfunctioning paper-tape reader and punch with a lamp!

A punch card can store information in 80 columns; each column can store 1 bit of information at 12 different levels. In other words, a punch card can store totally $\ln_2 (12 \times 80) = 9.9$ bit of information. The row location of the holes is an additional information on the punch card. Punch cards are also used for the storage of instructions to control a machine or similar applications where one card can contain a complete program.

7.7 Digital memories

Developments in the semiconductor industry during the past years have resulted in a major shift in the type of storage technology used in digital systems. Semiconductor memories are divided into three categories: random-access memories (RAM), read-only memories (ROM), and serial memories. A nonvolatile memory is presented with a core matrix. Core memory has as major advantage that it contains the stored data without energy supply.

A magnetic-core memory consists of small ferrite toroids, each with several wires through its centre to control the polarity of the magnet and to sense its polarity. Each core element represents one bit of information. A current passing through the core in one direction polarizes the core in a certain state. Reversing the current direction sets the polarity of the core in the opposite state. Storage of data is obtained if a certain core element in the matrix is addressed through an x and a y wire carrying each half the amount of current needed for polarization of the core. To retrieve information from ferrite cores a read operation is initiated through which read currents are sensed through the ferrite cores.

7.8 Videotape recording

The recording of high frequencies requires high head-to-tape speeds and problems arise if the low frequency extends to zero. A high head-to-tape speed is achieved by moving the tape at a relatively slow speed and rotating the video-record/playback head or heads, mounted on a wheel or drum at a high velocity to provide the necessary high head-to-tape speed. The action of the head is to scan the tape and the resultant track layout is called the record format. Basically the tape can be held across its width and scanned transversly from edge to edge or it can be wrapped longitudinally around a drum in the form of a helix to produce a helical format. The problem of recording zero frequency can be solved by modulating a high-frequency carrier and shifting the DC components up the band. If a video signal, with bandwidth from DC to 5.0 MHz modulates an 8-MHz carrier then, the modulated signal will have a bandwidth ranging from 3 MHz to 13 MHz, a frequency span slightly greater than 2 octaves.

As already mentioned, frequency modulation is nearly always used because of its tolerance to amplitude variations. The configuration of the tape deck differs considerably between manufacturers and formats, but the basic requirements are the same. It has a feed spool and take-up spool to contain the tape. A capstan is required to control the longitudinal speed of the tape. A rotating head wheel or drum scanner with a separate motor drive is also required to provide the head speed. This assembly has some way of coupling the R.F. signal to and from the heads. The audio heads are stationary and audio tracks are recorded longitudinally in the conventional manner with an audio erase head preceeding the record/playback head.

Erasure for the video track is also achieved with a stationary head upstream to the video heads. The rotational speed and phase of the video heads are electrically controlled by means of a servo mechanism on the head motor. During recording, the servo systems determines the video-track spacing whilst during playback it ensures accurate alignment of the video head with a recorded track. The recording electronics produce the frequency-modulated R.F. signal; the playback electronics amplifies the very small signal, equalizes for playback losses, switches between heads and finally demodulates the FM signal back to video. Playback equalization is normally adjusted for a flat video frequency response. Some sophisticated systems also provide automatic equalization on color signals and compensate for tape dropouts. The video on the output of the signal system has timing instability owing to the mechanics of the head-scanning process. These can be removed by further electronics.

The audio electronics are similar to conventional audio recording with high-frequency bias, pre-emphasis and de-emphasis. The audio quality is very often inferior to audio recorders of similar standard owing to the proximity of stray fields, grain orientation being on the wrong axes, poor tape contact on tape edges and reduced track width. Most video tape recorders incorporate an electronic editor which performs a controlled switch from the playback of a previously recorded scene to the recording of a new scene.

One advantage of helical scan is that stop-motion playback is possible by stopping the tape and keeping the head rotating at field rate. The scan angle is now slightly different owing to the lack of tape motion. If the scan starts on the centre of a track, it ends on the centre of an adjacent track, a longitudinal distance equal to the distance the tape has moved. In crossing the guard band a noise bar is produced in the centre of the picture. A slight adjustment of the track position moves the bar to the top or bottom of the picture, thus starting the scan in the guard band and ending in an adjacent guard band. The same recorded field is repeated on playback giving the effect of still frame.

Video recorders using a magnetic disc instead of the tape offer several applications and special effects. Still frame, slow motion and fast motion in forward or reverse are quite feasable. The storage disc is one where the recording medium is nickle cobalt which is coated with rodium to provide a hard protective surface. The tracks, unlike an audio record, are concentric rings and the heads are moved stepwise. The disc rotates once per T.V. field which gives head-to-disc speeds of approximately 7500 cm sec^{-1}, for the outer track and approximately 3000 cm sec^{-1} for the inner track. The tracks can be recorded from the outer edge to the inner limit sequentially, recording odd fields on one service and even fields oz the other, moving the head during the field it is not recording. If a greater storage capacity is required, at least two discs must be used, which adds some complexity to the recording and playback circuitry. A major part of video-disc technology is digital logic. Video discs are becoming more and more useful as tools for frame-by-frame animation, assembly of edited sequences, pre-programmed playback sequences at various speeds and in various directions.

References

1. Bycer, Bernard B. (1965): Digital Magnetic Tape Recording: Principles and Computer Applications, Hayden Book Company, p. 14–17 New York.
2. Davies, Gomer L. (1961): Magnetic Tape Instrumentation, McGraw Hill Book Company, New York.
3. Lowman, C.E. and Anjerbauer, G.J. (1963): Magnetic Recording Theory for Instrumentation, Ampex Corporation.
4. Robinson, J.F. (1975): Videotape Recording, Focal Press, London and New York.
5. Strong, P. (1970): Biophysical Measurements, Tektronix Inc., Beaverton.
6. Thompson, A.R. (1972): Recorder Linearity: Fact or Fantasy, Machine Design, April 20.
7. Yanof, Howard M. (1972): Biomedical Electronics, F.A. Davis Company Philadelphia.

8
AUTOMATION

John D. Laird

8.1 Introduction

Automation is a subject which rivals perhaps only taxes in its ability to evoke a strong response from nearly every member of society. Almost everyone, including the author, has an opinion on the subject. As we approach 1984 it would be difficult to think of a field of human endeavour which has not in some way been touched by both the promise and the problems offered under the flag of automation. The field of clinical medicine has certainly not escaped this turbulent process of revolutionary achievement and blatant failure.

Automation is neither good nor bad. It is not a panacea for the many problems challenging the practitioner, neither is it capable of itself of dehumanizing medical practice by increasing the distance between the physician and his patient. Automation is simply a modern concept encompassing many techniques which when used wisely in medicine or any other field offers a wonder of achievement or, when the application is ill considered, it results quite simply in a mess.

My goal in writing the chapter is neither to impress the reader with all the wonderful things which can be done to-day using automated techniques nor to recite a list of failures. I hope to illustrate to the reader some of the features which characterise applications of automation techniques in clinical medicine. It would be unreasonable in the short space of one chapter to attempt to make an expert of the uninitiated reader, however I can aim for an improved insight and judgement, a heightened sense of smell for whether a particular problem is ripe for the use of modern techniques of automation. Additional information on automation can be found by consulting the reading list at the end of this chapter (1–6).

How often have you stood at the bedside and drawn the conclusion that the patient just doesn't look good, is a worry, even though all the vital signs of the chart are within acceptable limits. This clinical judgement, so highly prized in medicine, is some poorly understood blend of education, experience and intuition. The extension of this subtle concept to encompass the measurement techniques and their automation is the goal of this chapter.

208

J.D. LAIRD

But, what is automation?

A very good question is that! Since the Oxford English Dictionary (shorter) does not yet have a definition of the word it is clear that we are concerned here with a fairly new concept, and each of us can come up with his own working definition. To some it may imply the use of computers while to others it may mean a process brought to completion without intervention by a human being. The trouble is that the concept is so new to our culture that we simply have not yet discovered either the limits of the application of the technique or a definition acceptable to everyone.

A feature of all applications of automation techniques is the minimization of the role of people in the direct accomplishment of the task itself.

The paradox of automation can be illustrated by considering the more limited area of computation. A survey conducted among our first year medical students revealed that the majority possessed a modern scientific pocket calculator. We all have within reach a level of computational power which only a few years ago was reserved for only the specialized researcher in a given field. The student of to-day can solve numerical problems which 30 years ago were simply beyond practical reach.

At the same time we have all had some frustrating "encounter with a computer" over some seemingly simple problem. Anyone who has moved from one house to another and has tried to convince the computers of the government, the gas supplier, the water, publishers of all the magazines to which one subscribes, etc. of this simple normal action will appreciate that there is apparently more to using computers than at first meets the eye.

8.2 Analysis of automation of measurements in clinical medicine

8.2.1 *Introduction*

The process of measuring in medicine is neither a begin nor an end in itself. It must be seen in the context in which it happens and, as importantly, it must also be seen in the light of the reason for attempting the measurement.

As has been extensively discussed elsewhere in this volume, the entire process can be dissected into a number of stages. This is shown schematically in Fig. 8.1. The reader may at first glance find this figure an unnecessarily formal abstraction of the problems presented by clinical medicine, so let us consider a concrete, earthy problem encountered daily at the bedside. Consider, for a moment the trivial task of determining the heart rate or more accurately the pulse rate of a hospitalized patient.

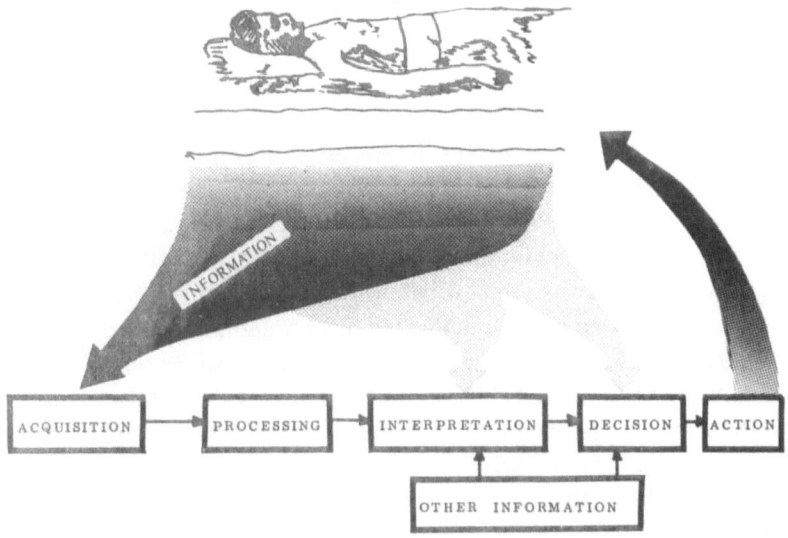

Fig. 8.1   Schematic diagram of steps in clinical measurements.

8.2.2 *The data acquisition phase*

Firstly the nurse or physician goes to the bedside and after, or while, having a chat with the patient locates the radial artery. The number of pulsations in a period of 15 seconds is then counted. In this formal analysis of this singular measurement task the data acquisition phase is now completed.

8.2.3 *The data processing phase*

In this trivial example the processing of the data consists simply of multiplying the number of beats counted in 15 seconds by 4, yielding the pulse rate in beats per minute.

8.2.4 *The interpretive phase*

Suddenly even this trivial example becomes more difficult to describe but might consist of the following steps or thought processes. Before attempting to incorporate the measured heart rate into the clinical picture one usually either consciously or unconsciously checks the data for its reliability. In this example this might consist simply of checking the plausibility of the answer. If the "measured" heart rate were 250 per minute it is almost certainly in error. This error could either have occurred during the counting (data acquisition) or when multiplying by four (processing). In either case one is forced to go back to square one.

It would be unthinkable in clinical medicine to decide on a course of action solely on the basis of a single measurement of, in this case, heart rate. To illustrate the nature of the interpretive process let us suppose the measured heart rate in this case was 120 per minute. This value would have two quite different implications in a 2 year old child or in a 60 year old man. If we are dealing with an adult a heart rate of 120 is a call for action of a very different nature in a man with fever just back from a stay in the tropics versus a patient with an uncomplicated history and unstable angina pectoris. Moreover, the *current* value of the heart rate is not enough. What was it an hour ago or yesterday? Is the trend up or down? I'm sure you'll agree that the decision making process in clinical medicine is complex, and made up of a blend of hard and soft facets. Some are easily quantifiable, expressed in numbers while other, equally important, factors are not.

8.2.5 *The decision phase*

It is evident from the above that the dividing line between interpretation of clinical findings and the making of a decision to act is not a sharp one. It is important to realize that the resolution to "wait and see," "let's look at it again in an hour," etc. is just as much a decision as deciding to start an infusion or rush the patient to the operating room.

8.2.6 *The action phase*

The action phase takes on a relative simplicity in contrast to the interpretive and decision making phase. It usually consists of initiating or modifying a particular therapy: starting an infusion, increasing or decreasing the speed of an already running intravenous drip, prescribing a medication, calling the surgeon, talking to the patient or his family, or waiting until 'to-morrow, etc.

8.3 Application of automation to measurements in clinical medicine

8.3.1 *Introduction*

In principle, automation could be applied to any or several of the phases of the clinical measurement process described in Fig. 8.1. Automation consists simply of accomplishing the particular task through the application of technology in such a way that the intervention by human beings is either eliminated or at least minimized.

When thinking back over the various phases involved in the heart rate problem discussed above it is evident that as one moves from the acquisition phase on to eventual action (processing, integrating, decision), the level of abstraction and complexity of the task increases significantly. This has important implications for the use of automatic techniques. In order to expect a box of electronics to accomplish a particular task it is essential that the task be capable of being reduced to a clear set of instructions. Boxes or computers do precisely as they are "told" and they do not have a supple, imaginative nature. On the other hand an automated system does not tire easily, nor run out of patience, nor require coffee, vacations or a pat on the shoulder. They may however require a sort of "love" in the form of periodic maintainance.

8.3.2 *Reasons for automating a particular task*

Modern clinical medicine is becoming both more complex and more costly. Trained paramedical personnel are in short supply and often overburdened. There are heavy pressures on a physicians facilities, faculties and time. Nowhere is this more evident than in the intensive care, coronary care, or emergency care unit of a modern hospital. Not surprisingly it is in these areas that considerable progress has been made with the introduction of automated techniques in clinical medicine. This development has not gone without much discussion and debate, some of it at times quite vehement. The use of automated techniques in the clinical chemistry laboratory, however, is by now so commonplace that one does not even give it a second thought. This latter application has become such a speciality that books have been written about it. It may be instructive to reflect on why the automated clinical chemistry lab has been introduced so quietly whereas similar attempts in bedside medicine must endure a much more stormy course.

The motivation to consider the introduction of automated techniques may consist of a blend of some of the following factors.
1. Inaccuracies in a manual method may be overcome.
2. A measurement may be repeated more often than with manual means.
3. Manpower costs may be reduced.
4. More sophisticated data processing may be possible.
5. A measurement may be possible with less disturbance to the patient.
Whether or not the verb "may be" can be replaced by "will be" is strongly determined by the specific application.

8.3.3 *Automating the acquisition/processing phase*

To illustrate the power and the pitfalls of automation, let us return to a specific problem. Let us consider the measurement of heart rate in an intensive care setting. In such a critical care unit the heart rate must be more or less continuously monitored. It will be obvious to everyone that a nurse cannot be expected to measure the pulse rate say once a minute. It just cannot be done manually!

Thus one must turn to automated techniques. The first step is of course to acquire the information required to determine the heart rate. As discussed in other chapters of this volume this consists of deriving an electrical signal corresponding to each heart beat. One could of course duplicate the standard clinical technique by devising a transducer to detect the radial pulse and turn it into an electrical signal. However the heart generates its own electrical signal (albeit small) in the form of the electrocardiogram, which when amplified is suitable for further analysis (see chapter 3). The ECG contains much more information (see chapter 9) than is required to detect the occurrence of a heart beat. For this purpose we need only to know the time of occurrence of the QRS complex. A textbook ECG is shown in Fig. 8.2.

8.3.4 *Detection strategies*

If we wish to detect the QRS complex in such a signal the most trivial of strategies would work. A simple threshold detector would work. We simply devise a circuit which goes beep (or quietly sends out a pulse) every time the voltage goes above 0.7 mV. But, of course not all ECG's look like Fig. 8.2. The simple strategy of a simple threshold would fail miserably on the ECG of Fig. 8.3.

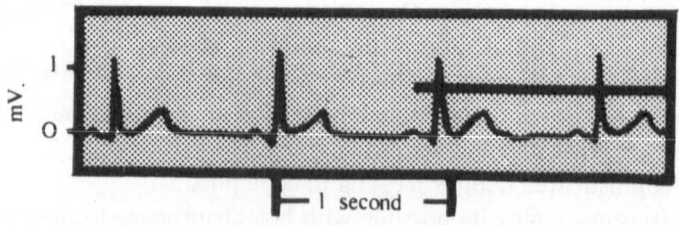

Fig. 8.2 Textbook example of a normal electrocardiogram. With a threshold set at 0.7 mV all QRS complexes in this example will be detected.

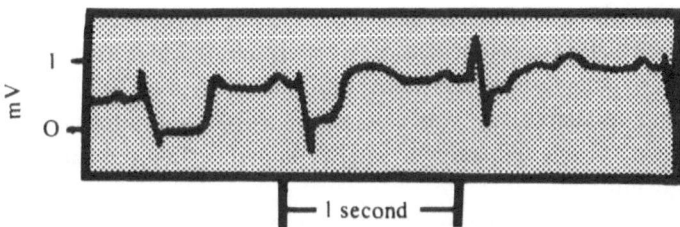

Fig. 8.3 Pathologic electrocardiogram. Due to changing morphology of the QRS and R-R interval variation a simple level detector would have difficulty with this ECG.

How do we decide for ourselves which is the QRS complex? Looking at Figs. 8.2 and 8.3 it's obviously the sharp spiky bit, that occurs regularly.

Perhaps we could more reliably detect the QRS complex if we were to enhance the "sharp spiky" bits. This, as the reader will appreciate from reading chapter 9 corresponds vaguely to looking at the higher frequency components of the signal. Thus, processing the data with such a high pass filter has the effect of selectively sorting out the faster moving parts of the signal.

But of course the ECG signal seldom looks like the textbook examples. Patients move, scratch their arms, put a hand under their head, reach for a glass of water, etc. All of this gives rise to sources of artifact or noise. A noisy signal such as show in Fig. 8.4 also contains a large number of spiky bits or high frequency components.

Does all this mean it's not possible to detect the QRS complex? Of course not! Witness the multitude of ECG monitors which are commercially available and in constant daily use in hospitals all over the world.

The object of the above discussion was simply to illustrate that such problems are much more complex than one might suspect.

It is often said that in order to understand the reactions of another person one should try and place oneself in the others shoes. Look at the problem

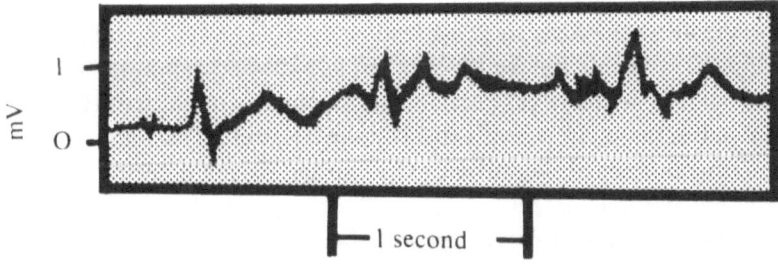

Fig. 8.4 Pathologic electrocardiogram contaminated by high frequency muscle artifact.

from his point of view. The same technique is applicable to trying to gain insight into the reactions of a piece of automated monitoring equipment. Try to imagine yourself as the circuits contained in an ECG monitor. The "actions" available to "you" consist of amplifying, filtering, "deciding" whether a voltage is bigger or smaller than a certain threshold, taking an average, etc.

A piece of automated apparatus cannot go about the task of trying to detect a QRS complex or a systolic pressure, etc. in the same way you or I would do it. If you look back to Fig. 8.4 you' d have no trouble in identifying the QRS complexes. Why is it then that even very sophisticated apparatus does have difficulties with such a task? The answer while poorly understood in the scientific sense, stems from the fact that a human being, or any animal for that matter has an uncanny ability to recognize patterns. We need only a fleeting glimpse of a face to know almost immediately if it's that of an old friend or loved one. And yet if one analyses a still photograph of a face its information content is enormous and would require a large computer in order to attempt a discrimination.

This is a very fundamental point and since medicine consists in no small part of recognizing patterns in fuzzy sets of data one can appreciate the background to so many disappointments.

8.4 Automation and the computer

8.4.1 *Introduction*

In designing an automation of a measurement the task of the automation has to be defined carefully. Let's return to the problem of the heart rate measurement. If we require a value of the heart rate measured on a time scale of minutes it is, in fact, not necessary to detect every single QRS complex with high precision. If over a period of minutes the detector misses a few beats (false negatives) or now and again triggers on an artefact or noise pulse, the average heart rate will not be seriously in error. This bit of leeway is in fact responsible for the efficacy of most cardiac monitors.

In a modern intensive care unit, however, one is as interested in the rythm of the pulse, the incidence of extra systolic beats etc. This slight change in requirements adds an important extra burden on the detector. One now needs to accurately detect the sequence of intervals. For example, a premature beat is followed by a compensatory pause. A decision logic in which a long interval is important should not be disturbed by a long interval corresponding to a beat missed by the detector. To make matters worse just in such a clinical problem the morphology of the QRS complex may differ significantly from beat to beat. The designer of the monitor has then no alternative but to try and build in automatically varying filters and self adjusting thresholds for the detector.

8.4.2 *If it's so complicated let's put it all in a computer and let it sort it out*

And how nice it would be if the solution were so simple. In order to see why on the one hand it's perfectly reasonable that todays students have nearly incomprehensible computing power in their pockets and so little seems practical in the clinic with a computer, put yourself now in the place of the computer.

Let us suppose once again that the problem is the same: the detection with high reliability of the occurrence of the QRS complex in the ECG.

8.4.3 *How do we "put it in the computer"?*

A digital computer does everything with numbers, and only with numbers. So the first-task is to convert the electrocardiogram into a sequence of numbers. The ECG of Fig. 8.2 is in fact a voltage which varies in time. The usual technique is to sample the magnitude of that voltage at frequent intervals and convert the measured voltage into a digital number. Hence the frequently heard word analog to digital converter (or in the jargon of computers, an A to D converter). How often this sampling should take place and how accurately it should be done is beyond the scope of this chapter (see also chapters 7 and 9). A computer must put these samples somewhere in order to be able to carry out a set of instructions designed to detect the occurrence of the QRS. Modern digital computers are characterized by a rather large memory capable of storing many numbers. A typical "present day" mini computer might have room for 32,000 words of storage. Is that large or small? A word is the basic currency used in a computer and represents a sequence of a certain number of bits. Typically it might take one word to store one sampled number. In order to accurately represent the ECG one might have to take 50 samples per second. Thus in one hour we would have accumulated 180,000 words worth of information. It is obvious that in that aspect a 32,000 word machine (32k in a computer jargon) is not so very large. On the other hand a 32k machine has a computational power which is enormous. It is simply a fact that a patient produces a very large flow of information, which *man* is relatively good at sorting out.

Thus, from the above it is clear that one must move quickly to analyse the incoming flood of information and then erase the "blackboard" to make room for more. That is, the analysis of the data must proceed more or less parallel with the imput of the data. This is referred to as "on-line" computing. If one collects the data and then takes the time later to "quietly" analyse it, one would call that off-line analysis.

In order to gain insight into the problem of signal processing and detection by use of the digital computer once again I ask you to place yourself in the shoes of the computer. The digitized ECG appears in the computer for analysis as a long string of numbers. To illustrate the problem I have digitized the ECG of Fig. 8.4 and the result is shown in Table 8.1. Although it's not easy, the trained observer has no trouble in spotting the QRS complex in the graphic form of Fig. 8.4 whereas in numerical form it's not at all obvious in Table 8.1. Fortunately, it is here that the computer has an edge over the human observer. The modern digital computer can do arithmetic operations $(+, -, \times, \div)$ at enormous speeds.

The problem now has to be solved by devising cunning schemes to on a numerical basis sort out the location of the QRS complex in such a string of numbers.

8.4.4 *Programming*

This is done by devising a sequence of operations to be carried out on the array of data typified by Table 8.1. This recipe of instructions is called a program, and by the way, it also must be stored in the computer. All the

Table 8.5 Result of digitizing the ECG of figure 9.4. The QRS complexes are located at approximately 0.095 and 1.047 seconds. Did you see it in the numbers?

Time	Volts	Time	Volts	Time	Volts	Time	Volts
0.000	0.333	0.357	0.542	0.714	0.625	1.071	0.979
0.024	0.271	0.381	0.604	0.737	0.604	1.094	0.813
0.047	0.313	0.404	0.750	0.761	0.667	1.118	0.313
0.071	0.396	0.428	0.771	0.785	0.792	1.142	0.792
0.095	1.063	0.452	0.688	0.809	0.750	1.166	0.833
0.119	0.708	0.476	0.667	0.833	0.750	1.189	0.917
0.143	0.208	0.500	0.563	0.856	0.771	1.213	1.063
0.166	−0.229	0.523	0.604	0.880	0.813	1.237	1.229
0.190	0.313	0.547	0.604	0.904	0.854	1.261	1.146
0.214	0.333	0.571	0.396	0.928	0.792	1.285	0.875
0.238	0.396	0.595	0.375	0.952	0.750	1.308	0.917
0.262	0.396	0.618	0.483	0.975	0.604	1.332	0.875
0.285	0.375	0.642	0.417	0.999	0.621	1.356	0.833
0.309	0.417	0.666	0.500	1.023	1.104	1.380	0.854
0.333	0.438	0.690	0.542	1.047	1.479	1.404	0.896

techniques of filtering which can be done with circuit hardware (see chapter 9) can also be accomplished numerically, along with much more.

The computer must be "told" (programmed) in minute detail how to go about solving the problem at hand. For this process a strategy is required and this recipe is called an algorithm. The art of writing highly sophisticated on-line programs is a specialty apart, and this when compled with the entirely different method of working used by a physician makes it difficult for him to specify with sufficient precision what is actually required. In addition the physician is used to handling exceptions to the rule whereas a computer program has very real difficulties with this sort of variability.

This means that if the program is to work effectively, all possible variants have to have been foreseen at the stage in which the program was written. Having done so it is usually possible to think up yet another variation which, while infrequent, is surely to occur during the use of the system.

I trust it is clear that in order to use a digital computer effectively the task must be capable of being defined in minute detail. The computer does not have a creative mind (or any other sort of mind, for that matter). It does precisely what it is told, not what one intended it to do. It can be thought of as a very hard working, dumb and fussy old lady.

8.5 Automation and the microprocessor

8.5.1 *Introduction*

Perhaps the best is the enemy of the good. Maybe it is more appropriate to use the computer for problems less complex than the ECG monitoring problem above, such as perhaps changing of the gain of a preamplifier when the signal is out of range, or every once in a while carrying out out the tedious, too seldom done job of calibrating a measurement system, etc. But, you say, a computer is too costly just to do such a simple job. But you are wrong. In the last few years there has been an enormous, even revolutionary progress in the development of smaller computers, so-called microprocessors. Even more importantly, the prices of microprocessors have fallen to a level ($< \$ 200$) where it is perfectly reasonable to incorporate such units in clinical measurement systems. Thus there are an increasing number of "smart" instruments incorporating micro-processors being introduced.

For example, blood gas measuring stations which are in essence "idiot proof" and self calibrating are now fairly commonplace. Microprocessor based pulmonary function testing is another example. In fact in all areas of modern medicine where repeated measurement of exactly the same sort must be done and where the subjective element in the interpretation is small, microprocessors can be usefully applied. In the case of the automated blood gas laboratory station the use of smart instrument design really does mean that blood gases can be determined at the treating unit by staff with no understanding of the intricacies of PO_2 electrodes etc. This stems from the fact that the calibration sequences, checking for drift, etc. can be accomplished automatically. In restricted application of function testing, cardiac catheterisation etc. automated techniques may be of some help.

Given how long it takes to develop software (algorithms, programs, etc.) one must however be assured that the medical requirements themselves are not likely to change importantly in the period of time required to develop the system. Thus attempts to apply automation on the frontiers of modern medicine are almost always doomed to failure in respect to their widespread applicability. The reader should note that this is not to say that selected research units should not attempt such applications since it is only by such attempts that one learns the limits of the application of automated techniques.

A word of caution is in order before the reader rushes off to buy a microprocessor to do some clever task in his research laboratory. A mircroprocessor consists of a few integrated circuit chips usually mounted on one printed circuit board. As it comes from the manufacturer it is capable of very very little. A great deal of effort is required in order to program it to accomplish the task you may wish it to perform. To illustrate this point let me describe an application with which I am familiar.

The thermodilution technique is a method quite commonly used to measure cardiac output in coronary and surgical intensive care units. The method consists of injecting cold saline into the right atrium and then measuring with a thermistor the temperature "washout" curve in the pulmonary artery. In order to compute the cardiac output one requires the injectate volume and temperature, the temperature of the patient and the area under the temperature vs. time curve as well as a constant. There are commercially available units which using analog circuit techniques accomplish the measurement quite satisfactorily. However, in practice errors involving the calibration as well as the integration were made with annoying frequency. These errors occurred largely when less skilled staff were on duty. With these considerations in mind along with a desire to gain experience with the application of microprocessors the decision was made to implement a microprocessor based design. The resulting prototype was very successful. The results were repeatedly very accurate and more importantly it was nearly impossible to get a wrong answer.

But what did the development cost. Before all the bugs were out of the system nearly one man-year's work by a master of science level electrical engineer with previous microprocessor experience was required. In addition there was approximately 6 man-months of technician work in the hardware design and development. If only one or two of such units were required and this was not an intentional learning experience, one could very seriously question the wisdom of the application for this particular problem. On the other hand, were this design time and cost to be amortized over hundreds of units it begins to become a reasonable proposition.

8.5.2 Tentative conclusions

Thus far we have focussed attention on the problem of automating the acquisition/interpretation phase of clinical measurements. In continuous, on-line applications typified by the intensive care unit and to a lesser extent the specialized function laboratories, it is not obvious that the current state of the art of large scale automation techniques together with the complex and poorly understood requirements of these clinics offer the prospect of real solutions to current problems. It would appear further that the degree of success depends on the extent to which the clinical problem can be specified in a concrete recipe, not susceptible to short term change. Few *continuous on-line* measurement problems are at this time amenable to such solutions.

In contrast it seems reasonable to expect that microprocessors combined with standard instrumentation techniques in "smart" instruments capable of self-calibration, diagnosis of measurement or operator errors, automatic scale changes etc., will become more and more a feature of clinical medicine in the near future.

8.6 Pocket calculators and computers as tools in the diagnosis

8.6.1 Introduction

At the beginning of this chapter I cited the fact that most medical students in our first year class have their own scientific pocket calculator and yet if one looks around a clinical departemnt in even a teaching hospital it is rare to find someone whipping out his pocket calculator to do a sum. Is this because there is no use for computation in clinical medicine? I am personally persuaded that here lies a great untapped potential, and the paradox cited above is more a gap in education and age than in need.

Let's face it. Most physicians are not giants in mathematics. Many, if not most, have scars to show for their encounters with mathematics in the pre-clinical science curriculum. A logarithm or exponential function meant little especially after the recipe for using log tables correctly had suffered a disuse atrophy. The student of today has the log tables plus much more at the push of a single button on a device he can privately afford. It remains to be seen how long educational institutions will require to realize that a revolution has happened in this area.

An anesthetist acquaintance of mine uses a pocket programmable calculator to work out acid base balance, circulation/appearance time and volume replacement in his daily clinical work. With some perseverance he learned to program the machine himself and has his own programs for each purpose on small magnetic cards which can be "read" by the calculator.

Fairly sophisticated statistical trend analysis and testing is another application which is now easily done with today's calculators. Many tasks which require the use of nomograms could easily be programmed on a pocket calculator. Importantly, these are applications which the physician has under his own control, he does not require a Phd. engineer or mathematician to do the job. The primary impediment to such innovation is the fact that those currently in clinical practice (the teachers) haven't had the personal "hands on" experience and those students who have are not yet in clinical work.

An intermediate solution which merits consideration is the desk top calculator. These units fall somewhere between pocket calculators and mini computers. Their computational power is large but even more important for the physician with a problem is their relative ease of programming and use. The language in which the programs are written is usually simple, less powerful, but easy to learn such as BASIC. The B in basic stands for "beginners" and the language was written for college students without a strong mathematical/sceintific background. A further advantage of desk top calculator systems is that they require mostly almost no knowldege of the details of how the system actually works. The keyboard is carefully designed to be self coaching. The minicomputer is designed to accomodate many configurations of peripheral apparatus, languages, etc., depending on the application. This is its great power, but it does require of the user much more knowledge in order to get it all together for one particular problem.

In desk top calculator systems this flexibility has been sacrificed to improve operating conveneince and to place less stringent demands on the experience of the user. Moreover, there are software packs available for specific tasks such as statistics, Radioimmunoassay, filter design, etc. One such system in a clinical department together with one person with a bit of a knack and knowledge of programming make it a very powerful tool.

Every system design represents a compromise between conflicting requirements and neither computers nor calculators are exempted from this rule. One can discern, however, that designers of calculator systems have made the design trade-offs based on a system of priorities different from those made by computer designers. Most computers, for example, can be programmed to "speak" any of several computer languages. This flexibility allows the user to write his programs in the language most suited to his application. This however also implies that the machine as it comes from the factory can do almost nothing. Most calculators "speak" only one language and in return for this compromise need only to be plugged into a wall socket in order to be made operational.

The same is true of peripheral equipment such as printers, plotters, graphic displays, etc. In general calculators are somewhat restricted in choice but what is gained is that those peripherals are much easier to use. These are important considerations for the physician thinking of using automatic processing in some part of his clinical work.

Equally important is the question of how much support in the form of clever people will be required to keep the system working at solving your own problems. Don't mistake acess to a computer for access to answers to your problems. In general, calculator systems require much less highly specialized people to be able to program them successfully. A physician with a special interest in or knack for programming can quite easily learn it, as can a number of his technical staff.

There are a number of problems which recur with sufficient frequency that the manufacturers of these calculators have found it to their advantage to offer customers packages of software (programs) for specific applications. The software packs require of the user no knowledge of the details of the program in order to use them effectively. A good example are the statistical packages offered by nearly all manufacturers of both computers and calculators. For a relatively low cost and occasionally for free one acquires a ready to go program for even very sophisticated statistical analysis and testing.

More specifically clinical examples of commercially available software packages are radiation, therapy and treatment planning, radio immunoassay (RIA), cardiac catheterization calculations, blood gas analysis, pulmonary function testing, etc. As is evident from these examples the list is large and the applications varied. If one's requirement is specifically covered, the chances are good that the commercially available software will offer a solution to your problem. On the other hand attempts to modify a program written by someone else to adapt it to a new requirement, even if the change is slight, usually results in disappointment, frustration and a waste of money and time.

8.6.2 *Off-line processing and interpretation*

The routine reading of the daily production of diagnostic electrocardiograms is typical of many tasks in a modern hospital which seem to grow and grow. Many hospitals have reached a through put which simply can't be surpassed with manual techniques and not surprisingly increasing use of automated analysis is being made.

On first considering the problem it may not appear essentially different from the on-line monitoring problem discussed earlier in this chapter. It is, however; and the difference is that there is time to carefully analyse the details of the ECG.

Basically two approaches have been used. The first consisted of trying to program into the computer the ECG reading recipes which physicians themselves use when manually reading the ECG. The second approach was fundamentally different. The computer was used to "measure" a large number of parameters (e.g. voltages, vector axes, etc.,) at different instants of time) and a large number of ECG's were processed from a training population of documented ECG's. A multivariate discriminate analysis was then performed on the results and a parameter subset derived to be used in practice. In practice use is made in this system of advanced statistical methods which has led on the one hand to a powerful and useful system but on the other some difficulties in user acceptance.

To illustrate this point, consider the following. The program developed by Pipberger (3) and colleagues makes use of the fact that the probability of encountering an ECG characteristic for acute anterior infarction is much higher in an ECG derived from a coronary care unit population than from the general hospital ward. Thus depending on the source of the ECG the discrimination criteria are different. This gives rise to the following paradox. Suppose a patient on a general ward develops symptoms of an acute myocardial infarction. An ECG is run and the computerised "diagnosis" gives a probability of infarction of 60%. The patient is transferred to the coronary care unit for further diagnostics and observations. On arrival in the ccu another ECG is made as part of the standard work up. Let us presume that the actual ECG is precisely identical with that taken in the general ward. It is possible now that the computer analysis will report a probability of infarction of 90% while the only thing that changed is the location of the patient. It takes some time for the staff to accomodate to such paradoxical results which stem from the difference in approach. In this application the computer is used to do the thing it's good at; calculation.

Another example of the difficulty of introducing such a system arises as a consequence of the reporting of the results in the form of "posterior probabilities." In the "old days" when a physician requested an ECG he got a specific diagnosis in the form of "signs of left ventricular hypertrophy" etc. In this automated system he receives a list of probabilities of a number of diagnoses: left ventricular hypertrophy 75%, anterior infarction 28%, normal 33%, etc. It is rather like the modern trend to report the weather forecast in terms of "probability of precipitation today is 65%." One is still faced with the decision of whether or not to take a raincoat.

The introduction of automated techniques may well result in the savings of manpower, money, or whatever was promised but it usually also requires some changes in thinking and ways of working.

8.7 Automating the decision/action phase

The closer one comes to closing the loop automatically with the patient the more controversial is the application. There are arguments both for and against. As with all applications of automation, the more circumscribed the task is, the greater the probability of success. One of the more ambitious applications is that developed by Kirklin and Sheppard (1). In this system for monitoring post cardiac surgical patients a carefully thought out decision tree was developed in order to specify and control blood transfusions following surgery.

As we have already discussed in this chapter one requires a very precise analysis of the problem and specification for action, before a program could be written. This is, in fact one of the most interesting aspects of the Kirklin system. They argue quite convincingly that in a working clinical unit there are substantial variations in the level of skill and experience of the clinical staff present in the unit in the course of the day and for that matter throughout the year. Moreover, when crises develop it is not unlikely that just at that point the staff is tired and less likely to make consistently well thought out wise decisions. Wouldn't it be better, he argues, to try and think it all through with the top clinical team when they are fresh and at their best. The number one team is not always in the hospital and they are not always at their best. Yet, their most carefully thought out strategies and therapeutic protocol is preserved in the programmed decision tree.

The fact that the loop is closed automatically is not the central point of their system, although because of the spectacular nature of a "machine treating the patient" much attention is devoted to that aspect. In fact it is the carefully thought out "therapy program," the best thoughts of the team which through the surrogate of the computer manages the treatment. It is, as it were, "still the physician who has written the orders."

More interesting is the fact that they very carefully tried to analyse the basis for their actions as one would have to do in order to write a program. This phase alone was probably very instructive in shedding light on how their unit really worked.

8.8 A last remark

To return to the example used at the very beginning of this chapter one may think that one has gone off to the bedside to measure the pulse rate on an hourly basis. Hourly because the form of the patient chart requires it. If one does a work study analysis of the nurse's actions in that time it could easily appear that let us say 4 minutes were devoted to this task. But is that really what she did during those 4 minutes? No not quite, she had a chat with the patient as well about the noise in the corridor, she moved the glass of water within easier reach of the patient, checked the intravenous drip which was running, etc.

Many tasks performed by people are multi-faceted, and only by careful, systematic analysis can we really learn what has contributed to the success (or failure) of recovery. It is partly through this sort of analysis that one will be able to identify those tasks where it is both desirable and reasonable to consider the place of automation in clinical medicine.

Few things are as they seem at first sight.

References

1. Kirklin, J. and Sheppard, L.: Personal Communication.
2. Ostrow, H.G., and Ripley, K.L. (1976): Computers in cardiology. 1976 Conference proceedings, IEEE catalogno. 76CN1160–1C, Institute of Electrical and Electronics Engineers Inc., Long Beach, California.
3. Sayers, B.McA., Swanson, S.A.V., and Watson, B.W. (1975): Engineering in medicine. Oxford University Press, London.

4. Van Bemmel, J.H., and Willems, J.L. (1977): Trends in computer processed electrocardiograms, North Holland Publishing Co., Amsterdam.
5. Special number devoted to Microelectronics, Scientific American, November 1977.
6. Proceedings of the 29th Annual Conference on Engineering in Medicine and Biology, 76CHII39–5EMB, Chevy Chase, Maryland (1976), The Alliance for Engineering in Medicine and Biology.

9
SIGNAL PROCESSING

Jan Strackee and Adriaan van Oosterom

9.1 Introduction

9.1.1 *Definition of a signal*

The concept signal nowadays is used in nearly every discipline. Here signal will be defined as an ordered set of values. Most often the ordening is thought of as ordening in time, but signal analysis with ordening according to position occurs nearly as often (picture analysis).

In this chapter we will use the term time signal for those cases where the natural flow of time is of relevance. The computer, however, knows neither time nor position but only order. In the concept of a signal as introduced here there is no room for analog signals. In 9.2 it is shown, however, that analog signals are adequately covered by the given definition.

A broad distinction can now be made of the types of signals. On the one hand one has the deterministic or non-random signals, on the other hand the non-deterministic or random signals. The latter are also called stochastic signals. The essence of a stochastic signal is that its behaviour in the future cannot be predicted exactly as would be the case for a deterministic signal, for instance a sine wave.

Among the stochastic signals the point processes take a somewhat distinct place. For these processes the time axis is marked by the occurrence of events and the time intervals between these events are analysed (9). Theoretically the duration of the event should be infinitely short; in practice it should be short compared to the smallest interval time. A good example of a biological point process is the occurrence of the peak of the R wave in electrocardiograms. No special techniques are needed for this type of process. The consecutive intervals can be interpreted as the subsequent samples of an artificial analog process; however, no natural sample time exists. The intervals are ordered according to their number of occurrence.

In practice one often encounters a mixture of deterministic and stochastic signals. Sometimes one is interested in the deterministic part only, for instance the fitting of curves. Here one tries to delete the stochastic component. At other times one is interested only in the stochastic component but a slow trend is also present; for instance when studying the irregularity of the interval between successive R waves, the subjects' rhythm may decrease, say from 75 to 65 beats per second, in 10 minutes. Although this latter phenomenon can be intersting in itself it might for the problem at hand hamper the analysis and detrending or filtering is indicated.

9.1.2 *Purpose of analysis*

In one way or another every scientist processes signals. In the biomedical domain the gamut covers fitting of a straight line as well as image processing; but to what purpose? The simple example of straight line fitting already contains most of the elements of the scope of the analysis. From the fact that the points do lie on a straight line one obtains on the one hand insight into the genesis of the signal, and on the other hand a substantial data reduction – only two values (slope and cut-off) describe the data. Furthermore it provides the possibility of manipulation – one can interpolate – as well as of prediction, which in this case amounts to extrapolation.

In this simple example manipulation is not mentioned. But detrending and filtering are among the many possibilities of signal handling.

9.2 Special techniques

9.2.1 *Sampling*

The signals that are the result of some fundamentally continuous process $f(t)$ have to be discretized before they can be handled in digital form. This means that readings of $f(t)$ are taken at discrete time instants t_n. The resulting series of values $\{f(t_n)\}$, or samples, now represent $f(t)$. For short $f(t_n) = f_n$.

The natural problem that arises here is the selection of the total number of samples required and the way in which they should be spaced over the interval studied. In most cases a regular spacing is employed, i.e. the samples are taken at regular time intervals of length $T = D/N$, in which D is the total length of the signal studied and N is the total number of samples taken. The corresponding sampling rate $\nu_s = 1/T$ can be unequivocally determined in cases where the process studied is known to be band limited. This means that no frequency component over say ν_q (Nyquist frequency) is present. For such band limited signals the entire signal is fully contained in the samples taken, provided $\nu_s > 2\nu_q$ (Shannon's sampling theorem) (23).

When the signal studied is band limited and at the same time it is known that the signal is contaminated by a noisy component containing frequencies in excess of v_q it should be realized that the sampling theorem applies to the complete signal. This means that either the value of v_s has to be adapted (increased) or that the frequencies in excess of v_q have to be removed prior to the sampling process (analog filters).

When v_q is unknown an iterative analysis can be carried out in order to find some estimate of v_q (20). For some cases regular spacing can with improvement be replaced by irregular spacing. This possibility should be considered particularly when dealing with continuous functions in which the natural flow of time is absent, i.e. functions of space. Computing an integral over some function of space can be performed far more efficiently when a non-regular sampling grid is applied (Gaussian integration) than by regular sampling. The technical realization of the sampling procedure of time signals is discussed in chapter 3.

9.2.2 *Digitizing*

The accuracy of all subsequent data processing is directly related to the accuracy of the sampled input data. When the input is a continuous function of time the resolution (and accuracy) of the analog to digital (A/D) conversion has to be related to the inherent precision of the input signal as measured by the signal to noise ratio. This ratio is usually expressed in dB (decibel), say k dB, defined by $k = 10 \cdot \log_{10} (P_s/P_n)$ with P_s and P_n the average power content of the signal and the noise respectively. The value of k is directly related to the required precision (resolution, accuracy), say n bits, of the analog to digital conversion through

$$\frac{k}{n} = 6 \,(\text{dB/bit})$$

(see chapter 7).

9.2.3 *Discrete Fourier Transform*

The Discrete Fourier Transform (DFT) of a set of numbers $\{x_n\}$, $n = 0$, $N - 1$ may be defined by

$$X_k = \frac{1}{N} \sum_{n=0}^{N-1} W^{-kn} \, x_n$$

or $X_k = \text{DFT} \{x_n\}$ \hfill (9.1)

with x_n and X_k complex,

$$k = 0, N - 1$$
$$W = \exp(2\pi i / N)$$

This transform may be used whenever Fourier types of analysis seem appropriate for the continuous signals studied, i.e. when the signal can be represented by a sum of a series of sinuses and cosinuses. Examples are the study of hidden periodicities in a noisy background, the determination of power spectra and the use of the convolution theorem of the Fourier transform.

The usage of the DFT has become quite common in recent years, particularly since the introduction of the Fast Fourier Transform (FFT) algorithm by Cooley and Tukey (8), which rapidly computes the DFT as defined in equation 9.1.

Besides relation 9.1, an inverse transform can be defined through

$$y_m = \sum_{k=0}^{N-1} W^{km} X_k \equiv DFT^{-1} \{X_k\} \tag{9.2}$$

When this inversed transform is applied to the X_k's of (1), $y_m = x_m$. The relations 9.1 and 9.2 thus form a closed transform pair. Other conventions exist regarding the exact definition of the transform pair, differing in either the normalizing factor (N) or in the sign of the exponent or both.

The theory of the transform pair can be fully developed on its own, i.e. without reference to the continuous Fourier transform, by using the properties of the W^{km} (7), (1). The results show a close likeness to the properties of the continuous Fourier transform (23).

A more direct insight into the properties of the DFT can be gained by relating it to the theory of Fourier series. The reason for this is that the inverse relation (9.2) allows for a direct extension outside the range of $m(0, N - 1)$ and results in a periodic repetition of the input data with period N.

This implicit periodicity of the input data whenever the DFT is used, is of major relevance when applied to various problems. The relation between the coefficients of the DFT and the continuous Fourier series expansion follows quite naturally as will now be demonstrated. Consider a signal $f(t)$ which is studied over a period of duration D. When the function is sampled at discrete points $t_n = nT$ the series $\{f_n\}$ results; $n = 0, N - 1$ and $D = NT$. When the DFT is applied to this series the implied input is periodic. This is

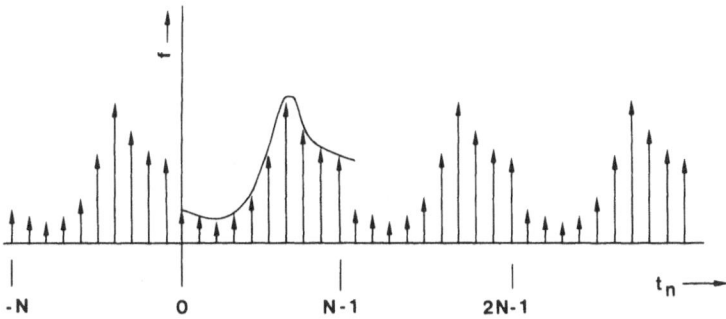

Fig. 9.1 Introduction of the implied periodicity of sampled data when using DFT.

depicted in Fig. 9.1, in which for convenience sake $f(t)$ is assumed to be real. Periodic extension of $f(t)$ outside the interval studied allows the following series representation (23):

$$f(t) = \sum_{-\infty}^{\infty} \alpha_k \, e^{2\pi ikt/D}$$

in which

$$\alpha_k = \frac{1}{D} \int_0^D f(t) \, e^{-2\pi ikt/D} dt \tag{9.3}$$

A straightforward numerical computation of this integral using the most primitive form of integration leads to:

$$\alpha_k \cong \frac{1}{NT} \sum_{n=0}^{N-1} f(nT) \, e^{-2\pi inkT/(NT)} \, T =$$

$$= \frac{1}{N} \sum_{n=0}^{N-1} f_n W^{-nk} = F_k ; \text{ with } F_k \text{ the } k^{th} \text{ DFT term of } f_n.$$

The nature of this approximation can be shown as follows.
Suppose

$$f(t) = \sum_{m=-\infty}^{\infty} \alpha_m \, e^{2\pi imt/D} \tag{9.4}$$

F_k as determined by N samples of this function can be found by substitution of equation 9.4 into equation 9.1:

$$F_k = \frac{1}{N} \sum_{n=0}^{N-1} \left\{ \sum_{m=-\infty}^{\infty} \alpha_m \, e^{2\pi imnT/(TN)} \right\} W^{-kn} =$$

$$= \sum_{m=-\infty}^{\infty} \alpha_m \left\{ \frac{1}{N} \sum_{n=0}^{N-1} W^{(m-k)n} \right\} = \tag{9.5}$$

$$= \sum_{m=-\infty}^{\infty} \alpha_{k+mN}$$

which follows directly from the properties of W.

Thus the DFT coefficients F_k are seen to be the sum of α_k and all α values that are shifted over multiples of N. The α_k values outside the range $-\frac{1}{2}N + 1 \leq k \leq \frac{1}{2}N$ are said to fold back into this range. N is assumed to be even as is usual in this type of work. When the Fourier series coefficients outside the range $(-\frac{1}{2}N + 1, \frac{1}{2}N)$ are zero it is immediately clear that the F_k's are identical to the corresponding α_k's, which are then simply repeated over a period N. This is a direct proof of the sampling theorem. The folding frequency is one half of the sampling frequency.

9.2.4 Least-squares approach

When a signal is known to be a member of a well defined class of analytical functions or assumed to be so, a common problem is the estimation of the parameters of the function involved. In practical circumstances the fundamental relationship is usually perturbed by an error term. This error term may be caused by the crudeness of the function (model) used for the description of the data or by the limited accuracy of the procedure measuring $f(t)$.

Quite generally the problem can be stated as follows: suppose y_n, $n = 0$, $N - 1$, are the readings on a function $f(x_n, p)$, with x_n the (K-dimensional) variable involved, e.g. time or a simple spatial coordinate, and p a M-dimensional parameter vector, i.e. vector $p = (p_1, p_2 \ldots p_M)$. The y_n values will be approximated by the f_n values that can be computed through $f(x_n, p)$ assuming certain values of the parameter vector p. Let ϵ_n be the difference between these values, i.e. $\epsilon_n = y_n - f_n$. The problem that has to be solved is the determination of that parameter vector that will minimize the errors involved. The series of errors form a N-dimensional vector.

Before the minimization problem can be tackled a suitable measure of the total error encountered has to be selected (4). In practice nearly always the sum of squares is used, i.e. $\sum_{i=1}^{N} \epsilon_n^2$. Minimizing this expression leads to the by far most commonly used procedure: the least squares approach. The popularity of this procedure stems from the fact that the resulting minimization procedure has an analytical solution for some important practical ·cases; amongst others the straight line.

Another reason lies in the fact that when the individual errors ϵ_n can be assumed to have a Gaussian distribution, confidence intervals of the estimated parameters can be computed; see 9.4.2. The resulting parameter vector p ensures a fair, overall fit of f to y. Individual large values of ϵ_n may, however, occur.

There are other measures for the error criterium. One of these leads to another well known procedure: the minimax principle. All individual errors now remain below the (minimized) maximum error found. The theory of this principle is less well developed than the least squares approach with the exception of the expansion using Chebyshev polynomials. For numerical procedures the minimax principle has the advantage of requiring less processor time.

We should like to emphasize here that a careful choice should be made before either method is selected for any application. In the sequel the sum of squares norm will generally be assumed.

The remaining minimization problem can be subdivided into three important cases:

1) f is linear in p.
2) f is linear in p, and the components p_m of p are the coefficients of an orthogonal base.
3) f is nonlinear in p.

Case 1.   Here f can be written as

$$f(x) = \sum_{m=1}^{M} g_m(x)p_m$$

if we take

$$x = x_n, n = 0, N - 1$$

then

$$f_n = \sum_{m=1}^{M} g_{nm}p_m \quad \text{or} \quad f = Gp$$

in which latter form f and p are N- and M-dimensional column vectors respectively and G is a N, M matrix. Solution of the minimization procedure now runs as:

$$\epsilon = y - Gp$$

$$E \triangleq \epsilon^t\epsilon = \text{total squared error} = (y - Gp)^t(y - Gp) = y^ty - 2y^tGp + p^tG^tGp$$

with the lable "t" referring to the transposed vector. For an extreme:

$$\frac{\partial E}{\partial p_j} = 0, \text{ which results in}$$

$$G^tGp = G^ty \text{ and finally } p = (G^tG)^{-1}G^ty \tag{9.6}$$

When G is square and non-singular the solution simplifies to $p = G^{-1}y$.

Case 2. In this case the functions g are such that

$$\sum_n g_j(x_n) \cdot g_k(x_n) = 1 \quad \text{if } j = k$$
$$= 0 \quad \text{otherwise}$$

As a consequence the matrix G^tG in equation 9.6 simplifies to the unit matrix (which can be omitted) so

$$p = G^ty \tag{9.7}$$

It must be stressed that the orthonormality conditions should hold for the x_n values considered. This limits the applicability of equation 9.7. When the x_n values are irregularly spaced and the g functions are simple powers of x a suitable orthonormal set can be set up for the particular set of sample points considered (12).

Case 3. When the function is nonlinear in its parameters no general analytical solution can be given. An iterative search procedure may now be started to find the M components of vector p so as to minimize $E = \sum_n(f_n - y_n)^2$ or some other error norm. An initial estimate of the parameter vector p is iteratively adjusted until no further significant decrease in E is encountered. The results of the optimization procedure will inevitably depend on the initial values provided.

Several algorithms have appeared in recent years. They all try to find ways to determine the size and direction (in M-dimensional space) of subsequent iteration steps, and to minimize the total number of iterations required. The direction of any subsequent iteration is chosen with reference to the local gradient at the present position in the iterative procedure. These gradients can either be worked out analytically or for more complex situations, by means of numerical differentiation (21), (11) and (6).

9.3 Stochastic signals

9.3.1 *Description in terms of statistical parameters*

In the following the stochastic signal is, in accordance with a digital approach, assumed to exist in the form of a discrete infinite series of values f_n; in practice one always has of course a finite subset $n = 0, N - 1$. Additionally the f_n's are assumed to be equispaced at distance T. For simplicity we assume that

$$\sum_{n=0}^{N-1} f_n = 0$$

When this is not true, making the assumption valid is one of the first steps in most forms of signal processing. Now, if the f_n's are mutually independent one speaks of a random signal or of (band limited) white noise. Under the additional assumption that the f_n's are normally distributed the description with one single parameter σ^2 – the variance – is exhaustive. However, when relationships between the f_n's exist – for instance when a positive value of a f_n is mostly followed by a negative value – more parameters come into play.

9.3.2 *Autocorrelation function*

The way to describe these interdependencies is by means of the auto-correlation function (ACF). To introduce this concept let us create from our original data file f_0, f_{N-1} two new ones: the series f_0, f_{N-2} and the series f_1, f_{N-1}. From these two series we compute the correlation coefficient (see chapter 10). Let us call this coefficient r_1, because the second series was created by a shift of one unit of the first series. In the same way we construct r_2 as the correlation coefficient of f_0, f_{N-3} and f_2, f_{N-1}. This procedure gives us exactly r_{N-1} values. To these we add r_0 which is the correlation of the series with itself and thus $r_0 = 1$. Under the assumption that the data are normally distributed we now have an exhaustive description. If the data are not normally distributed – and proving normality is rather difficult – an exhaustive description is nearly impossible. In practice therefore one always stops at the level of the ACF even if the material is not normally distributed.

For practical purposes let us remark that for white noise all ρ_k's–except ρ_0 which is by definition 1 – should be zero. To see how this comes about with actual data one has to take rather a long series (N > 1000) of data.

9.3.3 *Spectral density*

We now introduce a second method to compute the r_k's. We therefore put the data first in a frequency frame and for this we turn again to the tool of the Discrete Fourier Transform. Applying this transform to our data we are left with a set of real (R_k) and imaginary (I_k) values composing F_k; see equation 9.1 of 9.2.3. Thus

$$F_k = R_k + i\,I_k$$

We now define the periodogram as

$$P_k = R_k^2 + I_k^2 = F_k \overline{F}_k$$

This concept is still sometimes encountered in the literature but today the concept of spectral density C_k is nearly always used. The spectral density is a normed periodogram, i.e.

$$C_k = \frac{T}{N} P_k$$

with T the sample time. One should be aware of the fact that the norming T/N depends of the precise form of the applied DFT algorithm.

The essence of this frequency approach is the fact that the C_k's as defined here are directly related to the r_k's of the ACF. This relationship is again of the form of a Fourier transform and universally known as the Wiener Khinchin relation. The main point is that this frequency approach leads to the same descriptive parameters r_k. Why this devious way? This is rather a matter of taste. Both the ACF–the r_k's–and the spectral density–the C_k's– contain the same information. It is presented, however, in an entirely different format.

On the other hand if one hunts for hidden periodicities in the signal a different interpretation of the C_k's can be given. To that end one takes a resistor of $1\,\Omega$ and the C_k's can now be read as the energy which frequency $\nu_k = k/(NT)$ $(k = 0, \frac{1}{2}N)$ dissipates in this resistor. Before one actually can use the C_k's for a measure in the sense of energy contribution much more work still has to be done. The computed C_k's scatter namely wildly round their true – but helas unknown – values. To that end one starts to smooth the C_k's, i.e. one averages a number of values adjacent to C_k and takes this average as a new C_k value. There are many other techniques and these procedures are usually referred to as windowing.

For a thorough and practical analysis of this subject we refer to Jenkins and Watts (15), Cox and Lewis (9) and for a broad introduction to Vol. 3 of Kendall and Stuart (19).

9.3.4 Description in terms of an ARMA process

A fundamentally different technique of describing a stochastic signal is in terms of an autoregressive moving average representation (ARMA). Instead of estimating the r_k's as in the foregoing section (9.3.1) one tries to find a rule which converts the data at hand (f_n) to white noise, i.e. to a set of mutually independent values ϵ_n. To understand how this works, let us assume that the data obey the following model:

$$f_n - \lambda f_{n-1} = \epsilon_n$$

i.e. we know the f_n's – the data – and if we also knew λ we could construct the ϵ_n's and test for independency. This could be done by computing the ACF of the ϵ_n's. This ACF should have $r_k = 0$ for all k's (of course $r_0 = 1$).

If the latter is indeed true one says that the data result from a Markov process, i.e. from an autoregressive process of order 1 (AR(1)). In this case λ turns out to be equal to r_1 of the f_n values. For the constructed ϵ_n's we compute its variance (chapter 10) and we now again exhaustively have described our data by saying that they arise from a Markov process with a specified λ and a specified ϵ-variance. In quite a number of cases this type of description pays off; what to do if the ACF of the ϵ_n are not zero? One first could try an AR(2)

$$f_n - \lambda_1 f_{n-1} - \lambda_2 f_{n-2} = \epsilon_n$$

But a 1-st order moving average process might also be effective

$$f_n = \epsilon_n - \beta \epsilon_{n-1}$$

Finally a mixed representation might be the solution like an ARMA (1, 1)

$$f_n - \lambda f_{n-1} = \epsilon_n - \beta \epsilon_{n-1}$$

In the excellent book of Box and Jenkins (2) a systematic approach to this modelling is given. From practical and theoretical point of view the problem is by no means completely solved.

9.4 Deterministic signals

Quite often one obtains as the result of a measurement a set of data which should, according to some explicit theory, lie on a curve, the standard example being of course a straight line. In practice deviations due to errors occur but given the form of the analytical expression of the curve one tries to find its parameters. It is evident that if the measured values were the absolutely true ones, i.e. free of noise, one could limit oneself by taking as many data points as unknown parameters and solve the ensuing equations. However, data nearly always contain errors and although the basic signal is deterministic an additional stochastic component is present. It is this latter component which spoils this procedure.

The problem of having equations which are nonlinear in the parameters was already touched upon in 9.2.4.

9.4.1 *Model known* (as function, as differential equation)

In the preceding we assumed that the model was known in the form of an analytical function. Sometimes, however, the model is given in the form of a differential equation with unknown parameters. If one can solve the differential equation the best approach to the fitting problem is to do this and fit the analytical solution to the data. Most frequently, however, no general solution is possible. In 9.4.2 we shall mention some techniques for coping with this problem.

9.4.2 *Parameter estimation*

First some assumptions have to be made about the behaviour of the noise at each measuring point. The simplest case is that for which each data point is contaminated with an error ϵ_n for which $\text{Cov}(\epsilon_n, \epsilon_{n+k}) = \sigma^2 \delta_k$ and of course $E(\epsilon_n) = 0$. This means that at each measuring point the noise has the same variance and furthermore that the ϵ_n's are independent. In some cases the errors although independent have different variances for the different measuring points. Most techniques can be extended to situations where either the individual variances are known or can be estimated from multiple measurements, see Draper and Smith (10).

Let us thus assume that

$$\text{Cov}(\epsilon_n, \epsilon_{n+k}) = \sigma^2 \delta_k$$

and that the function is known in some form; in brief

$$y = f(x, \alpha); \text{vector } \alpha = (\alpha_1, \alpha_2, ..., \alpha_M)$$

The problem is to estimate the parameter vector α, given N readings (x_i, y_i).

In 9.2.4 this problem was discussed. Of main importance now is the question of what can be said about the accuracy of α. In general it is possible to give some crude measures for the accuracies with which the α_i's are estimated (4) and (10). Especially in comparing α_i for one set of data with an α_i estimated with another set of data one should be very careful in pronouncements about equality.

We now come to the case where the model is specified in the form of a differential equation or even a mixed differential integral equation. As already stated if an analytical solution can be obtained one should start from there. In some cases it is possible to perform a Laplace transformation of the differential equation or even of the mixed equation and if the result can be brought into the form of a ratio of two polynomials a rather simple procedure is again feasible.

Although one can theoretically expand the ratio of two polynomials into a series of rational fractions (Heaviside expansion theorem) and thus obtain an analytical solution, this can be rather difficult in practice. The difficulty lies in the fact that one has to find the roots of the denumerator and this repeatedly at each iteration. A simple technique to bypass this problem is based on the z-form as introduced by Boxer and Thaler (3). The z-form is a special technique for converting a representation in the s-domain to the z-domain with emphasis on the step response bypassing the Heaviside expansion theorem. Once a representation in the z-domain has been obtained a digital solution is possible.

The necessary differentiation to the parameters α_j must be performed numerically.

As an example we present a case where we applied this technique to the following integro-differential equation

$$\frac{dC}{dt} = -k_1 \, C(t) + k_2 \, \int_0^t C(t - \tau)g(\tau)d\tau$$

in which $g(t) = a(at)e^{-at}$. Estimates had to be obtained for $C_0 = C(0), k_1, k_2$ and a. Fig. 9.2 gives the data points as well as the fitted curve. As is shown the fit is excellent.

Of the more general methods of estimating the parameters of differential equations we would like to mention the technique of quasilinearization as given by Kagiwada (17). This is indeed a general method in which the basic idea consists of expressing the parameters α_j in terms of the solutions of the system of first order differential equations:

$$\frac{d\alpha_j}{dx} = 0 \text{ with } \alpha_j(0) = \alpha_j$$

The original differential equation is reduced to a set of first order equations and extended with the newly created equations. Since this set of equations is no longer linear a linearization process, i.e. quasilinearization, is introduced which makes an iterative approach possible. The method converges for a given set of initial parameter values either rather fast or not at all.

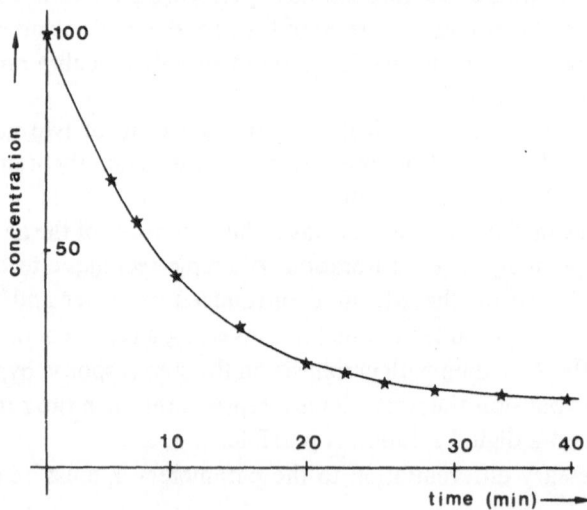

Fig. 9.2 Result – full line – of fitting a model with help of the z-form technique to give data points (asterisks).

The main drawback is the sensitivity to the initial parameter values. One should have proper initial estimates otherwise no convergence will occur. A second drawback is the fact that no simple possibility exists for obtaining a measure for the accuracy of the α_j's.

A straightforward method which is not afflicted with the above mentioned problems is the following rather crude technique. For a given set of initial parameters one solves the differential equation – any numerical technique will do. Differentiation to the parameter is done numerically and one proceeds as in any straightforward iteration technique. The drawback here is computing time. Convergence is slow even with automatically adjusted step size.

9.4.3 *Model unknown*

Quite often one is confronted with the following problem: a curve has been measured and without having a model, i.e. a function or otherwise, one still needs a description, for instance for manipulation (see 9.5). Assuming that the data have a deterministic genesis and are contaminated by noise there are three main approaches: polynomial, exponential and Fourier representation. Our choice will most often be guided by some vague a priori knowledge. For instance if the data stem from radio-active compartment analysis one certainly opts for an exponential approach. Nearly straight lines favour a polynomial approach etc.

Polynomial representation. Nearly everyone is familiar with fitting a straight line. Quite often the data points slowly curve up- and/or downward and it looks attractive to add a quadratic term or even higher powers to the representation of the data values d_n

$$d_n = f(t_n) + \epsilon_n \equiv a_0 + a_1 t_n + a_2 t_n^2 + a_3 t_n^3 + \epsilon_n \quad (n = 0, N - 1)$$

Minimizing $\{d_n - f(t_n)\}^2$ with respect to the a_j's (j = 0,3) results in 4 linear equations which can be solved in the standard way (see 9.2.4). There are two serious drawbacks to this technique. First of all: adding an extra term $(a_4 t^4)$ would mean that all coefficients $(a_0, a_1, a_2$ and $a_3)$ change and that the entire numerical procedure must be repeated. Secondly: even for rather low polynomial orders solving the equations may lead to large numerical errors. This is caused by the fact that the determinant of the system may easily assume a very small value. Both difficulties can be avoided by the introduction of the so called orthogonal polynomials (5). A practical algorithm is given by Forsythe (12). In 9.5.5 it is shown that raising the order of the polynomial does not always imply an increasingly better fit.

Exponential representation.   The representation is according to

$$d_n = f(t_n) + \epsilon_n \equiv A_1\,e^{-\lambda_1 t_n} + A_2\,e^{-\lambda_2 t_n} + \ldots \epsilon_n$$

For the problem of how many terms should be considered (system identification) we refer to Lanczos (20). The difficulty of this latter problem is also demonstrated by this author; he fits slightly truncated data beautifully – truncation here causing the experimental errors ϵ_n – generated by a third order system with a second order system. As long as one simply wants to represent the data no harm would have been done by using the second order system. Any physical interpretation, however, would lead to an erroneous conception of the system.

Fourier representation.   Let us stress again the fact that we merely want to represent our N data points on the interval 0, D with $D = t_{N-1} - t_0$. Now a Fourier representation makes use of orthogonal functions (sin and cos). The orthogonal property of sin and cos not only holds on a continuous interval but also for a discrete set of equidistant points.

In representing the data with a Fourier representation one implicitly introduces a periodicity of the data points outside their range. As the data values are prescribed only within the data interval we can also proceed as follows. On the interval $- D, 0$ we create artificial data points; either symmetrically according to $d_{-n} = d_n$ or antisymmetrically with $d_{-n} = - d_n$. The extended set of points d_n, $n = - N + 1, N - 1$ on the interval $- D, D$ are now used in the computation. This leads to a pure cosine representation or to a pure sine representation respectively. The idea behind this procedure stems from the following. The usefulness of the Fourier representation hinges on its speed of convergence. If the function and its first derivative are continuous at the beginning and end of the interval the convergence is at least as k^{-3}.

A symmetric continuation, leading to a cosine representation, is continuous at $t = 0$ but may or may not have a continuous derivative at this point.

A procedure which insures a continuous derivative as well is to subtract first a linear function from the data points in such a way that the first and last data point become zero. If one now extends the newly created data points d_n^* ($n = 0, N - 1$) in an antisymmetric ($d_{-n}^* = - d_n^*$) one gets a sine representation on the interval $- D, D$ with convergences at least as k^{-3}.

Other similar procedures are possible. Broome (5) gives numerical procedures which permit one to generate orthogonal sets of sequences. One of these sequences approaches the Laguerre functions and is therefore quite useful in the representation of impulse responses.

9.5 Manipulation

9.5.1 *Data reduction*

The aim of data reduction is to facilitate representation, handling e.g. mutual comprison and storage of a set of signal data. This reduction can be effected all the better the more a priori information about the process concerned is available. Thus the technique as discussed in the previous paragraphs also pertains to the problem of data reduction. Let us give some examples; consider
1. the very use of the sampling theorem for band limited signals,
2. the expansion of the signal in terms of elementary functions –which may or may not form an orthogonal set – and storage of the expansion co-efficients only,
3. direct estimation of the parameters of an actual physical model of the system generating the signal.

9.5.2 *Filtering*

In practice signals may frequently be considered as the sum of a certain "wanted" signal and some additional component obscuring the first one. Note the use of the term additional; more complex forms of contamination (e.g. multiplicative effects) will not be considered here. The general problem of separating these two kinds of signals is called filtering. The term stems from a well established practice both in electrical engineering and in optics. The traditional domain in which the effectiveness of any filter may be judged is the frequency domain.

The response $y(t)$ of a linear time invariant filter H operating upon a signal $x(t)$ can be found through:

$$y(t) = H \cdot \{x(t)\} = \int_{-\infty}^{\infty} x(\tau) \cdot h \, (t - \tau) d\tau \equiv x(t) \times h(t)$$

a convolution of the input signal $x(t)$ and the response $h(t)$ of the filter on the application of an impulse function $\delta(t)$. The application of the convolution theorem of the Fourier transform theory now leads to

$$Y(v) = X(v) \cdot H(v)$$

a direct multiplication of the Fourier transform of the signals concerned. For linear filters the reason for the excursion into the frequency domain is thus obvious: a multiplication is somewhat easier to grasp than a convolution. Application of a filter to the sum of two signals x(t) and n(t) shows directly that the success of the desired separation of x(t) and n(t) will depend on the amount of overlapping of their respective Fourier transforms.

When filtering is applied to (real-) time signals the class of filters that can be applied is restricted to those having an impulse response h(t) such that h(t) = 0 for t < 0. The associated limitations of this "causality" requirement are absent when the natural flow of time is absent, as for functions of space or for effectively disrupted functions, i.e. for signals which are recorded first and filtered subsequently. This possibility can be fully exploited in performing filtering procedures numerically (digital filtering) on sampled input data.

We will now briefly discuss some aspects of digital filtering as applied to equally spaced data. Two main lines of approach are possible: the filtering can be performed either in the frequency domain or in the time domain.

When filtering in the frequency domain the entire sequence of N points must be recorded first. Next the discrete Fourier transform $X_k = \mathrm{DFT}\{x_n\}$ is computed and the filter H_k is specified as N complex numbers. Finally the inverse discrete Fourier transform is applied to $H_k X_k$, resulting in the filtered output:

$$\{y_n\} = \mathrm{DFT}^{-1}\{H_k X_k\}$$

As stated this procedure is restricted to prerecorded signals. Using the FFT algorithm the entire procedure takes roughly $2N_2 \log N$ multiplications. This number should be compared with the figures on the time domain approach to follow.

The N filter coefficients ($\frac{1}{2}N + 1$ for the real part and $\frac{1}{2}N - 1$ for the imaginary part) can be specified at will allowing a great freedom of choice for the filters. The fact that no equivalent analog counterpart of the discrete filter may exist, is irrelevant. However, when specifying the filter coefficients, which may have all the desired properties for separation in the frequency domain, it is often essential that the associated time behaviour of the filter is according to some norm. This may be a requirement on the

rise time, the amount of overshoot or the amount of ripple of the step response of the filter. To this end (using DFT) one should study the effect of
the filtering procedure on the series (step response)

$$\{x_n\} = \tfrac{1}{2} \quad n = 0 \text{ and } n = \tfrac{1}{2}N$$
$$\quad\quad = 1 \quad n = 1, \tfrac{1}{2}N - 1$$
$$\quad\quad = 0 \quad n = \tfrac{1}{2}N + 1, N - 1$$

Another check on the time behaviour of the specified filter coefficients
H_k is to compute directly

$$\{h_n\} = DFT^{-1}\{H_k\}$$

being the impulse response of the filter.

We now turn to filtering in the time domain. Let the series $\{x_n\}$ be the
ordered set of observations on some variable. From this series we can
construct another series $\{y_n\}$ as a linear combination of the $\{x_n\}$ and $\{y_n\}$
values through:

$$\sum_{l=0}^{L} b_l y_{n-l} = \sum_{k=0}^{K} a_k x_{n-k} \tag{9.8}$$

The set of coefficients

$$a_k, k = 0, K$$
$$b_l, l = 0, L$$

are the filter coefficients and determine the properties of the filtering procedure. These properties may, as for the continuous (analog) case
be studied by looking at the impulse response of the filter. This is the series
$\{h_n\}$ arising from the application to the filter of the series: $\{x_n\} = \delta_n$. The
series $\{h_n\}$ can be generated directly through the generating equation 9.8.

Alternatively the step response can be worked out by computing the
effect of the filter specified by the series

$$u_n = 0 \quad n < 0$$
$$\quad = \tfrac{1}{2} \quad n = 0$$
$$\quad = 1 \quad n > 0$$

The criteria to be met by the time behaviour of a filter are usually interpreted and expressed best in terms of conditions of this step response (rise
time, overshoot etc.) rather than in terms of the impulse response. Having
specified the filter in the time domain the next point of interest of course
lies in the determination of its frequency behaviour. To this end we consider the output of the filter to the input series:

$$x_n = e^{2\pi i\nu nT}$$

with ν the frequency of interest and T the sampling interval. We introduce here

$$s = 2\pi i\nu$$

resulting in

$$x_n = e^{nsT}$$

The ensuing output series, being a linear combination of such terms,
will have the same appearance as x_n, differing only in a multiplicative
factor $A(\nu)$ and a phase term $\varphi(\nu)$. This approach is equivalent to the
electrical engineer's determination of amplitude and phase characteristic
of an analog filter. The output series

$$\{y_n\} = A(\nu)e^{i\{2\pi\nu nT + \varphi(\nu)\}}$$

can now be found through:

$$\sum_{l=0}^{L} b_l A(\nu)e^{2\pi i\nu(n-l)T+i\phi(\nu)} = \sum_{k=0}^{K} a_k e^{2\pi i\nu(n-k)T}$$

Hence

$$H(\nu) \triangleq A(\nu)e^{i\phi(\nu)} = \frac{\sum\limits_{k=0}^{K} a_k e^{-2\pi ik\nu T}}{\sum\limits_{l=0}^{L} b_l e^{-2\pi il\nu T}}$$

or

$$H(v) = \left[\frac{\displaystyle\sum_{k=0}^{K} a_k e^{-ksT}}{\displaystyle\sum_{l=0}^{L} b_l e^{-lsT}} \right]_{s = 2\pi i v} \tag{9.9}$$

This transfer function $H(v)$ of the filter is a continuous periodic function of frequency v with period $1/T$, reflecting the sampled nature of the data. The Fourier series expansion of the periodic function $H(v)$ forms the impulse response of the filter. This provides an indirect determination of $\{h_n\}$.

When computing $H(v)$ for a discrete set of frequencies it may be computationally efficient to use the FFT algorithm.

Two special filter classes can be distinguished depending on the range of the filter coefficients (24)

1) non-recursive filtering:

$$b_0 y_n = \sum_{k=0}^{K} a_k x_{n-k}$$

Here the output is a simple moving average of the applied input data. The impulse response is:

$$\{h_n\} = \frac{1}{b_0} \{a_n\} \text{, i.e. the}$$

impulse response is a direct replica of the filter coefficients. As such the impulse response is of finite length: FIR filters (finite impulse response).

2) (pure) recursive filtering:

$$\sum_{l=0}^{L} b_l y_{n-l} = x_n$$

The filtering is effected on the output signal itself. This is also known as an autoregressive procedure. The impulse response in this case is of infinite length: IIR filters (infinite impulse response). As compared to non-recursive filters the recursive filters can be shown to require fewer terms to obtain a given cut-off slope of the filter. In contrast the phase characteristics of the autoregressive procedure is inferior to that of non-recursive filtering. However, in case of prerecorded signals this shortcoming can easily by bypassed by subsequently performing an identical filtering, in which the series $\{y_{N-n}\}$ is taken as the input (25).

The design of digital filters frequently (18) makes use of the vast amount of literature available on analog filters. Some possible procedures are outlined below.

Let us suppose that an analog filter has the desired transfer function

$$H(v) = \{H(s)\}_{s=2\pi i v}$$

in which H(s) is a quotient of two polynomials:

$$H(v) = \left[\frac{\sum\limits_{m=0}^{M} \alpha_m s^m}{\sum\limits_{l=0}^{L} \beta_l s_l} \right]_{s = 2\pi i v} \tag{9.10}$$

We now look for ways to establish the coefficients a_k and b_l of the digital filter of equation 9.8 such that the associated frequency transfer function (equation 9.9) resembles equation 9.10.

A first method makes use of the rational fraction expansion of equation 9.10 in which H(v) is expressed as

$$H(v) = \sum\limits_{p=0}^{L} \left(\frac{w_p}{s - s_p} \right)_{s = 2\pi i v} \tag{9.11}$$

with s_l the poles of H(s). Consider one such term, represented here as $\alpha/(s + \alpha)$. The rational fraction expansion of equation 9.9, being a similar quotient of polynomials in e^{-sT}, gives rise to terms of the form

$$\frac{c_1}{e^{-sT} + c_2} \tag{9.12}$$

To relate equation 9.11 to 9.12 it is thus required to express s in terms of e^{sT}. The method at hand uses the identities

$$s + \alpha \equiv \frac{1}{T} \ln e^{(s + \alpha)T} \text{ and } \alpha = \frac{1}{T} \ln e^{\alpha T}$$

as well as the first term of the expansion

$$\ln v = \sum_{n=1}^{\infty} \frac{1}{n}\left(\frac{v-1}{v}\right)^n$$

This results in:

$$\frac{\alpha}{s+\alpha} = \frac{\frac{1}{T}\ln e^{\alpha T}}{\frac{1}{T}\ln e^{(s+\alpha)T}} \cong \frac{e^{\alpha T}-1}{e^{\alpha T}-e^{-sT}}$$

The parameters of the digital filter section equation 9.12 corresponding to $\alpha/(s + \alpha)$ found in this way are:

$$c_1 = 1 - e^{-\alpha T} \quad \text{and} \quad c_2 = -e^{\alpha T}$$

If the H(s) of the analog filter is solely $\alpha/(s + \alpha)$, i.e. a simple low pass filter, the corresponding digital filter is

$$b_0 y_n + b_1 y_{n-1} = a_0 x_n$$

with

$$a_0 = e^{\alpha T} - 1$$
$$b_0 = e^{\alpha T}$$
$$b_1 = -1$$

The impulse response of this digital filter element is

$$\{h_n\} = (1 - e^{-\alpha T})e^{-n\alpha T}$$

as can be verified by substitution in equation 9.8. This discrete impulse response is therefore proportional to the impulse response of the analog filter section $\alpha/(s + \alpha)$ at time instants $t = nT$

$$\{h(t)\}_{t=nT} = \{\alpha e^{-\alpha t}\}_{t=nT} = \alpha e^{-n\alpha T}$$

The resulting frequency transfer function of the digital filter element is similar to that of the analog filter, and more so, as αT becomes smaller.

This is demonstrated in Fig. 9.3.2a, where the amplitude transfer function

$$A(v) = \{H(v) \, \overline{H(v)}\}^{1/2}$$

of the continuous filter section:

$$\alpha/\{(2\pi v)^2 + \alpha^2\}^{1/2}$$

is compared to that of the digital filter section:

$$(e^{\alpha T} - 1)/(e^{2\alpha T} + 1 - 2e^{-\alpha T} \cos 2\pi v T)^{1/2} \tag{9.13}$$

for different values of α.

To summarize, when this transformation technique is applied to all individual terms of the Heaviside expansion of the (continuous) $H(v)$ and the resulting discrete filter terms are summed, the resulting digital filter has an identical impulse response and a similar frequency transfer function (extended periodically). However, for the method to be applicable the location of the individual poles of $H(s)$ has to be known.

The second method uses the first term of the expansion

$$\ln v = 2 \sum_{n=1}^{\infty} \frac{1}{2n - 1} \left(\frac{v - 1}{v + 1}\right)^{2n-1}$$

applied directly to the identity:

$$s = \frac{1}{T} \ln(e^{sT})$$

The resulting substitution is

$$s \rightarrow \frac{2}{T} \frac{1 - e^{-sT}}{1 + e^{-sT}}$$

which does not require a known pole location. This substitution is carried out for all s terms in $H(s)$.

As an example we consider the application of this method to the same low pass filter

$$H(s) = \alpha/(\alpha + s)$$

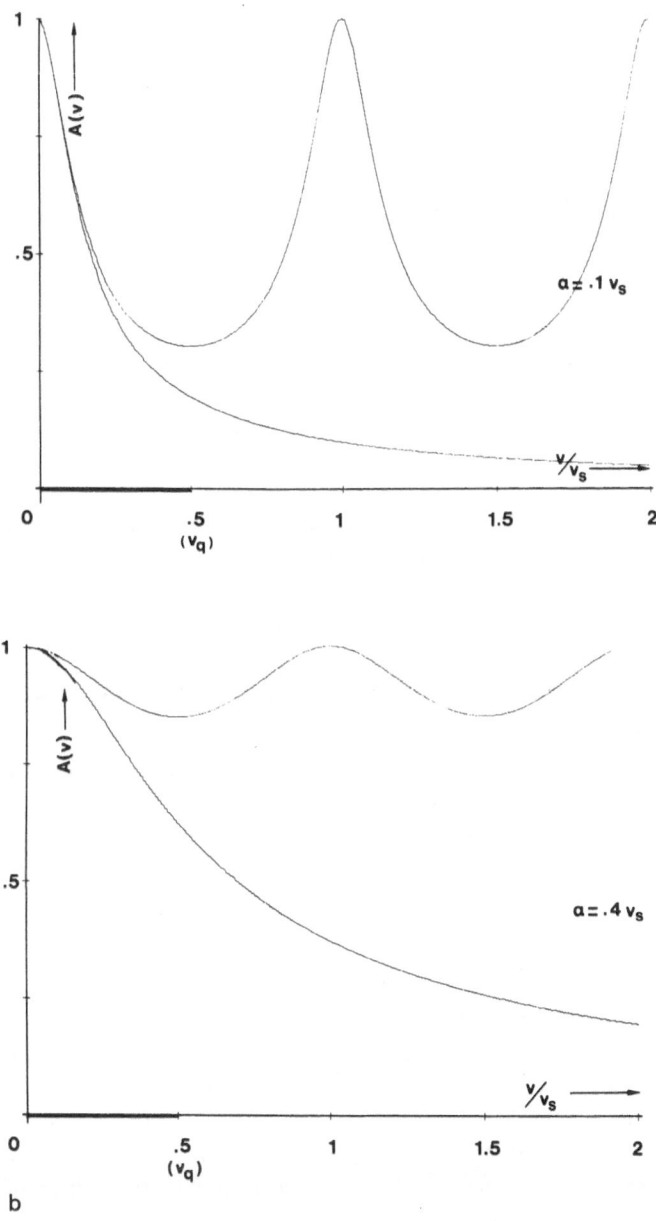

Fig. 9.3a, b Amplitude transformation of a simple low pass filter. The periodic curves represent the behaviour of the digital filter according to equation 9.13. The other curves represent the original analog filter. Values of α are indicated. Note that the sampling theorem restricts the v values to those for which $v < v_q = 0.5\, v_s$.

The corresponding digital filter transfer function is in this case

$$\frac{\alpha}{s + \alpha} \rightarrow \frac{\alpha}{\dfrac{2}{T}\dfrac{1 - e^{-sT}}{1 + e^{-sT}} + \alpha} = \frac{\alpha T(1 + e^{-sT})}{2 + \alpha T + (\alpha T - 2)e^{-sT}}$$

Here the digital filter becomes:

$$b_0 y_n + b_1 y_{n-1} = a_0 x_n + a_1 x_{n-1}$$

with

$$a_0 = \alpha T \quad b_0 = (\alpha T + 2)$$
$$a_1 = \alpha T \quad b_1 = (\alpha T - 2)$$

The amplitude transfer function $A(v)$ in this case can be worked out as:

$$A(v) = \alpha T \left\{ \frac{1 + \cos 2\pi vT}{4 + \alpha^2 T^2 - (4 - \alpha^2 T^2)\cos 2\pi vT} \right\}^{1/2} \tag{9.14}$$

This function is plotted in Fig. 9.4 for different values of α. These curves are to be compared to those of Fig. 9.3.

More sophisticated procedures still are possible, e.g. utilizing the already mentioned (9.4.2) z-forms (3) (16). For filters that should have a sharp cut-off special care has to be taken in working out the coefficients of the digital filter in order to ensure stability of the filtering procedure (24).

9.5.3 *Differentiation*

Various techniques for differentiating a signal can be given according to, again, the a priori available information about the signals, whereas, as before, a distinction has to be made between real-time signals and other signals. When a signal is known to be a member of a certain class of functions and the available data are prerecorded the best strategy would be to estimate the parameters and differentiate analytically. When no model is available local estimation of the derivatives can be obtained using implied analytical differentiation of local estimates of the signal (14).

In real-time application the procedure necessarily has to be applied to previously recorded values. This generally reduces the associated accuracy. Finally, differentiation may be performed in the frequency domain using

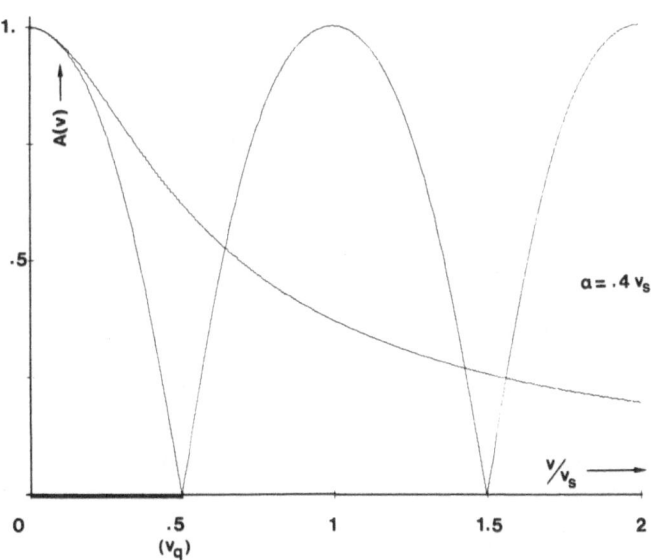

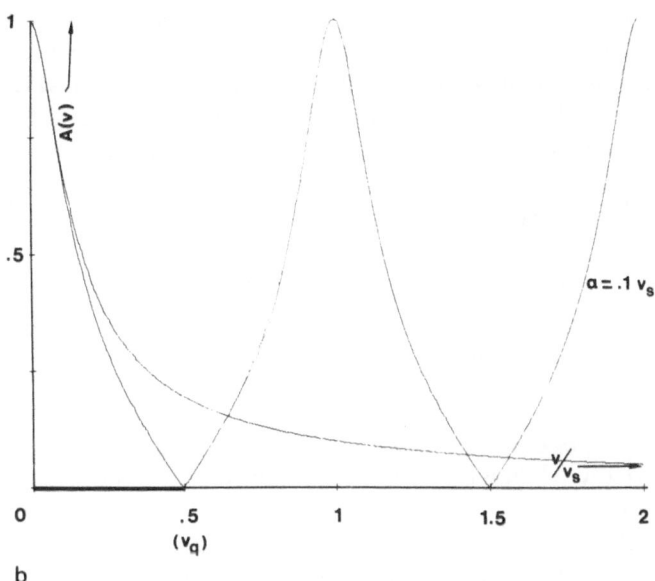

b

Fig. 9.4a, b Amplitude transformation of a simple low pass filter. The periodic curves represent the behaviour of the digital filter according to equation 9.14. The other curves represent the original analog filter. Values of α are indicated. Note that the sampling theorem restricts the v values to those for which $v < v_q = 0.5 v_s$.

equidistant samples since

$$F \cdot \left\{ \frac{d}{dt} f(v) \right\} = 2\pi i v F \cdot \{f(t)\}$$

In the actual application of this technique using the DFT the function $2\pi i v$ is implied to be periodic (a saw tooth results) and its spectrum to be unbounded. In order to avoid undesired side effects of this property the teeth of the saw should be blunted slightly by selecting some smooth approximating function (18).

9.5.4 *Integration*

The computation of an integral of a signal as specified by sampled data can be performed using a weighted sum of the samples:

$$\int_a^b x(t)dt \cong \sum_{n=0}^{N-1} w_n \, x(t_n)$$

As in the previous sections the weights may be derived using the amount of a priori information available. Some algorithms that are widely used will be listed. The error involved can be shown to be directly proportional to the maximum of the k-th derivative of the signal over the interval studied (k being related to the complexity of the formula used) and to the k-th power of the sampling interval. Thus the accuracy of any procedure is greater the smoother the signal involved. For equally spaced samples an equally weighted sum can be used when the signal is known to be band limited and the sample density is – conform the sampling theorem – at least twice the highest frequency present.

$$\int_a^b x(t)at = \frac{b-a}{N} \sum_{n=0}^{N-1} x(t_n)$$

In this case the relation is exact as can be seen from section 9.2.3 realizing that $\alpha_0 = F_0$.

In all other cases more elaborate procedures have to be recommended. The next algorithm uses a linear expansion in between sample points:

$$\int_a^b x(t)dt = \frac{b-a}{N} (\tfrac{1}{2}x_0 + x_1 + x_2 + \ldots x_{N-1} + \tfrac{1}{2}x_N)$$

The formula follows from repeated use of the trapezium rule. A slightly more complex formula is Simpson's rule:

$$\int_a^b x(t)dt = \frac{T}{3}(x_1 + 4x_2 + 2x_3 + 4x_4 + \ldots 2x_{N-2} + 4x_{N-1} + x_N)$$

with T the sample distance.

More complex algorithms can still be worked out using even higher order local polynomials. However, when the procedure is applied to measured data with some associated limited accuracy the more complex formulas may give poorer rather than better results when compared to repeated application of some lower order expansion (14); see also section 9.5.5. When the sample points can be chosen at will extra freedom is introduced into the approximating procedure of the integral, requiring fewer sample points for a given accuracy compared to equally spaced samples. This method is known as Gaussian quadrature. However, it can be fully employed only when the signal to be integrated is known to be a member of a certain class of functions (14).

9.5.5 Interpolation

The technique of assigning values to signals in between sample points using measured data nearby is known as interpolation. Clearly when nothing is known about the signal but the sampled data any suggested procedure is as good as another. It is when looking at the implementation of some such procedure that one may decide to give preference to one technique rather than the other; this usually implies some further knowledge about the process concerned.

One may thus, again, express the signal in some set of functions and estimate the expansion coefficients or parameters involved using the techniques as discussed in the previous sections. The resulting analytical representation of the signal may be used in turn to obtain the values between sample points.

Some useful techniques are the following:

1) *Lagrangian interpolation.* This technique employs the unique $(N-1)$th order polynomial form N (not necessarily equally spaced) data points y_n taken at the points x_n. The polynomial passes exactly through all recorded data points, which should have complete accuracy.

The need for this latter requirement is demonstrated in Fig. 9.5. Here part of a 5-th order polynomial is shown. The six points on this curve,

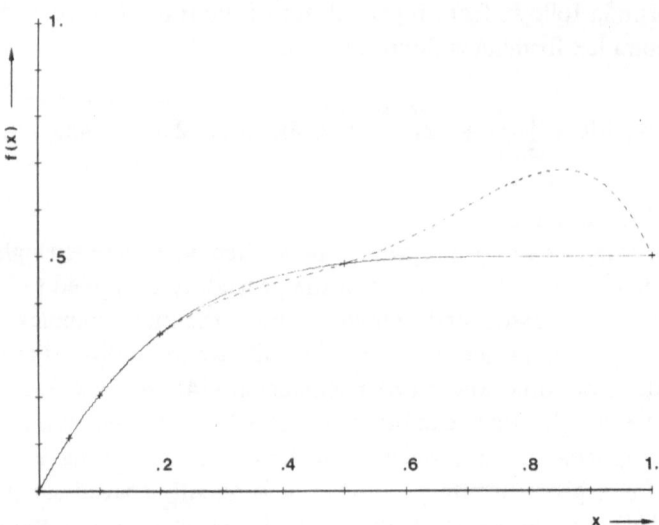

Fig. 9.5 Solid line: the polynomial $y = 0.5(x - 1)^5 + 0.5$. The 6 asterisks (note the one near the origin) were computed from this expression with 3 digit accuracy. Dotted line: the Lagrangian interpolation polynomial.

indicated by asterisks, have been computed using some arbitrarily chosen 5-th order polynomial. The dashed line indicates the fifth order polynomial that arises from Lagrangian interpolation applied to these six points using 3 digit accuracy.

2) *Spline interpolation.* This is a local fitting procedure using cubics and minimizing the local curvature (13).

3) *Fourier interpolation.* For band limited signals a completely accurate interpolating procedure can be given based on the sampling theorem. Using the N, and sufficiently and closely and equally spaced data points taken at $t = t_n$, the procedure under the DFT is as follows: the DFT of $x(t_n)$ is computed with the original N samples. The DFT coefficients are then augmented by as many zero entries as is required for the desired inter-polation, say M. These zero's are entered in the middle of the original series. Applying the inverse DFT to the augmented series results in a completely accurate time representation of the signal specified by a number of points equal to N + M.

These techniques have been generalized to work in more dimensions. The sample points then usually have to be taken from a rectangular grid.

The general case in which samples are taken randomly from multi-dimensional space is extremely difficult to handle. In recent years some algorithms have appeared for dealing with this problem (22).

References

1. Bergland, G.D. (1969): A guided tour of the Fast Fourier Transform. IEEE Spectrum, 6, 41–51.
2. Box, E.P., and Jenkins, G.M. (1971): Time series analysis. Holden-day, San Francisco, Cambridge, London, Amsterdam.
3. Boxer, R. and Thaler, S.: Extension of numerical transform theory. Report RADC-TR-56-115, Rome Air Development Center, New York.
4. Brandt, S. (1970): Statistical and computational methods in data analysis. North-Holland Publishing Company, Amsterdam, London.
5. Broome, P.W. (1965): Discrete orthonormal sequences. J.A.C.M., 12, 2, 151–168.
6. Chambers, J.M. (1973): Fitting nonlinear models: numerical techniques. Biometrika, 60, 1–13.
7. Cochran, W.T., Cooley, J.W., Favin, D.L., Helms, H.D., Kaenel, R.A., Lang, W.W., Maling, G.C., Nelson, D.E., Rader, C.M., and Welch, P.D. (1967): What is the fast Fourier transform. IEEE Trans. Audio Electroacoust., Au-15, 45–55.
8. Cooley, J.W. and Tukey, J.W. (1965): An algorithm for the machine calculation of complex Fourier series. Mathematics of Computation, 19, 90, 297–301.
9. Cox, D.R. and Lewis, P.A.W. (1966): The statistical analysis of series of events. Methuen and Co. Ltd., London.
10. Draper, N. R. and Smith, H. (1966): Applied regression analysis. John Wiley and Sons, Inc., New York.
11. Fletcher, R. and Powell, M.J.D. (1963): A rapidly convergent descent method for minimization. The Computer Journal, 6, 163–168.
12. Forsythe, G.E. (1957): Generation and use of orthogonal polynomials for datafitting with a digital computer. J. Soc. Indust. Appl. Math., 5, 2, 74–88.
13. Gueville, T.N.E. (1967): Spline functions, interpolation, and numerical quadrature. Mathematical Methods for Digital Computers 2. Ralston, A. and Wilf, H.S. (Eds.), Wiley, New York.
14. Hamming, R.W. (1962): Numerical methods for scientists and engineers. McGraw-Hill Book Company, Inc., New York.
15. Jenkins, G.M. and Watts, D.G. (1969): Spectral analysis. Holden-day, San Francisco, Cambridge, London, Amsterdam.
16. Jury, E.I. (1964): Theory and application of the z-transform method. John Wiley and Sons, Inc., New York.
17. Kagiwada, H.H. (1974): System identification. Addison-Wesley Publishing Company, Inc. Advanced Book Program, Reading, Massachusetts.
18. Kaiser, J.F. (1963): Design methods for sampled data filters. Proc. 1st Annu. Allerton Conf. Circuit System Theory, 221–236.
19. Kendall, M.G. and Stuart, A. (1966): The advanced theory of statistics vol. 3. Charles Griffin and Company Limited, London.

20. Lanczos, G. (1961): Applied analysis. Prentice Hall, Inc., Englewood Cliffs, N.J.
21. Marquardt, D.W. (1963): An algorithm for least-squares estimation of non-linear parameters. J. Soc. Indust. Appl. Math., *11*, 2, 431–440.
22. McLain, D.H. (1976): Twodimensional interpolation from random data. The Computer Journal, *19*, 2, 178–181.
23. Papoulis, A. (1962): The Fourier integral and its applications. McGraw-Hill Book Company, Inc., New York.
24. Rader, Ch.M. and Gold, B. (1967): Digital filter design techniques in the frequency domain. Proc. IEEE, *55*, 149–171.
25. Verburg, J. and Strackee, J. (1974): Phaseless recursive filtering applied to chestwall displacements and velocities, using accelerometers. Medical and Biological Engineering, *12*, 483–488.

10
STATISTICAL ASPECTS

John H. Annegers

10.1 Introduction

The object of this chapter is to introduce some concepts relative to the logic of applied statistics, and to suggest the usefulness of statistics in the design of experiments that can yield maximum information. Many excellent text-books make available statistical tables, formulas for computations, more complete discussions and a wider selection of topics. The cited references (1, 2) have been most helpful to the present writer.

10.2 The scope of applied statistics

10.2.1 *Introduction*

Statistics provides a tool which can aid the user in the interpretation of experimental data. Consider, for example, that twenty subjects are available for testing the effectiveness of a therapeutic agent in the treatment of obesity. Ten subjects are given the test agent and ten serve as "controls." Both groups are allowed to eat *ad lib*. Body weight is measured before and at weekly intervals after daily medication is started. The results of this hypothetical study are summarized in Table 10.1. This simple example raises several questions about the adequacy of the experimental design and about the interpretation of the findings.

In the first place, what population does the initial twenty subjects represent? How were the subjects assigned to the control and treated subgroups? What constitutes valid control? Such questions must be considered by anyone who conducts or evaluates an experimental study.

In the second place, the results may seem to be puzzling. The control group gained an average of 0.04 pounds per week whereas the treated group lost 0.09 pounds. Few investigators would expect that the weight changes would be identical, even if the treatment were without any effect in a controlled study. How does statistical analysis aid in the evaluation of the apparent effect of the treatment?

Table 10.1 Weight change in pounds per week.

	control	treated
	0.4	−0.1
	0.2	0.1
	−0.1	0.2
	0.1	−0.4
	−0.2	−0.2
	0.1	0.2
	0.2	−0.5
	−0.3	0.1
	0.2	−0.1
	−0.2	−0.2
means	0.04	−0.09
range	−0.3 to 0.4	−0.4 to 0.2

standard error of the two means = 0.105
$\underline{t} = 0.13/0.105 = 1.24$ $P > 20\%$

To begin, one recognizes the variability *within* each subgroup of ten subjects. This variability is not caused by the treatment; rather, it is due to such uncontrolled influences as total caloric intake, energy expenditure, state of hydration and the like. Such variability is presumed to be random and hence to affect both subgroups nearly equally in a properly controlled study. Superimposed on the random variability within the subgroups is a systematic difference *between* them, identified by the average weight difference of 0.13 pounds.

Statistical analysis tests whether the systematic difference can be accounted for by the degree of random variability that is present. If so, then the treatment has not been shown to have an effect. If not, then the treatment probably has an effect. Thus, statistical analysis enables one to cope with random variability when looking for possible systemic differences between groups of like measurements.

Theoretical statistics is a branch of mathematics which formulates hypothetical events into a model, and then analyzes the behavior of the model. Applied statistics borrows a model from statistical theory in order to handle observations that have been made in the real world. Usually, the borrower must assume that the available data conform to the assumptions that are implied by the model. There are a number of useful statistical applications.

10.2.2 *Description of results*

Literally pages of data can be reduced to a few descriptive statistics. For example, the mean and standard deviation concisely and objectively summarize a sample of any size that comes from a normally-distributed population. Sample standard deviation quantitates the distribution of members of a sample about the mean of the sample. The standard deviation of the mean, commonly called the standard error, quantitates the probable locus of the true population mean relative to a sample mean. Accordingly, standard deviations are given when one wishes to describe the distribution of the individual measurements which make up a sample. Standard errors are stated when one wishes to draw attention to differences between two or more means, or when one wishes to specify the range within which the true population mean probably falls. The latter application is a "confidence interval estimate," and as such is a generalization beyond available sample data. Many bivariate samples are described appropriately by either a correlation coefficient or a regression equation.

10.2.3 *Tests of significance*

The logic of most statistical tests can be illustrated by reference to Table 10.1. The object is to answer whether the difference in weight change between the subgroups, 0.13 pounds, is "significant." One formulates an appropriate "null hypothesis." In this case, the hypothesis states that there is no real difference between the subgroup means, assuming that the apparent difference is no greater than would be expected if only random variability is present. The null hypothesis is tested by calculating a "t" statistics, where t equals (observed difference – theoretical difference) divided by standard error, or $(0.13 - 0)/0.105 = 1.24$. The t statistic is evaluated by asking: *if* the null hypothesis is correct, what is the probability of finding a $t = 1.24$ or larger? Available t tables indicate that such would occur in over 20% of cases. Since null hypotheses ordinarily are not rejected if the probability of their being correct is over 5%, the difference between the subgroup means is not statistically significant.

The statistic, t, can be considered to depend on the ratio of systematic to random variability. The numerator of the ratio is the difference between means. The standard error term of the denominator is derived from the variability of each subgroup around its mean and the size of the subgroups. Statistical theory identifies the tabular t values that would occur at specified probabilities for samples of a given size, if the null hypothesis is correct. It is customary for a null hypothesis to be rejected at either 1% or 5% probability, or not to be rejected.

In this fashion a test of significance leads to a conclusion about a treatment. The probability of the conclusion being in error is stated where the treatment is judged to have an effect. For example, if a null hypothesis were rejected at a probability of 1%, one would conclude that the treatment is effective, but that there is a 1% chance that random sampling variation accounts for the apparent effect.

10.2.4 *Populations and samples*

The statistician evaluates data which have been obtained from unbiased samples in order to make inferences about the population(s) from which the samples were drawn. A population consists of a large number of individuals that are homogeneous with respect to some measured characteristic. The members are alike, except for random variability. Since random variability needs to be minimized if small systematic effects are to be demonstrated, it is desirable to sample from a population that is as homogeneous as is practical. Thus, all pure-bred beagles in the world constitute a population that is more homogeneous than is the larger total canine population. On the other hand, litter-mate beagles are more homogeneous than are beagles in general.

Theoretically a population should be indefinately large, so that after a sample has been taken from it, the mean value of the characteristic under study has not changed in the remaining population. Usually, one must settle for "very large" populations in the real world. Populations are presumed to be stable. It is most important that the mean value for a characteristic does not change between the time that a sample is taken and inferences are made about the population from analysis of the sample. Certain populations are unstable because the natural incidence of a disease, diagnostic criteria, and so on, change with time. Samples of opinions are commonly drawn from populations that are notoriously unstable with respect to the incidence of an opinion. Statistical inferences assume the stability of populations, even though yesterday's sample may not reflect correctly tomorrow's population.

A sample consists of an available fraction of a real or a theoretical population. The members of a sample are presumed to be chosen "without bias". This means that each member of the population has an equal chance of appearing in the sample. Since it is impractical to number each member of the population and to draw randomly the desired sample according to well-mixed numbers in a large bowl, usually the investigator must take his sample where he can. Thus in the weight control example, Table 10.1, the initial twenty subjects could have been volunteer students, consenting outpatients with a common obesity problem, and so on. However, even though a sample which is truly representative of a specified population may not be available, the further division of the sample into control and test subgroups must be unbiased. This might be done by a coin-toss or by assigning successively available subjects to control and test procedures until the planned subgroups were filled. Although the statistical analysis itself is unaffected by biased sampling, no valid conclusion can be made regarding the effect of the test agent when a bias in the sampling can be identified. Suppose, for example, that in the present case (Table 10.1) subjects that wished to lose weight were given the therapeutic agent while subjects with no weight problem served as "controls". Under such circumstances any difference between the outcomes of the subgroups could be due to the treatment, to differences in motivation, or to both.

Control experiments are carried out to prevent bias, not to add bias. Adequate control means that in so far as is possible the only *systematic* difference between the control and treated subgroups is the inclusion of a test agent in the latter. Thus, in a "double blind" study of an agent in humans, neither the operator nor the subjects know the recipients of test agent or placebo until the data from the study have been collected. Other test procedures may be controlled by performing "sham" operations, by injecting an inert material, and so on. When the procedures that are carried out on control and test groups differ in some way that is evident to the subjects, or when controls and tests are conducted at different times or in different locations, the possibility is present of confounding the effects of a test agent with some other concomitant systematic effect.

Measurements made on one member of a sample should be independent from measurements made on other members. For example, three determinations on subject A plus five determinations on subject B are *not* eight independent measurements. The significance of this distinction is that replications obtained from a single subject tend to vary less than the same number of measurements obtained from independent subjects. An incorrect reduction in random variability can result in a conclusion that errs in the direction of claiming significant effects where there may be none. In simple experimental designs the "independent" measurements might come from single tests done on individuals or from averages of replicates obtained on each subject. More complex designs can utilize replicate measurements, with an appropriate assignment of "degrees of freedom" to be associated with the numbers of individual subjects, numbers of replications, and so on.

All null hypotheses are evaluated with attention to the correct number of degrees of freedom. Sample size is the determinant of degrees freedom in simple experimental designs. The basic idea is that a larger sample tends to yield a more nearly correct estimate of the population mean and variability of the characteristic under study. Therefore in general, the difference between the means of two small samples must be greater than that of large samples in order to show a statistically significant difference. Positive conclusions (i.e. significant effects) based on small samples are as convincing as are those from large samples where differences are significant at, say 1% in both cases, when the experimental data conform to the chosen theoretical model. On the other hand, failure to reject a null hypothesis upon analysis of data from small samples is not very convincing. Regardless of sample size, failure to demonstrate a difference between sample means does not prove that the means come from a common population. Thus, statistical conclusions may have two types of error. In one case, what actually is an unlikely sampling variation is interpreted as being the effect of some test procedure. The probability of this type of error is stated. In the second case, a true treatment effect is not demonstrated, often because of a paucity of data. The probability of this second type of error is not revealed by the conventional statistical analysis.

Sometimes the major purpose of a statistical analysis is to draw inferences about the sampled population. It is important to identify correctly the population in such cases. For example, where a sample consists of ten measurements from a single subject, A, the theoretical population consists of an indefinately large number of possible measurements on subject A. No valid inferences can be made about the different population of which A is one member. The question often arises whether a given number of measurements, dictated by the available facilities, should be carried out with many replications on a few subjects or on many subjects. When the investigator wishes to make inferences about a specific population, usually it is more efficient to maximize the number of individuals from that population in the sample. In other instances there may be less interest in a specific large population. For example, the pituitary glands from ten animals might be pooled, homoginized and sets of replicate aliquots used to identify the effects of given treatments.

10.2.5 *Measurements and attributes*

Thus far the discussion has focused on measurement data wherein some characteristic, which may have any valued dependent upon the range and sensitivity of the recording device, is obtained on tested subjects. The mean of the sample measurements, a statistic, provides an estimate of the population mean, a parameter. The true population mean seldom if ever is known, but inferences about it can be made from the sample. Other observations yield attribute data. Each member of an unbiased sample is examined for the presence of some specified attribute. The attribute might be survival 48 hours after treatment, the presence of a urinary bladder tumor of specified size at a specified time interval, and so on. Thus, the sample yields a sample frequency of the attribute which is an estimate of the frequency of the attribute in the sampled population.

The computations for statistical analysis of measurement data differ from those of attribute data. However, the statistical inferences from samples of both kinds of data are similar. Thus, given a sample frequency and sample size, one can estimate the range within which the population frequency probably falls, a confidence interval estimate. Also, one can state the probability that a given sample frequency comes from a population in which the frequency of the attribute is known, or the probability that two or more sample frequencies come from a common population. Such null hypotheses which are appropriate for the analysis of attribute data are tested by calculating and evaluating the statistic, chi-square. The probability of a chi-square value that has been obtained under a null hypothesis is given in available tables. As with the evaluation of t values chi-square is rejected at a probability of 1% or 5%, or is not rejected.

Chi-square analyses are based on a statistical model which has a binomial or multinomial distribution. This model generally is appropriate except where one is dealing with sample and population frequencies that are near zero. Since the binomial distribution is symmetrical around the population frequency, distributions around a value close to zero might deviate from the model.

10.2.6 *Non-parametric methods*

The theoretical model for many of the statistical methods that are applied to measurement data assumes a normal distribution of measurements in the sampled population. The chief obvious characteristics of a normal distribution are symmetry and a central tendency; the familiar "bell-shaped" curve. The parameters of a normal distribution are the mean and the standard deviation. Thus in the normal distribution, 68% of the population is within ±one standard deviation from the mean; 95% is within ±two standard deviations, and so on. In so far as measurement data come from a normally-distributed population, this model is appropriate.

Frequently, measurements obtained on a sample appear to deviate from a normal distribution; some deviation in a small sample is to be expected because of random variability. The degree of deviation from symmetry and central tendency can be observed by making a histogram of the data. Distributions that are flat, badly skewed, and so on, do not look like bell-shaped curves. A non-parametric method may be appropriate for analysis of such data.

Non-parametric methods make no assumptions about the population distribution. However, assumptions regarding unbiased, independent sample measurements and adequate control apply as with parametric methods. Currently, nonparametric tests are widely used in the simpler experimental designs. The logic is similar to that of the parametric $\underline{t}$ tests. A statistic which is based on the ranking rather than on the numerical measurement values is calculated. The probability of the calculated statistic belonging to some region – being larger than a given value – is determined under a stated null hypothesis. The hypothesis is rejected at a stated probability, or is not rejected.

10.3 Some experimental designs and statistical applications

The following examples assume normal distribution of measurement data.

10.3.1 *Paired data*

Frequently, control and test procedures are administered to each member of a sample of individuals in order to identify a significant effect of a test agent. Such a design is summarized in Table 10.2 wherein each subject contributes a difference between control and test measurements. The mean difference may be tested by a null hypothesis which states that the test agent has no effect, the apparent effect being due to random variation. This hypothesis is tested and by calculating and evaluating a $\underline{t}$ statistic, mean difference/standard error of the mean difference. The probability

Table 10.2 Paired data.

subject	X_1	X_2	$X_1 - X_2$
1	100	110	10
2	85	90	5
3	90	92	2
4	120	124	4
5	98	99	1
6	93	96	3
7	80	85	5
8	105	109	4
9	98	103	5
10	110	113	3
means	97.9	102.1	4.2
st. dev.	11.9	12.0	2.45
st. error			0.78

$\underline{t} = 4.2/0.78 = 5.4$ $t_{01} = 3.25$, for 9 degrees freedom $P < .01$

of the t statistic is obtained from tables, using $n - 1$ degrees of freedom, where n is the number of paired measurements. In the present example a t statistic with a numerical value of 3.25 (or larger) would occur in 1% (or fewer) of cases where the null hypothesis is correct. Accordingly, the null hypothesis is rejected. One concludes that the treatment had a significant effect since there is less than a 1% probability that the mean difference was due to random variation. In other cases where the null hypothesis is not rejected, i.e., the probability of the calculated t is greater than 5%, there are several alternative explanations. *One* possibility is that the test agent truly was without effect. Accordingly, null hypotheses are never "accepted;" test procedures are not proven satistically to have no effect.

The basics for the paired test is that random subjects tend to differ consistently. The consistent differences persist whether replicate or different test procedures are applied. This tendency is evident in the data on Table 10.2. The X_1 and X_2 values could be measurements on paired organs from different animals, control and test measurements on different subjects, and so on. The range of variability in the X_1 measurements is from 80 to 120, or 40 units, and that of the X_2 measurements is 39 units. However, the range of variability in the sample of ten differences is only nine units. Thus, most of the variability in the data is associated with the consistent differences among the test subjects and is independent of the effect of the treatment. the pairing procedure removes the variability which is unassociated with the test agent, and yields a test of the mean difference with its relatively small standard error.

In performing a paired test the assumption is made that the variability around the mean is not changed significantly by the test procedure. Thus in Table 10.2 the standard deviations are similar for the X_1 and X_2 measurements; the *variances* (standard deviation squared) are homogeneous When, in a paired experiment, the variability between replicates is as great as the variability among subjects, or when the sample measurements markedly deviate from a normal distribution, or the test procedure introduces further variability into the measurements, the paired t test may not be appropriate. Non-parametric tests for analyzing paired measurements are available.

10.3.2 *Non-paired data*

When data are obtained from samples of different size, or when there is no valid basis for pairing sets of measurements, a group comparison can be made. For example, the data from the obesity study of Table 10.1 was subjected to a non-paired analysis because the order of the appearance of

Table 10.3 Data in two groups.

	number	mean	variance	st. deviation
group one, X_1's 2, 3, 3, 2, 4, 1, 3, 3	8	2.63	0.84	0.92
group two, X_2's 2, 6, 3, 5, 6, 4, 6, 5, 8, 3	10	4.80	3.29	1.81
difference between means		2.17		

pooled standard error = 0.71
t = 2.17/0.71 = 3.05
t_{01} = 2.92, 16 degrees freedom
$P < 0.01$
F = 3.29/0.84 = 3.92
$F_{.05}$ = 4.83, for 9 and 7 degrees freedom (two-tailed)
$P > .05$

subjects in the control and treated groups was random. Similarly, Table 10.3 summarizes a statistical analysis of measurements from eight subjects from one group and ten subjects from a second group. The object is to determine whether the two group means differ significantly. A null hypothesis that the means come from a common population is tested by calculating and evaluating a t value in which the numerator is the difference between the means and the denominator is a pooled standard error.

The standard error is calculated from the variability in each group around its mean and the number of independent measurements in each group. The two groups each contribute $n - 1$ degrees of freedom for evaluating the t statistic. Since a t of 2.92 (or larger) would occur in 1% (or fewer) of trials where the null hypothesis is correct, one concludes that the group means differ at a probability of less than 1%. Recall that a similar analysis failed to reject the null hypothesis in the example of Table 10.1 because the probability of the t was greater than 20%.

The variances in the groups should be homogeneous if a pooled standard error term is to be used. This assumption is tested by evaluating the ratio of the larger to the smaller variance by means of available F-ratio tables. Thus, a null hypothesis is stated that the variances are equal; i.e., that they differ no more than would be expected when two samples are drawn from a common population. If the null hypothesis were correct in the present case, Table 10.3, an F-ratio lower than 4.83 would occur in more than 5% of trials. Therefore the variances are not shown to differ and may be considered to be homogeneous.

The group t test is biased when the variances of the groups differ. Sometimes the bias operates to cause the rejection of a true null hypothesis whereas at other times one fails to reject a false null hypothesis. Thus, *if* the variances around the means are similar, the present test is appropriate for evaluating the difference between the means.

Since data from a paired experiment can be analyzed by using either the paired or the non-paired t test, as well as a non-parametric test, it is important to recognize the consequences of the use of an inappropriate test. Generally, sensitivity is decreased when group-comparison analysis is applied to validly-paired measurements. Also, when either grouped or paired data deviate from a normal distribution, non-parametric tests tend to be more sensitive for showing differences between the groups or the pairs.

10.3.3 *Analysis of variance, single classification*

Recall that variance is the standard deviation squared and depends on the differences between each measurement in a sample and the mean of the sample. The variance analysis is useful when the means of more than two groups are to be compared. In the example which is shown in Table 10.4, 18 subjects have been assigned to three test conditions. The statistical question is whether the column means differ more than probably would occur if there were no real differences in the effects of the treatments.

The total variance around the mean of all 18 measurements is partitioned between systematic differences among the column means with two degrees of freedom and random variability with 15 degrees of freedom. The variance ratio, or F ratio, tests the null hypothesis that the column means differ only randomly. If this hypothesis were correct, variance due to differences among column means and random variance both would be estimates of random variability. The F ratio usually would have a value near 1.0, but would be as much as 6.36 in 1% of trials in which three groups of six subjects were measured. Thus, there are differences among the column means at a probability of less than 1% in the present case. The assumption is made that the variances around each column mean are homogeneous; i.e., although the column means may differ, the variability around each mean differs only randomly.

When the variance analysis reveals differences among more than two means, as in the present case, a further test can estimate which mean differs from other means at a specified probability. The experimental design, Table 10.4, is singly classified according to the test condition. The variability within each column is considered to be random and to be normally distributed. Usually, equal numbers of measurements are obtained for each test condition.

Table 10.4 Analysis of variance, single classification.

	Test 1	Test 2	Test 3
	33	35	36
	30	34	36
	31	33	37
	32	34	37
	28	32	34
	32	34	38
means	30.8	33.7	36.3

Source of variation	Degreees freedom	Variance
Tests	2	45.5
Random	15	2.0

$F = 45.5/2 = 22.8$
$F_{01} = 6.36$, for 2 and 15 degrees freedom (one-tailed)
$p < 0.01$

10.3.4 *Multiple classification*

An additional source of possibly systematic variation is included in the experimental designa of Table 10.5. Thus, five subjects *each* receive three test procedures. The individual subjects tend to differ consistently during all tests in this example. Similar subject variability is assumed to be present in the paired-comparison design, and is removed by analyzing the differences between two tests which have been performed on each subject. The present design extends the pairing to more than two procedures, tests whether difference between subjects is significant, and controls statistically such differences so that a more sensitive test can be applied to the possible effects of the experimental procedures.

The variance due to the test procedures (plus random effects) is derived from the column means and has two degrees of freedom. Similarly, the variance due to different subjects (plus random effects) is derived from the row means and has four degrees of freedom. The remaining eight degrees of freedom are associated with random variability. The F ratios examine whether test plus random variance exceeds random variance, and whether subject plus random variance exceeds random variance. Thus, the two null hypotheses are, respectively, that test and subject effects are zero. Both hypotheses are rejected at probabilities of less than 1.% in the present example. Therefore, there are statistically significant differences among the three test means and among the five column means. Further tests can show which column or row mean differs from others.

The experimental design and analysis, Table 10.5, assumes homogeneity of variance around the row means as well as around the column means. The analysis also assumes that there is no significant interaction between row and column classifications. In the present case, this means that there should be a consistent effect of treatments on subjects rather than a situation in which the effect of a procedure depends upon which subject is treated. Significant interaction in this example is unlikely because random subjects from a population tend to respond similarly to test procedures. However, if the column classification, Table 10.5, were dose levels of substance A and the row classification were dose levels of substance B, the response to substance A might depend on the concomitant dose of substance B. Thus, positive or negative "synergism," or "potentiation," or "inactivation" might occur.

Were there to be interaction between row and column classifications, or were the variances around row or column means to be non-homogeneous, such would tend to make the estimate of random variance excessively large. This would tend to mask possibly significant effects of the column or the row agents.

More elaborate designs and analyses permit the testing for significant interaction. Replicate measurements are required for each combination of

Table 10.5 Analysis of variance; multiple classification.

	Test 1	Test 2	Test 3	Means
subj. 1	23	29	30	27.3
2	19	21	31	23.7
3	25	26	40	30.0
4	22	32	30	28.0
5	20	22	27	23.0
means	21.8	26.0	31.6	

Source of variation	Degrees freedom	Variance
tests	2	121.0
subjects	4	28.5
random	8	1.15

for tests: $F = 121/1.15 = 105$
  $F_{01} = 8.65$, for 1 and 8 degrees freedom (one-tailed)
for subjects: $F = 28.5/1.15 = 25$.
  $F_{01} = 7.01$, for 4 and 8 degrees freedom (one-tailed)

row and column agents. Statistical analysis may demonstrate significant
synergism between two agents. In addition, one can test whether the effect
of an agent is greater than can be accounted for by interaction with a second
agent; i.e., whether an agent has a significant independent effect.

10.3.5 *Multivariate samples*

The preceding experimental designs and statistical tests have dealt with
samples in which a single variable was measured. Additional variables either
were assumed to have only random effects or were kept under experimental
or statistical control. Here are presented some examples wherein each
member of a sample contributes two or more measured variables.

10.3.6 *Correlation*

Commonly, members of a sample furnish concomitant measurements such
as plasma cholesterol and plasma thyroxine concentrations, test scores in
biology and in chemistry, and so on. Upon arbitrary designation of one
variable as X_1 and the other as X_2, a plot of the measurements might
resemble figure 1. Each point in this figure represents X_1 and X_2 measures
on single subjects, all of whom are presumed to come from some specified
population. Both X_1 and X_2 vary randomly and are presumed to be normally distributed around mean X_1 and mean X_2 in the sampled population.

The statistical question is whether, in a sample, increments (or decre-
ments) in X_2 are associated with increments in X_1 to a degree that is greater
than is likely to happen by chance when the variable are unrelated in the
sampled population. If the variables appear to be nearly linearly related as
in Fig. 10.1, linear correlation analysis is the appropriate statistical method.
This analysis yields a correlation coefficient, $\underline{r}$, which is a pure number in
value between -1.0 and 1.0. The numerical value will be the same whichever variable is designated as X_1. The correlation coefficient describes the
tendency for the plotted points to fall near a straight line that can have any
slope except zero or ninety degrees. The coefficient is near zero when the
plotted points fall within a circle, or fall on a vertical or horizontal line.
Coefficients near 1.0 mean that X_1 and X_2 increase or decrease together,
and that plotted points fall quite close to a line. Coefficients near -1.0
mean that one variable decreases while the second variable increases and
the points fall near a line.

Whether a sample correlation coefficient differs significantly from a
theoretical population coefficient of zero depends on the numerical
value of the coefficient and on sample size. For example, it would not

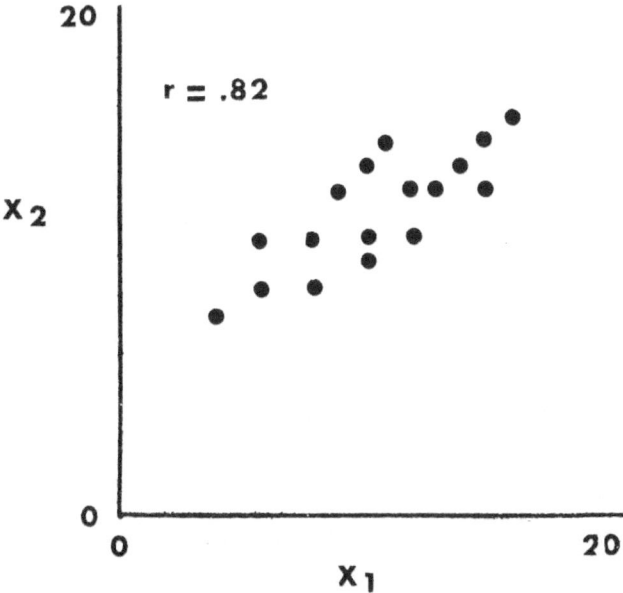

Fig. 10.1 Linear Correlation.

be unusual to obtain a coefficient near ±1.0 by chance in a sample of three from a population in which X_1 and X_2 vary randomly but are unrelated. On the other hand, one would be unlikely to find a coefficient as large as 0.25 in a very large sample from a population in which the variable are unrelated. Available tables cite the numerical values for correlation coefficients which are significant at probabilities of 5% and 1% for samples ranging in size from three up to one thousand. There is less than one chance in 100, probability less than 1%, that a value of 0.82 or larger would occur in a sample from a population where the variables are unrelated. A highly significant correlation coefficient does not establish a *causal* relationship between the variables.

10.3.7 *Regression*

The statistical models for linear correlation and for linear regression differ in one important respect. Two measures are made on each member of a sample in both models. However in the regression model, one of the variables is measured "without error." This variable is identified conventionally as the independent variable, X, and is plotted along the horizontal axis. The second variable, Y, is assumed to be normally distributed around its mean value at each value of X. Data which conform to the regression

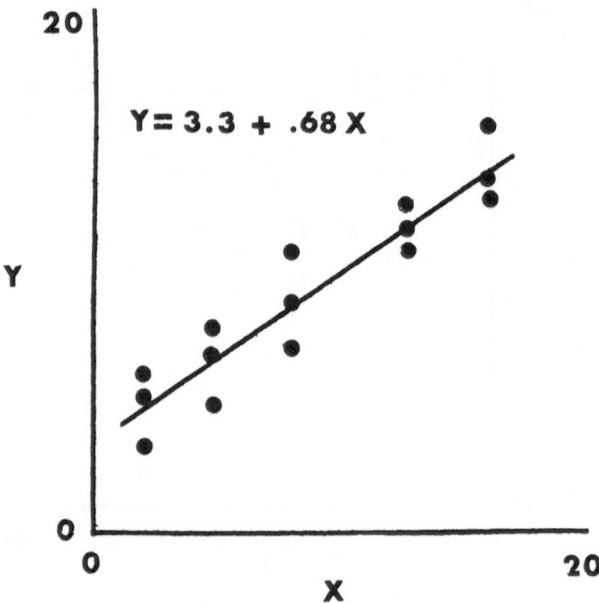

Fig. 10.2 Linear Regression.

model are shown in Fig. 10.2. Other features of the regression model are that Y increases or decreases linearly with increasing X, if X and Y are related, and that the variance in Y is homogeneous for all values for X. Thus in Fig. 10.2 the scatter of plotted points from the regression line is about the same throughout the range of X.

The analysis yields a regression equation by means of which a straight line can be fitted to the plotted points. The fit is by the method of least squares, which means that the vertical deviations from the line, squared, summate to a smaller total than would be obtained if measured from any other straight line drawn through the data. The regression equation describes the relationship of the dependent variable to the independent variable; it includes the value for Y at $X = 0$ and it quantitates via the regression coefficient the change in Y, positive or negative, for unit increases in the value for X. Unlike the correlation coefficient, the regression coefficient can have any numerical value which is appropriate for units of measuring the variables; i-e. $\Delta X / \Delta Y$.

The usual statistical test which is applied in regression analysis answers whether the apparent change in Y with increasing values for X is greater than would occur by change in a sample from a population wherein there is no significant regression of Y and X. Thus, one tests whether the sample regression coefficient, $|\underline{b}|$, is greater than zero. The coefficient divided by its standard error, s_b, yields a $\underline{t}$ ratio that is evaluated at $\underline{n} - 2$ degrees of freedom, where $\underline{n} = $ the number of X, Y measurements. The analysis can yield further inferences about the sampled population. For example, sometimes it is useful to test for a significant departure from a direct proportion. In such cases the sample intercept value, 3.3 in Fig. 10.2 would differ significantly from a population intercept value of zero. This test requires the calculation of the standard error in Y at the value of X equals zero.

The correlation and regression models cannot both be valid for given experimental data. However, the preliminary computations are the same for correlation and regression analyses and many computer programs conveniently yield numerical values for $\underline{r}$ and for $\underline{b}$. Thus, one frequently sees a straight line "fitted" to correlation data as if one of the variables were measured without error. Also, a correlation coefficient often is cited as evidence that a regression relationship is significant.

10.3.8 *Covariance*

Linear regression methods can be extended to the evaluation of two or more bivariate samples. For example, consider an experimental situation wherein Y and X have been measured on multiple samples and the regression lines for the samples do not superimpose because of differences in slopes (regressing coefficients, $\underline{b}$), differences in position (parallel lines) or differences in both slope and position. One statistical question is whether the regression coefficients for the several samples differ more than probably would occur by chance if the samples came from a common-slope population. The second statistical question is whether the value for Y in the samples would differ, if each was measured at an appropriate common value for X, more than could be accounted for by sampling variability from a single population. Thus, covariance analysis tests for significant differences among regression coefficients and for significant differences among "adjusted mean values" for Y. In the latter test the values for Y in all samples are compared at the average of all X values. Obviously, if the several regression lines have different slopes, the values for Y in at least two of the samples are the same at some value for X. The further testing for differences in adjusted mean Y values may be inappropriate in such cases.

Recall that regression analysis can test whether a sample regression coefficient differs from a theoretical, errorless population coefficient of zero; or 1.0. When two or more linear regression coefficients, each with its standard error, are to be compared, covariance analysis is useful.

10.3.9 *Multiple linear regression*

Regression analysis can be extended to an evaluation of the effects of two or more independent variables on a dependent variable. Suppose, for example, that one wished to measure the rate of glucose absorption from the intestine as the dependent variable, Y, as influenced by different intestinal lumen concentrations of glucose, X_1, and of a possible competing hexose, X_2. While X_1 and X_2 could be controlled experimentally, it is likely that the net rate of fluid absorption would vary and might affect glucose absorption; accordingly, net fluid absorption could be measured as X_3. Similarly, the lumen concentration of sodium might vary as the different combinations of hexose concentrations were applied. Thus, average lumen sodium concentration could be calculated and considered to be another meaningful independent variable, X_4. One would like to quantitate the effects, if any, of each independent variable on the rate of glucose absorption. The multiple linear regression model requires that the effects of the independent variables, if significant, are linear and that the variance in the dependent variable is homogeneous at all values for each independent variable. Non-linearity or non-homogeneity might result in failure to identify a significant effect of an independent variable.

The analysis applied to the present example would describe the results of many combinations of glucose concentrations, inhibitor hexose concentrations, fluid absorption rates and sodium concentrations by a multiple regression equation of the following form:

$$Y = a + b_{1 \cdot 234}X_1 + b_{2 \cdot 134}X_2 + b_{3 \cdot 124}X_3 + b_{4 \cdot 123}X_x$$

Y is the estimated rate of glucose absorption at given values for each independent variable. The coefficient $b_{1 \cdot 234}$ is that for the effect of glucose concentration when the other independent variables are held constant at their average values. Similarly, the other coefficients quantitate the independent effects of inhibitor hexose concentration, net fluid absorption rate, and sodium concentration, respectively.

A completed analysis also reveals the degree of correlation between pairs of independent variables, X_1 vs X_2, X_1 vs X_3, and so on. It is important that these pairs of variables not be highly correlated if significant independent effects are to be found. Also, a standard error can be calculated for each regression coefficient so that one can test whether each differs significantly from zero. Finally, by calculation of a multiple correlation coefficient, R, one may estimate the fraction of the total variability in Y that can be accounted for by the effects of the measured independent variables. The numerical value of R ranges from zero to 1.0. Low values indicate that much of the variability in Y is random or is due to unmeasured, additional independent variables.

References

1. Hald, A. (1952): Statistical Theory with Engineering Applications. John Wiley and Sons, New York, N.Y., U.S.A.
2. Snedecor, G.W. and Cochran, W.G. (1967): Statistical Methods. Iowa State College Press, Ames, Iowa, U.S.A.

11
PRESENTATION OF INFORMATION FOR PRINTED PUBLICATIONS, SLIDES AND POSTERS

Linda Reynolds, Herbert Spencer and Robert S. Reneman

11.1 Introduction

A vitally important part of any research activity is the dissemination of results to other research workers, to students, and to those in a position to implement the findings in a practical way. This dissemination process can take place in a number of ways, but in most cases some kind of visual presentation of all or part of the information is necessary. Printed material, in the form of journal articles, reports and books, is perhaps the most commonly used medium for publicizing research results, and here the presentation is entirely visual. At conferences, visual aids such as slides are used as an essential complement to the spoken word. More recently posters have begun to become popular as a means of presenting summaries of research results. In all three cases, the quality of the visual presentation will affect the ease with which potential users of the results are able to read, understand and remember the information given.

Authors often have very little control over the typography and layout of text in journals and books, though in many cases they can determine the presentation of tables and figures by providing the publisher with good quality artwork. In some instances the author or editor of a book may have sufficiently close contact with the publisher to be able to influence the presentation of the text also. Authors of reports are often faced with the task of designing the entire document themselves. Similarly, the presentation of information on slides and posters is generally the direct responsibility of the research worker, and having decided on the content he may often find it necessary to prepare the artwork himself.

The effective communication of information in a visual form requires an appreciation of the reader's needs and his information processing capacities in a given situation, an understanding of some of the factors which affect the legibility of printed materials, slides and posters, and a knowledge of basic design principles. In general research workers in the biomedical sciences have no formal training on these subjects. The aim of the present chapter, therefore, is to provide relevant theoretical and practical information which will be of use in the preparation of visual materials.

11.2 Investigating legibility

11.2.1 *The reading process*

A knowledge of the perceptual processes involved in reading is essential
for an understanding of the way in which typographic factors affect reading
performance, or legibility. A number of early studies were concerned with
the way in which words are recognised and these have been summarised by
Spencer (34). As long ago as 1885 Cattell (8) demonstrated that in a normal
reading situation each word is recognised as a whole rather than as a series
of individual letters, and Erdmann and Dodge (11) concluded in 1898 that
it is the length and characteristic shape of a word that are important for
recognition. Javal (17) showed that the upper half of a word is more easily
recognized than the lower half (Fig. 11.1), and Messmer (19) suggested that
letters with ascenders contribute most to word recognition. Similarly,
Goldscheider and Müller (14) found that consonants, with their ascenders
and descenders, contribute to characteristic word shapes more than vowels.
Tinker, who spent many years investigating legibility, has stressed that it is
the overall structure of the word – the internal pattern as well as the outline – that is important for recognition. It has been demonstrated recently,
however, that rearrangement of the middle letters of words has very little
effect on word recognition. This suggests that accepted theories about the
way in which words are recognized are very much oversimplified (28).

Eye movement studies have contributed much to our knowledge of the
reading process (34). The eyes move along a line of print in a series of
jerks, or saccadic movements, during which no clear vision is possible. At
the end of each movement there is a fixation pause lasting approximately
200 to 250 milliseconds, during which perception occurs. On average, 94%
of reading time is devoted to fixation pauses and the remainder to inter-
fixation movements. Sometimes the eyes make a backward movement or
regression to re-examine material not clearly perceived or understood, and
at the end of each line they make a return sweep to the beginning of the next.

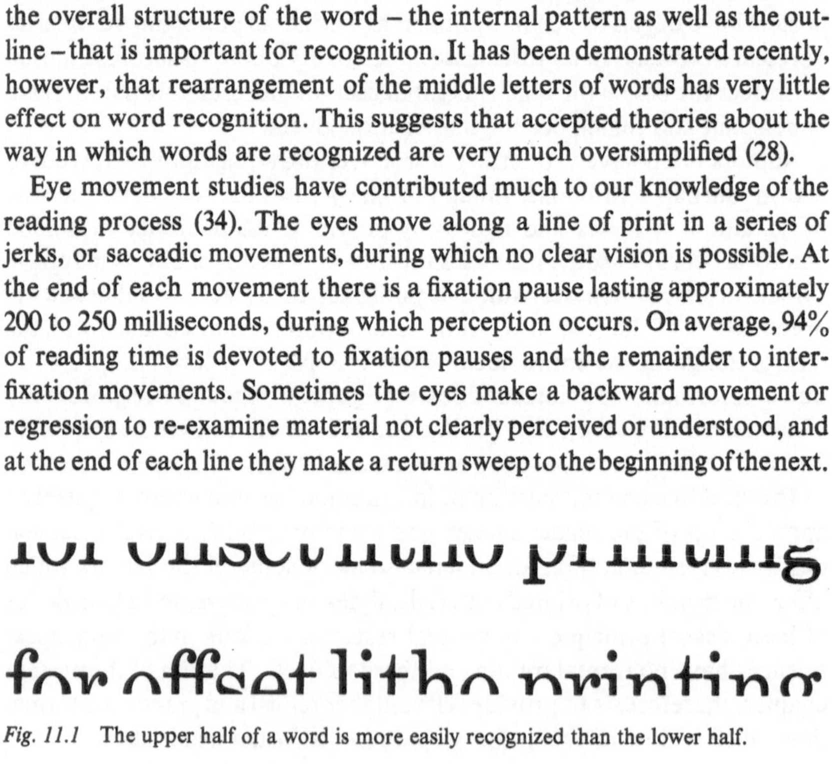

for offset litho printing

Fig. 11.1 The upper half of a word is more easily recognized than the lower half.

The frequency and duration of fixations and regressions are increased when typography is non-optimal and when the content of the reading material is complex.

At a normal reading distance of 30 or 35 cm, only about 4 letters of normal size print fall within the zone of maximum clearness or foveal vision. The field of peripheral vision usually encompasses about 12 or 15 letters on either side of the fixation point. During a single fixation it may therefore be possible to read up to 30 letters in word form, as opposed to only 3 or 4 unrelated letters. Short exposure experiments have shown, however, that most adults can perceive in their field of vision about 3 or 4 words in 1/50th of a second. Allowing for the time occupied by movements between fixations, this means that the eyes are capable of perceiving printed images at ten times the rate at which they do in a normal reading situation. This suggests that the reader is ultimately limited by his rate of comprehension.

11.2.2 *Methods of research*

There are numerous definitions of the term "legibility"; it is used in relation to studies ranging from the visibility and perceptibility of individual characters and words to the ease and speed of reading of continuous text. A precise definition of legibility can only be given in relation to the technique used for measuring it. A number of methods have been used, and these vary in their validity, that is, the extent to which they measure that which they are intended to measure, and in their reliability. Different techniques are appropriate for different reading situations and for different types of reading materials, and results are rarely relevant to conditions other than those tested. Tinker (41) and Zachrisson (48) have published detailed descriptions of some of the principal techniques, and some of them are summarised below.

Speed of perception.   Speed of perception is measured by means of a tachistoscope, which is used to give very short exposures of individual characters, words or phrases. This method is useful for determining the relative legibility of different characters and of different character designs, but results have little relevance to the reading of continuous text.

Perceptibility at a distance.   Perceptibility is often measured in terms of the maximum distance from the eye at which printed characters can be recognised. This method is appropriate when used to assess the relative legibility of characters which are intended to be read at a distance as on slides and posters, but can be misleading if applied to continuous text intended to be read under normal reading conditions. A device called a focal variator is sometimes used to throw the test characters out of focus by a measured amount. The distance threshold is then established in the usual way.

Perceptibility in peripheral vision.   This is assessed by measurement of the horizontal distance from the fixation point at which a printed character can be recognized. It has been used to compare the relative legibility of single characters and to compare the legibility of black letters on white with white on black.

Visibility.   Visibility is usually measured by means of the Luckiesh-Moss Visibility Meter (41). This piece of equipment allows the contrast between image and background to be varied by means of filters. It has been used to determine the threshold contrast, and hence the relative legibility, of different type faces, sizes and weights, and to examine the effects of variation in contrast between image and background.

Blink rate.   Reflex blink rate has been used as a measure of legibility by Luckiesh and Moss (41), but the validity of the method is questionable. The assumption is that the blink rate will increase as legibility decreases. The same authors also advocated heart rate as a measure of legibility, but this method is extremely suspect.

Visual fatigue.   This has been suggested as a criterion of legibility, but no satisfactory objective method of measurement has been devised. Subjective assessments of fatigue are subject to modification by a great many factors which may be totally unrelated to the experimental situation.

Eye movement studies.   Eye movement records provide a very useful insight into the effects of typographic factors on reading performance. They indicate whether a slower rate of reading resulting from non-optimal typography is associated with more frequent fixations, longer fixations, more regressions, or some combination of these. Total perception time, which is the time spent in fixation pauses, and fixation frequency have been found to be valid and highly reliable measures of legibility. The duration and frequency of regressions have been found to be less reliable.

Rate of work.   Rate of work is by far the most satisfactory measure of legibility. This can be assessed either in terms of the time required to complete a given reading task, or in terms of the amount of work done within a set period. A number of methods of measurement have been used, including speed of reading continuous text (with or without a comprehension check), speed of scanning text in search of target words, and look-up tasks of various kinds on structured information such as indexes and directories. Rate of work is undoubtedly one of the most valid measures of legibility because the task can be directly related to the normal use of the material under test. Most of the methods used are also highly reliable when carefully administered. There has been some dispute as to the time periods necessary for these methods. Tinker (41) maintains that differences which are significant after 1 minute 45 seconds will continue to be so after 10 minutes, but he accepts that significant differences may emerge after 10 minutes which would not be apparent during a short reading period.

Readers' preferences.   Readers' preferences cannot be relied upon as an accurate indication of the relative legibility of printed materials. Tinker carried out an experiment in which he compared measurements of visibility, perceptibility at a distance, reading speed and judged legibility for ten different typefaces. He found that judged legibility corresponded more closely with visibility and perceptibility than with speed of reading. Tinker and Paterson (41) studied the relation between judgements of "pleasingness" and judgements of legibility for a number of typographical variables. They found that in general the agreement between the two sets of judgements was high. Tinker concluded from these studies that readers prefer those typefaces or typographical arrangements which they can read most easily. Burt (6), however, put forward the argument that readers are able to read most easily those typefaces which they find most pleasing.

11.3 Books, journals and reports

11.3.1 *Introduction*

The typography and layout of text in books and journals is often outside the control of authors and contributors but, nevertheless, an awareness of the results of research on the legibility of continuous text will be helpful in cases where the author is able to influence the publisher and printer to some degree. Many of the general principles which apply to conventionally type-set text can also be applied to the design of reports, where the author generally has much greater freedom in determining the final appearance of the document.

The author usually has much more control over the design of tables, graphs and diagrams. For some journals, and for reports, he will often be responsible for the artwork itself. Research has shown that the efficiency of tables and figures depends very much on the relation between the logic used by the author in preparing them and the questions which the reader will be asking in using them. An appreciation of this important relation-ship is therefore essential in the design of good tables and figures. Similarly, research has shown that the comprehensibility and legibility of tables and figures can be greatly influenced by certain factors in the layout and typo-graphy of the artwork.

Much of the research on text, tables and figures described below relates to letterpress and good quality litho printing, but other printing processes are being used increasingly to reproduce original documents. Electrostatic and other office copying processes used for "on demand" printing tend to impair the quality of the original image in certain ways. Background "noise", loss of contrast and image degradation are typical effects and the likelihood of their occurrence needs to be taken into account in the design of scientific and technical information.

The problems which arise in transferring information from one medium to another must also be carefully considered. If it is intended that art-work prepared for a printed document should subsequently be used for the production of slides or posters, this must be taken into account in the initial design. The increasing use of microfilm as a publishing medium also has considerable implications for the design of printed materials (35). In some cases a single master must serve for a printed edition and for a microfilm edition. This means that factors such as type size, style and weight must be selected with the needs of both publications in mind. Typography which is suitable for a printed image may not be suitable for a negative image projected onto a screen. Similarly, a different ap-proach to the organisation of information and to its layout is required for microfilm. Conversely, data from many computer systems are output directly onto microfilm and full-size copies are produced on demand. Prints from source document microfilm are also widely used. Much re-search is needed to determine how legibility can be maximised for simul-taneous print and microfilm publications and microfilm prints, and for computer output microfilm and prints.

11.3.2 *Type forms*

Typeface.   Much early research was concerned with establishing the relative legibility of different typefaces for continuous text, though many of these studies compared types of the same body size rather than of the same visual size (Fig. 11.2). The effect of letter form on legibility could not, therefore, be separated from the effects of letter size. It would seem, however, that there is very little difference in the legibility of typefaces in common use when well printed, and that familiarity and aesthetic preference have much to do with any observed differences in reading speed. The relative legibility of seriffed and sans serif typefaces has been a particularly controversial issue. It has been argued that serifs contribute to the individuality of the letters, that they are responsible for the coherence of the letters into easily recognisable word shapes, and that they guide the eye along each line. None of these claims has been conclusively proved however, and it would seem that any differences in legibility in favour of seriffed faces may be due to familiarity rather than to any intrinsic superiority. On the other hand, there is evidence to suggest that sans serif faces are the more legible for children and poor vision readers, and they are widely accepted as being suitable for display purposes, because of their simple form.

The choice of typeface is likely to influence considerably the legibility of any paper or microfilm copies of the information. Recent research (40) has shown that some typefaces are better able to withstand extremes of image degradation than others (Fig. 11.3). The legibility of typefaces such as Baskerville is markedly impaired by any "thinning down" of the image, whereas typefaces such as Rockwell are severely affected by "thickening up". Faces such as Times and Univers are less affected by these extremes of degradation.

A number of studies has been made of the relationship, in terms of

Fig. 11.2 Five typefaces of the same body size are here juxtaposed to demonstrate the importance of measuring type in terms of its visual size rather than its body size if valid comparisons of legibility are to be made.

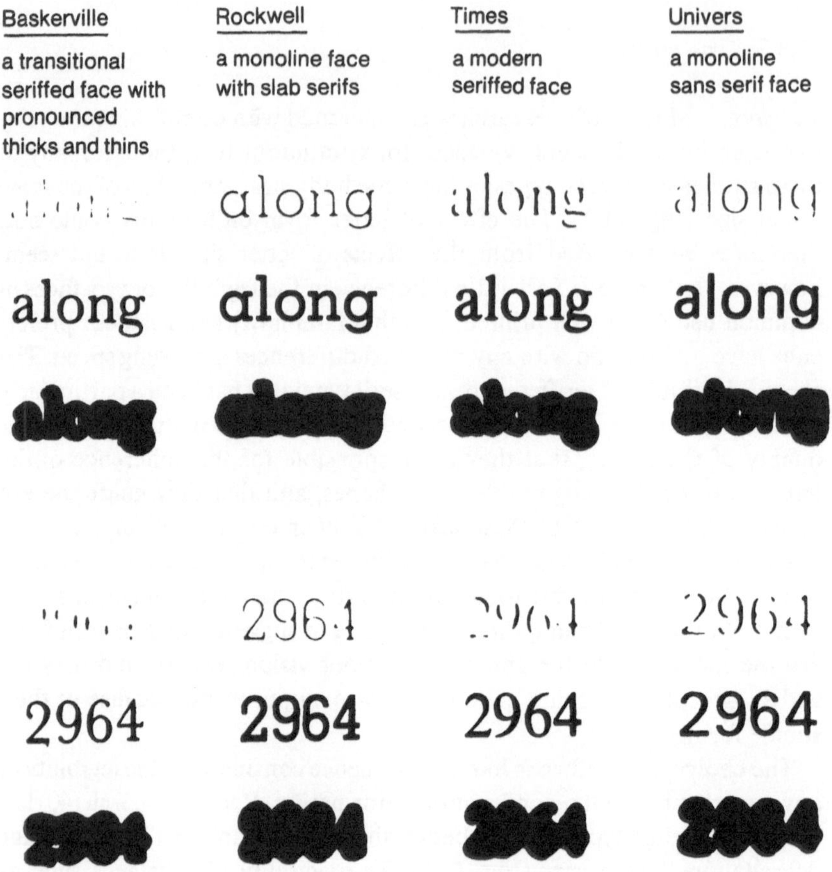

Baskerville	Rockwell	Times	Univers
a transitional seriffed face with pronounced thicks and thins	a monoline face with slab serifs	a modern seriffed face	a monoline sans serif face

along along along along

along along along along

along along along along

''''·: 2964 ·)(,·| 2964

2964 **2964** 2964 **2964**

Fig. 11.3 Enlargements of 9 point type which has been photographically thinned-down and thickened-up; in each case a "normal" image is shown for comparison. Differences in in the legibility of the typefaces are largely accounted for by differences in certain design characteristics such as x-height, normal stroke thickness, variation in stroke width within characters, the openness of counters, the presence or absence of heavy serifs, and the spacing between characters.

appropriateness or congeniality, between typefaces and the content of the printed message. Several investigators have attempted to categorise typefaces according to their "atmosphere value". This consideration is particularly relevant to persuasive material such as advertisements, but has rather less significance in information publishing. Nevertheless, Burt (6) found that readers of serious publications do have preferences with respect to typeface. He found that some readers had a tendency to prefer old faces while others preferred modern faces. The larger group who preferred old faces comprised mostly students and lecturers in the faculty of arts, whereas those who preferred modern faces included regular readers of scientific and technical materials. The latter group would almost certainly have been more familiar with modern faces in their everyday reading than the former group.

Type weight. It has been reported that readers prefer typefaces "approaching the appearance of bold face type" and that letters in bold face are perceived at a greater distance than letters in normal lower case. Speed of reading tests, however, revealed no difference between bold face and normal lower case, and 70% of readers preferred the latter. This suggests that bold face should not be used for continuous text, but there is no reason why it should not be used for emphasis (41).

Italics. Experiments have shown that the use of italics for continuous text retards reading, especially in small typesizes·and under non-optimal illumination. Further, readers apparently do not like italics; 96% prefer roman lower case. It is unlikely, however, that the occasional use of italics for emphasis significantly retards reading (41).

Capitals versus lower case. All-capital printing has been shown to markedly reduce the speed of reading of continuous text. Reductions of 13.9% over a 20 minute reading period have been recorded (41). This is partly because words in lower case have more distinctive shapes than words in capitals and may therefore be easier to recognise, and partly because text set entirely in capitals occupies about 40 to 45% more space than text in lower case of the same body size. Eye movement studies have shown that all-capital printing increases the number of fixation pauses and that, because the text occupies a larger area, the number of words perceived at each fixation is reduced. These findings are very significant with respect to upper case computer printout and computer output on microfilm. As a general rule, sections of upper case only computer printout should not be reproduced directly for inclusion in a printed document. The data should be typeset or typed.

Only under exceptional circumstances are capitals likely to be more legible than lower case. This may happen, for example, where very small type sizes approaching the threshold of visibility are used. Experiments have shown that capitals may then be more easily discriminated than corresponding lower case letters because of their greater size. For the same reason they may be more legible than lower case letters when the threshold for distance vision is approached.

Numerals. The legibility of numerals is particularly important because each individual character must be identifiable and contextual cues are usually weak or absent. Tests on Modern and Old Style numerals showed that Old Style numerals, which vary in height and alignment, were more easily recognised at a distance than Modern numerals which are all the same height (Fig. 11.4), but in normal reading situations the two kinds of numeral were read equally fast and with equal accuracy (41).

The legibility of certain numerals, such as 6 9 3 5 might be improved by accentuation of the open parts. Use of optimal height to width and stroke width to height ratios, which have been shown to be 10:7.5 and 1:10 respectively, might also be beneficial (41).

Arabic numerals are read significantly faster and much more accurately than Roman numerals because the former are so much more familiar. The differences are so marked, particularly for numbers greater than ten, that there is little justification for using Roman numerals in most cases.

Punctuation marks. Relatively little work has been done on the legibility of punctuation marks. Prince (27) has reported on the difficulties of poor vision readers in distinguishing between the comma and the full point. He suggested that the full point should be 30% and the comma 55% of the height of the lower case "o". Similar difficulties are often encountered with computer line printer output, particularly on carbon copies, where commas and full points, and semi-colons and colons, may be virtually indistinguishable.

11.3.3 *Typography and layout for text*

Type size. Much research has been devoted to the relative legibility of different type sizes. Type size is usually specified in terms of the body depth

Tinker (41) ranked isolated numerals in this order of legibility:

Modern: 7 4 1 6 9 0 2 3 8 5
Old Style: 7 4 6 0 1 9 3 5 2 8

When the numerals were arranged in groups the order was:

Modern: 7 1 4 0 2 9 8 5 6 3
Old Style: 8 7 6 1 9 4 0 5 3 2

Fig. 11.4 Tinker's (41) ranking of numerals.

in points (see Fig. 11.2). One point is approximately equal to 0.33 mm. The most reliable investigations suggest that the more commonly used sizes, 9 to 12 point, are of about equal legibility at normal reading distance and 10 or 11 point is generally preferred. The legibility of smaller sizes, such as 6 point, is impaired by reduced visibility, and larger sizes, such as 14 point, reduce the number of words which can be perceived at each fixation (41). Optimum type size cannot, however, be determined independently of leading and line length, since these three factors interact very strongly with one another in their effects on legibility.

Line length. Line length may be varied within broad limits without diminishing legibility, but research has shown that very short lines prevent maximal use of peripheral vision and thus increase the number and duration of fixations and hence total perception time. Very long lines cause difficulty in locating the beginning of each successive line and the number of regressions after the backsweep to the beginning of each line is greatly increased. The effect of short lines is particularly severe with large type sizes, as is the effect of long lines with small type sizes. The optimal line length seems to be one which accommodates about ten to twelve words, or 60 to 70 characters. In general readers prefer lines of moderate length (41). The effects of non-optimal line lengths are especially noticeable in the very short lines of some newspapers and the very long lines of up to 132 characters which are typical of certain types of computer output.

Leading. It would appear from the experimental evidence available that the use of leading (additional space) between lines improves the legibility of all sizes of type, but that smaller sizes benefit most. The additional space is helpful in locating the beginning of successive lines, and the effect is enhanced for smaller sizes because of the greater number of characters in any given line length. Leading appreciably modifies the effect of line length on legibility. Tinker (41) found that 2 points of leading allowed the line length of 8 and 10 point type to be extended without loss of legibility. Experiments suggest that the optimum amount of leading is 1 or 2 points for 8 or 10 point type (34). The results of these studies show clearly that type size, line length and leading interact in a complex way and must therefore be selected in relation to one another.

Justified versus unjustified setting. For reasons of convenience and economy, the use of unjustified setting which results in a ragged right hand margin is increasing with the use of typesetting processes other than conventional typesetting. It has the advantages of consistent word spacing and no hyphenation, and is being adopted for a wide variety of printed materials including books, catalogues and reference works, magazines and newspapers. Experiments by several researchers have shown no significant differences in reading time or eye movement patterns for justified and unjustified text (34). In some cases there has been a tendency for unjustified text to be read more quickly, particularly by less skilled readers who may be confused by hyphenation and uneven word spacing.

Paragraphs. Paragraphs were originally indicated by paragraph marks within the text, but when the practice of beginning each paragraph on a new line was adopted, the mark became redundant. It then became conventional to indent the first line of every paragraph. An alternative to indentation which is increasingly used is the insertion of extra space between paragraphs as well as or instead of indentation.

The division of text into paragraphs containing related information undoubtedly improves readability (41). It has also been shown that when the same text is divided into a small number of long paragraphs or a larger number of shorter paragraphs, readers judge the text with shorter paragraphs to be easier to read (32). This is attributable to the larger amount of white space on the page which results from more frequent paragraph breaks. The use of additional white space between paragraphs is therefore likely to increase the apparent "accessibility" of long text passages even further.

Headings. Assuming that the text itself is arranged logically, the headings should clearly reflect this logical structure. In many cases a hierarchical system of headings with several levels will be needed. It is important to establish at the outset how many levels are required. If more than 3 or 4 are used it may be necessary to rewrite portions of the text in order to avoid confusing the reader. The chosen number of levels must then be represented in a consistent fashion so that the relationship between successive sections of text is clear.

Headings at different levels can be "coded" both by variations in type form and by the use of space. In conventionally printed materials, variations in type size, type weight or type style are often used. Changes in type weight are apparently more obvious to the average reader than changes in type style or face (37). In typewritten materials, underlining certain levels of heading may be helpful, but double underlining is aesthetically displeasing. The use of capitals for headings should be limited if possible, for the reasons given in section 11.3.2. Headings can also be coded by means of a logical system of spacing. The amount of space left before and after each level of heading and paragraph must be consistent and must clearly indicate the relation between the different levels (3). Another alternative is to place superior headings in the left hand margin. This device is sometimes useful in typewritten materials where the possibilities for coding the headings typographically are very limited. Centred headings may also be used, but as the eyes tend to move automatically to the left hand margin at the end of each line, centred headings may well be less efficient than side headings.

11.3.4 *Tables*

Structure. The ease and speed with which tables can be understood depends very much on the tabulation logic. The author must ask himself what information the reader already has when he enters a table, and what information he is seeking from it. The row and column headings should relate to the information he already has, thus leading him to the information he seeks in the body of the table (46). The importance of using appropriate tabulation logic is illustrated in Table 11.1a and b.

Table 11.1 a and b These two tables show alternative ways of presenting the same information. The first table (*a*) will be more useful for readers wishing to know which raw materials have certain characteristics; the second table (*b*) is more suitable when they wish to find out the characteristics of particular raw materials (after Wright, 46).

a)

| | | Consumer Acceptability | |
		Good	Poor
High transportation costs	Fast processing time	B	
	Slow processing time	A	[E]
Low transportation costs	Fast processing time		F
	Slow processing time	[C]	G

b)

Raw material	Transportation costs	Processing time	Consumer acceptability
A	High	Fast	Good
B	High	Slow	Good
C	Low	(?)	Good
E	High	(?)	Poor
F	Low	Fast	Poor
G	Low	Slow	Poor

The structure of the table should be as simple as possible. Many members of the general public apparently experience difficulty in using two dimensional tables, or two way matrices (Table 11.2), and a one dimensional presentation is therefore preferable (46). Research suggests that a vertical sequence of entries is preferable to a horizontal one. Tinker (42) has shown that searching for a single item in a random list is quicker if the items are arranged vertically rather than horizontally. Wright and Fox (47) carried out a series of tests on tables for converting pre-decimalization British currency in pounds, shillings and pence (£ s d) to decimal currency in pounds and new pence (£ p). They found that conversions were quicker and errors fewer with a vertically arranged table which was essentially one long column, than with a horizontal arrangement where each row gave the values for each multiple of a shilling and the columns gave the values for 0 to 11d from left to right (47). This latter arrangement was possibly too much like a two-way matrix. It has also been shown that when subjects are required to compare pairs of numbers in a table and identify them as the same or different, they are able to make the comparison more quickly if the numbers are printed one above the other than if they are printed end to end (45).

Frase (12) has stressed the importance of presenting the information in a sequence which will be compatible with the reader's strategy of processing. In most cases this strategy will operate from left to right and from top to bottom. Wright and Fox (47) found that conversions from £p to £sd were faster if £p values were on the left, and vice versa. Earlier studies had shown, however, that if conversion is always in the same direction it makes

Table 11.2   A £p to £sd conversion table in a two-way matrix format (after Wright and Fox, 47). £sd = British currency in pounds, shillings and pence. £p = decimal currency in pounds and new pence.

Tens of new pence	Units of new pence									
	0	1	2	3	4	5	6	7	8	9
0		2d	5d	7d	10d	1/-	1/2	1/5	1/7	1/10
10	2/-	2/2	2/5	2/7	2/10	3/-	3/2	3/5	3/7	3/10
20	4/-	4/2	4/5	4/7	4/10	5/-	5/2	5/5	5/7	5/10
30	6/-	6/2	6/5	6/7	6/10	7/-	7/2	7/5	7/7	7/10
40	8/-	8/2	8/5	8/7	8/10	9/-	9/2	9/5	9/7	9/10
50	10/-	10/2	10/5	10/7	10/10	11/-	11/2	11/5	11/7	11/10
60	12/-	12/2	12/5	12/7	12/10	13/-	13/2	13/5	13/7	13/10
70	14/-	14/2	14/5	14/7	14/10	15/-	15/2	15/5	15/7	15/10
80	16/-	16/2	16/5	16/7	16/10	17/-	17/2	17/5	17/7	17/10
90	18/-	18/2	18/5	18/7	18/10	19/-	19/2	19/5	19/7	19/10
100	£1									

Table 11.3a An explicit conversion table. (After Wright, 46.)

Old s d	New Pence	Old s d	New Pence	Old s d	New Pence	Old s d	New Pence	Old s d	New Pence
1d	½p	2/1d	10½p	4/1d	20½p	6/1d	30½p	8/1d	40½p
2d	1p	2/2d	11p	4/2d	21p	6/2d	31p	8/2d	41p
3d	1p	2/3d	11p	4/3d	21p	6/3d	31p	8/3d	41p
4d	1½p	2/4d	11½p	4/4d	21½p	6/4d	31½p	8/4d	41½p
5d	2p	2/5d	12p	4/5d	22p	6/5d	32p	8/5d	42p
6d	2½p	2/6d	12½p	4/6d	22½p	6/6d	32½p	8/6d	42½p
7d	3p	2/7d	13p	4/7d	23p	6/7d	33p	8/7d	43p
8d	3½p	2/8d	13½p	4/8d	23½p	6/8d	33½p	8/8d	43½p
9d	4p	2/9d	14p	4/9d	24p	6/9d	34p	8/9d	44p
10d	4p	2/10d	14p	4/10d	24p	6/10d	34p	8/10d	44p
11d	4½p	2/11d	14½p	4/11d	24½p	6/11d	34½p	8/11d	44½p
1/-d	5p	3/-d	15p	5/-d	25p	7/-d	35p	9/-d	45p
1/1d	5½p	3/1d	15½p	5/1d	25½p	7/1d	35½p	9/1d	45½p
1/2d	6p	3/2d	16p	5/2d	26p	7/2d	36p	9/2d	46p
1/3d	6p	3/3d	16p	5/3d	26p	7/3d	36p	9/3d	46p
1/4d	6½p	3/4d	16½p	5/4d	26½p	7/4d	36½p	9/4d	46½p
1/5d	7p	3/5d	17p	5/5d	27p	7/5d	37p	9/5d	47p
1/6d	7½p	3/6d	17½p	5/6d	27½p	7/6d	37½p	9/6d	47½p
1/7d	8p	3/7d	18p	5/7d	28p	7/7d	38p	9/7d	48p
1/8d	8½p	3/8d	18½p	5/8d	28½p	7/8d	38½p	9/8d	48½p
1/9d	9p	3/9d	19p	5/9d	29p	7/9d	39p	9/9d	49p
1/10d	9p	3/10d	19p	5/10d	29p	7/10d	39p	9/10d	49p
1/11d	9½p	3/11d	19½p	5/11d	29½p	7/11d	39½p	9/11d	49½p
2/-d	10p	4/-d	20p	6/-d	30p	8/-d	40p	10/-d	50p

no difference which way round the values are given, so long as the arrangement is constant.

It is important that information in numerical tables should be given in full, that is, the table should be explicit rather than implicit (Table 11.3). In an implicit table the reader may be required, for example, to add together two values in order to obtain a third which is not explicitly stated in the table. Implicit tables save space, but require more effort on the part of the reader and may cause confusion (47).

Typography and layout. Some aspects of the fine detail of typography and layout for tables have been investigated by Tinker (41) and by Wright and Fox (47). Tinker carried out two studies on mathematical tables and concluded that one set of columns only should be used per page, with about 50 entries per page. He found that grouping entries vertically in fives or tens was helpful in reading across from column to column. Grouping in

Table 11.3b An implicit conversion table (after Wright 46).

Old s d	New Pence	Old s d	New Pence
1d	½p	1/-d	5p
2d	1p	2/-d	10p
3d	1p	3/-d	15p
4d	1½p	4/-d	20p
5d	2p	5/-d	25p
6d	2½p	6/-d	30p
7d	3p	7/-d	35p
8d	3½p	8/-d	40p
9d	4p	9/-d	45p
10d	4p	10/-d	50p
11d	4½p		
1/-d	5p		
1/1d	5½p		
1/2d	6p		
1/3d	6p		
1/4d	6½p		
1/5d	7p		
1/6d	7½p		
1/7d	8p		
1/8d	8½p		
1/9d	9p		
1/10d	9p		
1/11d	9½p		
2/-d	10p		

fives might be by means of space and grouping in tens might be by the use of a rule. In general, grouping in fives tended to be more helpful than grouping in tens.

Tinker suggested that a space of at least 1 pica (approximately 4 mm) should be used between columns, but he found little difference between a 1 pica space and a 1 pica space plus a rule. Unnecessary rules between columns may be distracting however, tending to draw the eye down the page instead of across. Similarly, the space between columns should be just sufficient to clearly separate them, but no more. Wide column spacing will increase the risk of error in locating related items in adjacent columns.

Some form of typographic differentiation between columns of information may also be helpful. Tinker (41) suggested the use of bold for the first column of mathematical tables. Wright and Fox (47) found that the use of light and bold type to distinguish between the two columns in their conversion table reduced the number of errors in making conversions, though it made no difference to the speed at which conversions were made. They concluded that such differentiation would be particularly helpful in multi-column tables. They found that if typographic distinctions were made between the columns, conversions were made faster without abbreviations after each item in the body of the table. The use of bold to emphasize items in the body of the table which might serve as "landmarks" was found to be confusing until readers became familiar with the convention. It may, therefore, be preferable for information in any subheadings to be redundant.

Wright and Fox (47) concluded that tables should be set in a type size in the range 8 point to 12 point. Tinker (41) found no difference between 8 and 9 point when entries were grouped in fives. As a general rule it is not advisable to set tables in a smaller type size than the accompanying text, particularly if the information is likely to be copied or microfilmed subsequently. If printed tables are likely to be used for the preparation of slides, this must be taken into account in choosing the type size (see section 11.4).

11.3.5 Graphs, charts and diagrams

Graphs.   When should a graph be used as well as, or instead of, a table? Carter (7) suggests that if the reader needs precise numerical information, this can be obtained more rapidly and accurately from an explicit table than from a graph. Graphs are helpful where an interpretation of the relation between the variables is important, but they should not be used alone unless speed and accuracy of reading are relatively unimportant.

Graphs should be kept as simple as possible. The number of lines which can reasonably be placed on one graph will, of course, depend on the amount of overlap, but 3 or 4 lines is likely to be the maximum. In some cases there may be an argument for multiple graphs as opposed to multiple lines on one graph, but Schutz (29) has shown that if the reader wishes to compare the lines the latter arrangement is superior. There was no difference between the two presentations in terms of reading off the values of specific points however.

If a multiple line presentation is used, the lines and points should be very clearly distinguished from one another. This can be achieved by the use of different geometric shapes for the points and by the use of lines of different thicknesses, dotted lines, or lines broken in a variety of patterns. Schutz found that colour coding slightly improved performance on multiple line graphs, but this is not a realistic proposition for printed materials, particularly when they **are** likely to be copied in black and white subsequently.

The choice of scales for the x and y axes should be very carefully considered. An ill-advised choice of scales can give a very misleading impression of the data. For the same reason it is preferable that the ordinates should meet at zero and that the axes should not be shortened (31). Any departures from this rule should be indicated by a break in the axis.

The relation between the scale values and the grid squares significantly affects the accuracy with which graphs are used. Experiments by Beeby and Taylor (1) showed that when subjects were required to enter a graph on the vertical axis, read across to a line and read off a value from the horizontal axis, they made mistakes in interpolating scale values on both axes. Their accuracy was greatest when 1 grid square corresponded to a 1 unit increase in the scale. Scales with 0.5 or 2 units per square were little different from one another, but were not used as accurately as the scale with 1 unit per square, and scales with 4 squares per unit were used 50% less accurately. Small grids (with a larger number of rulings) were not found to be advantageous. When subjects were asked to plot points on a 10 mm and a 1 mm grid, the 1 mm grid increased accuracy only by a factor of 2 or less. The closely spaced lines were often misread. Beeby and Taylor concluded that the best accuracy would be given by squares of 5 mm, but they suggest that there are few applications where accuracy greater than that given by 10 mm is needed.

The labelling of points on the scales is also very important. Ideally every unit increase on the scale should be labelled. As little interpolation as possible should be required of the user, in order to minimize errors. If single units cannot be marked, it has been suggested that multiples of 2, 5 or 10 should be used (46). The axis should not extend beyond the last marked scale point.

Errors in relating points on a curve to scale values can also be reduced by repeating the x and y axes on the top and right hand sides of the graph respectively (21). Thickening of every fifth line of the grid is also likely to be helpful in guiding the eye accurately between the curve and the scale values (1).

Charts. Charts of various kinds are sometimes a useful alternative to tables or graphs. Wright (46) reports that for the general public bar charts may be more comprehensible than tables for certain purposes. She also cites studies which suggest that bar graphs may be preferable to line graphs where the data can be subdivided in a number of different ways and the reader is required to make static comparisons. It would seem, however, that line graphs are superior where dynamic comparisons of data plotted against time are required. Schutz (30) found that where extrapolation of a trend was required, line graphs were better than vertical bar charts, and vertical bar charts were better than horizontal bar charts.

Where a number of different variables are to be presented in a bar chart, the information can usually be arranged in several different ways (Fig. 11.5). The ease of comparison of specific variables will depend on the arrangement chosen. The author must therefore decide which comparisons the reader is most likely to want to make and design the chart accordingly.

Pie charts may also be useful in some instances. Wright cites a study carried out in 1929 which suggested that a pie chart may be superior to a bar chart where three or four comparisons are being made, but for accuracy a table would be superior to both. Hailstone (15) has suggested that charts or symbols which rely on proportional area to denote quantity may not be entirely satisfactory as most people tend to make comparisons on a linear basis.

It would seem that more research is needed on the relative effectiveness of tables, graphs and charts, and of charts of different kinds.

Diagrams. Relatively little research has been carried out on the effectiveness of different kinds of diagrams. Dwyer (10) has found that although photographs may be superior in terms of immediate retention of information, line diagrams are better for delayed retention. He found that simple line diagrams in colour were the most effective form of illustration (in a text on the functioning of the heart). The addition of realistic detail to the diagram did not automatically improve it, and excessive detail tended to reduce its effectiveness. Students preferred coloured illustrations, even when the colour conveyed no additional information.

Artwork. A4 is usually a convenient size for artwork, whether it be for a book, journal or report.

Simmonds (31) reports that the majority of journals ask that the capital letter height of lettering on figures when reduced to final printed size should not be less than 2 mm. Assuming that the figure will be reduced to fit a column width of 75 mm, artwork on a vertical A4 page will need to be reduced to approximately half its original size. The lettering on the original must therefore be at least 4 mm high. A number of typewriter faces would satisfy this criterion. If the A4 original is horizontal rather than vertical a greater reduction will be required however, and the lettering must then be at least 6 mm in height.

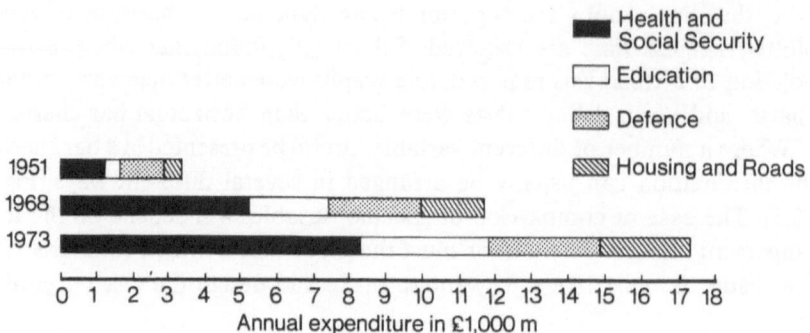

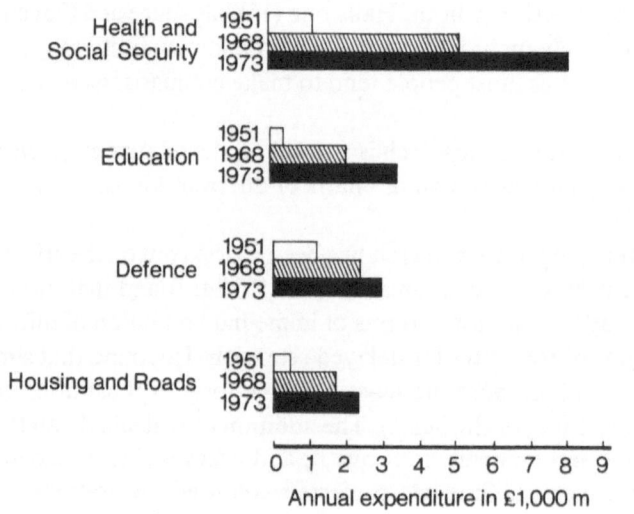

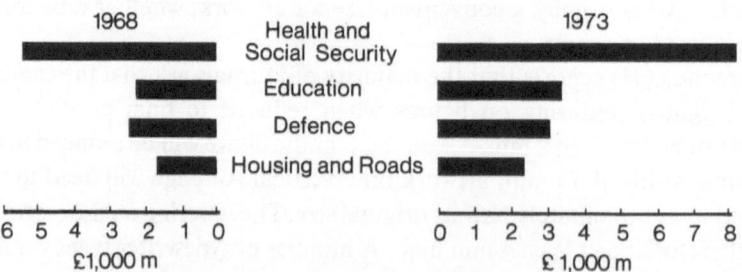

Fig. 11.5 Three alternative presentations of the same information using horizontal bar charts. Different arrangements of the information facilitate different comparisons (after Wright, 46).

The line thickness on original artwork must also be carefully controlled. Simmonds (31) suggests that no line on an A4 original should be thinner than 0.2 and he recommends a thickness of 0.35 or 0.4 mm. A thickness of 0.4 can be achieved with a Rotring pen which can be used for both drawing and stencilling. This thickness will also coordinate well with typewritten lettering.

All artwork should be prepared in jet black ink on good quality paper. The use of colour is rarely appropriate for printed or typewritten materials, partly because of the added expense of colour printing, and partly because if the material is copied or microfilmed the colour will be lost.

In some cases shading or tones may be required in charts and diagrams. Any lettering superimposed on a tone should be placed in a white surround. Some background patterns (such as regular dot screens) can severely reduce the legibility of any superimposed lettering. It is also necessary to bear in mind that some kinds of shading will reproduce better than others when they are copied electrostatically. Large dark areas generally do not copy well. A shading made up of discrete points will copy better than a continuous tone. If several tones are used, they need to be clearly distinguishable. Subtle distinctions may be lost in copying or microfilming. Ideally shading should be avoided altogether in publications which will be microfilmed.

11.3.6 *Indexes*

Very little research has been carried out on the layout of indexes for books. In 1977 Burnhill et al. (4) reported on an experiment in which they tested three styles of index. These were: (a) traditional, with sub-headings indented and listed, page numbers ranged right; (b) the same but with page numbers immediately following headings; and (c) with sub-headings running on and page numbers immediately following headings. No significant differences were found. Style (a) has the practical disadvantage of increasing setting time because of the right ranged numbers. As styles (b) and (c) were no more difficult to use, there would seem to be no serious disadvantage in adopting them.

In contents lists there is a strong argument for positioning the page numbers to the left of the headings. This eliminates the problem of reading across to the right hand side of the page to locate the numbers, and Wright's work on tables suggests that placing the numbers on the left would cause no difficulty. Such an arrangement would simplify typesetting, and Hartley et al. (16) have shown that typists are able to produce this format more quickly than the traditional right ranged numbers.

11.3.7 *Bibliographies*

The efficiency with which bibliographies are used can be maximised by the judicious use of spatial and typographic coding devices (38). Assuming that the references in the bibliography are ordered alphabetically by author, the location of specific entries will be facilitated if the entries are clearly distinguished from one another and if the author's name is clearly distinguished from the rest of the entry. This can be achieved simply by indenting all but the first line of each entry, by leaving a line space between entries, or by successively indenting each element of each entry (author, title, journal, etc.), or some combination of these.

By convention a certain amount of typographic coding is often used. Authors' names may be given in small capitals, book titles and journal titles are often in italics, and journal volume numbers are often in bold. These conventions are unlikely to help or hinder in searching of the list for an author's name if the information is well coded spatially. If, however, the reader wishes to scan the titles of books and of journal articles, these conventions will not help him. Successive indentation would be of some assistance, but experiments have shown that ideally the titles should be in bold for this kind of search.

In typewritten bibliographies the possibilities for typographic coding are extremely limited. Research has shown that underlining and the use of capitals for authors' names are not especially helpful. One of the most satisfactory coding devices would appear to be the indentation of all but the first line of each entry. This serves to distinguish the author's name very clearly. Distinctions between other parts of the entries such as the titles could be achieved by beginning each part on a new line.

11.3.8 *Page layout*

Single versus double column layouts. The area of the page can either be treated as a single wide column or it can be divided into two or more narrower columns. A study by Poulton (25) on the comprehension of scientific journals suggests that in the majority of cases a larger type size in a single column is preferable to a smaller type size in a double column layout. With an A4 page however a single column of typeset text would be too wide (unless part of the page area was unused), so double column layouts are usually adopted. In certain other situations there may be advantages in using two or three columns because a smaller type size can then be used without the lines being excessively long. For typewritten text on an A4 page a single column layout is the most suitable. This will result in lines of a reasonable length and adequate margins.

The position of tables and figures must be considered in relation to the layout of the text. A study of the effect of centred tables on the legibility of single and double column formats in scientific journals suggests that text is more easily scanned when tabular matter is set within the width of the column; two column layouts with tables cutting across both columns were found to be confusing, and 74% of readers preferred the single column layout (5).

Margins. Margins conventionally used in books reduce the printed area to approximately half the total page area. Many book designers insist that margins of this order are desirable on aesthetic grounds, and that because of a "part-whole proportion illusion," an area of type matter occupying 50 per cent of the total page area will seem much larger. Some writers have suggested that margins help to keep out peripheral colour stimuli and prevent the eye from swinging off the page at the end of a return sweep. Experiments with margins of 2.2 cm and 0.16 cm suggested that the wider margin gave no advantage, and the authors therefore concluded that margins could only be justified in aesthetic terms (41). Margins do, however, have a practical function in providing space for the fingers to hold the book without obscuring any of the text (which is why outer and foot margins are conventionally wider than inner and head margins), and in providing space for making notes.

Page size. A survey conducted in 1940 of 1000 text books and 500 American and foreign journals revealed that nearly 90% of the text books were 20 × 12.5 cm or smaller, but that journals were produced in a wide range of sizes from 17.5 × 10 cm up to 35.5 × 28 cm (41). It is probable that a similar survey conducted today would reveal larger average page sizes and perhaps still greater diversity in both text books and journals. This great diversity of sizes causes a number of practical difficulties in storing the materials, in the use of office copying processes, and in microfilming. The gradual adoption of International Paper Sizes by publishers and printers in many countries during recent years is, however, alleviating the situation.

Page size can have a significant, though indirect, effect on legibility. Choices of page size, number of columns, line length, type size and leading should be carefully considered in relation to one another. This is especially true for printed materials which must serve as microform masters, where the page size and type size must be carefully chosen in relation to the proposed film width and reduction ratio.

Large pages may also affect legibility in that users may tend to hold a large book in such a way that the angle between the page and the line of sight is greater than 90 degrees. This diminishes the optical size of the type. A 12 point type held at an angle of 135 degrees is approximately equivalent to 8 point. Large flimsy publications such as newspapers which are bent or folded are subject to more complex distortions as a result of page curvature and reading angle (41).

11.3.9 *Paper and ink*

Maximum paper reflectance undoubtedly promotes maximum reading efficiency because it provides the greatest contrast between background and image. Tinker (41) has concluded, however, that for text printed in black ink all paper surfaces, whether white or tinted, are equally legible if they have a reflectance of at least 70%.

The effects of glossy papers on legibility have been examined in several studies. High gloss papers often result in troublesome reflections if they are read under direct lighting, but experiments by Paterson and Tinker (41) showed that with well diffused illumination, text printed on a high gloss paper with 86% glare was read as fast as on paper with less than 23% glare. Operbeck (24) found, however, that readers preferred matt papers to medium or glossy papers.

Paper opacity is also an important factor. Spencer et al. (39) found that the "show-through" of print on subsequent pages which occurs with papers of low opacity can reduce legibility very markedly. The effects were particularly severe when papers of low opacity were printed on both sides and when the show-through occurred between the lines of type actually being read. Such papers should therefore be printed on one side only, and the lines of type on successive pages should be aligned.

Ink density will also affect legibility. Experiments by Poulton (26) with inks of different densities suggested that ink density should be at least 0.4, giving a contrast ratio of at least 60%. The question of ink density is particularly important in relation to typewriter ribbons. Typescripts for reproduction should be typed with a carbon ribbon for maximum clarity. Operbeck (24) studied ink gloss, and found that legibility was reduced by glossy ink, and particularly when it was printed on glossy paper.

The use of negative images (white type on a dark background) is not recommended. The irradiation which occurs around the edges of bright images viewed against a dark ground will tend to blur and thicken the type.

If negative images are used, this phenomenon must be taken into account and larger type with open counters and adequate character spacing should be used.

The use of coloured inks or papers is likely to reduce legibility if the contrast ratio falls below 70%. The quality of any subsequent copies or microfilm versions will also be affected.

11.4 Slides

11.4.1 *Introduction*

Slides have become an important communication medium for the exchange of ideas and data among research workers, but in spite of their extensive use, their quality is often disappointing. An increasing number of research workers in the biomedical sciences prepare the artwork of their slides themselves, and in doing so they tend to give too little thought to the fundamental differences between slides and printed books as ways of communicating information. As a result, slides often contain far too much information presented in too complex a manner, and the artwork is often prepared in ignorance of some of the basic rules which would ensure maximum legibility.

Relatively little research has been carried out on the legibility of slides, and there is a great deal of scope for further work. Some of the findings which do exist are discussed below, together with suggestions derived from research in other media and from experience.

11.4.2 *Information content*

A significant difference between slides and the equivalent information presented in printed form is that in the former case the time available for reading each slide is determined by the speaker, whereas in the latter case the reader is self-paced. This means that it is essential that a slide should only contain as much information as the audience can reasonably be expected to be able to assimilate in the time the speaker allows. The information presented should be simple and concise, whether it be text or tables, graphs or diagrams. Galer (13) quotes from a Transportation Research Board publication, in which the Board advises that no slide should contain more information than can be assimilated in 30 seconds, elaborate captions and details should be avoided, no table should contain more than 10–15 words, no graph more than two datum lines, nor any histogram more than two or three bars. The Ergonomics Society suggests that a slide should ideally contain one fact, and at the most two, and that graphs should not contain more than three curves.

In many cases however the strict application of these rules, thus dividing data between slides, might well cause even greater confusion. It is difficult or impossible to lay down any hard and fast rules, particularly as so little research has been done on the ability of audiences to assimilate information from slides which differ in the quantity and complexity of the information presented. What can be said, however, is that the speaker should clearly define the point he is trying to make on each slide, and should include only the information which makes this point. The selected information should be presented as simply as possible.

In planning a sequence of slides, it is often helpful to begin with a slide which lists the main topics to be covered in the talk. This will enable the audience to understand the logical structure of the information presented to them. Similarly, a final slide reviewing the main conclusions of the talk is likely to clarify the issues discussed and to help the audience to remember the salient points.

11.4.3 *Sources of material*

Printed information. Tables, graphs and diagrams photographed from books, reports and papers often do not make good slides for two main reasons. Firstly, as we have seen above, printed information often contains far too much detail to make a good slide. Secondly, the quality of the printing is often not adequate for photographing and displaying in slide form. Lettering will often be too small in relation to the overall image area, and, particularly in typed reports, the image quality may not be adequate for good legibility on the screen unless clean type and a carbon ribbon have been used. Printed material also has limitations in the use of colour which need not apply to slides.

Photographs. Galer (13) points out that in presenting a description of an experiment, the need for photographic evidence must be considered at an early stage, as it is often impossible to go back afterwards and take the necessary photographs. Ideally he suggests that each photograph should be taken to illustrate a particular point, and extraneous detail should be excluded as far as possible. If necessary, editing techniques can be used during processing to eliminate any unnecessary detail which might be confusing.

Artwork. The use of original artwork prepared specifically for slides eliminates the problems of using printed materials which are unsuitable in content and form, and it gives the speaker the freedom to design the presentation to suit his own particular purposes.

11.4.4 *Size of originals*

A4 (210 × 297 mm) is a convenient size for original artwork, because its proportions are similar to those of the standard film aperture (24 × 36 mm). If some other size is used, this proportion of 2:3 should be maintained so that best use is made of the available film area. Ideally all artwork should be of a standard size, as this requires only one camera setting and photography is therefore easier and cheaper.

The page may be photographed horizontally (landscape format) or vertically (portrait format). It is often recommended that slide orientation be standardized to the landscape format, but this can have disadvantages if printed materials are photographed. Graphs and diagrams in reports, for example, are often in the A4 portrait format. This means that a higher reduction than would normally be necessary for A4 will be required to fit the material onto a landscape slide. As a result the information will be smaller than it need be, and part of the slide area will be wasted.

A margin should be left between the information and the slide frame. Simmonds (31) suggests an information area of 160 × 240 mm for an A4 original. Blaauw (2) suggests an information area of 231 × 353 mm for an A3 page (requiring a reduction ratio of 10.5) and an area of 163 × 249 mm for an A4 page (reduction ratio 7.4).

11.4.5 *Character size*

Simmonds (31) suggests that the minimum character size for legibility at the back of a large theatre is determined by the formula:

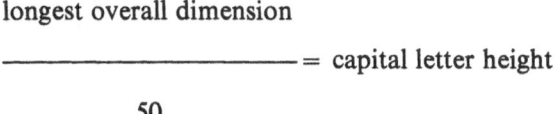

$$\frac{\text{longest overall dimension}}{50} = \text{capital letter height}$$

This is a minimum for good projection conditions. Under sub-optimal conditions, he suggests a ratio of 1:40 would be desirable. This would limit the maximum number of lines of text to 10 for a landscape page and 15 for a portrait page. Simmonds points out however that although 15 lines of text may be legible, this amount of information may not necessarily be desirable.

Galer (13) quotes a formula for determining the minimum size of numerals and letters for instrument panels which are to be read under a variety of illumination and other conditions:

$$H = 0.0022 \, D + K$$

where D is the viewing distance in mm and H is the character height. K is a constant whose value depends on the illumination of the work and the viewing conditions. A K value of 1.524 would represent very good conditions, and a value of 6.604 would represent very poor conditions. This means, says Galer, that if a slide is viewed from approximately 9.75 m under good conditions, the characters on the screen should be no less than 23 mm high.

Character height is often quoted in terms of its ratio to image height. Assuming the reader is 9.75 m from the screen, the image is 1.22 m high, and the lettering is 23 mm high, then the ratio of character height to image height would be 23:1220, or 1:53. For A4 artwork with a height of 210 mm, a ratio of 1:50 would mean characters of 4.2 mm. Galer quotes ratios of 1:17–1:13, 1:32 and 1:15 which have been recommended by various organizations. These would give a minimum character height of between 4 and 16 mm on A4 originals.

In the above example of characters 23 mm in height viewed from a distance of 9.75 m, the visual-angle* subtended by the character height would be about 8'. Van Cott and Kincade (43) recommend a minimum visual angle of 10' for slides. Experiments by Wilkinson (44) showed that as the visual angle was decreased from 15'21" to 7'20" and then to 3'50", there was a significant reduction in legibility at each stage. Metcalf (20) varied the visual angle from 10'26" to 23'50" and showed that visual angles of less than 11'53" were significantly less visible than larger angles. He concluded that the visual angle should be at least 16'19" for adequate visibility, and that the ratio of character height to image height should be a minimum of 1:27 under good viewing conditions.

Galer (13) concluded that a workable standard is to use lettering 6 mm in height (24 point) for text, and lettering 4.5 mm high (16 point) for labelling graphs and diagrams. He is assuming however that capitals will be used. This is not desirable, for the reasons given in section 11.3.2. Character size should therefore be based on x-height rather than on capital height.

---

*The visual angle is the angle subtended at the nodal point in the lens of the eye by the height of the object. Visual acuity may be expressed in terms of visual angle. The average subject can resolve two points and recognize their duality when the angle which they subtend is one minute or more.

As a final check on artwork, Galer suggests that if the artwork is readable from a distance 8 times its height, then provided no-one sits a greater distance from the projected image than 8 times the height of the screen image, the lettering should be legible.

11.4.6 *Lettering*

Methods of lettering for slides include typesetting, transfer lettering, stencilled lettering (stencils suitable for use with a Rotring pen), typewriting, or in exceptional cases, handwriting. Typesetting tends to be expensive however, so transfer lettering, stencilled and typewritten lettering are generally the most suitable.

Transfer lettering gives a wide choice of styles and sizes, and of special symbols. Stencilling is a relatively cheap, simple and rapid method, though stencils do not provide the same range of styles as transfer lettering. Typewritten lettering is adequate provided that a carbon ribbon is used and the type is clean and sharp, but it may be necessary to use an original smaller than A4. Galer recommends a size of 76 × 114 mm for artwork with typewritten lettering. If a golfball typewriter or an IBM composer is used, a variety of type faces will be available.

An important factor in the choice of type style is the width to height ratio of the characters. McCormick (18) has suggested that the ratio should be 1:1, but that this can be reduced to 1:1.66 "without serious loss in legibility". It is difficult to make firm recommendations on the basis of the evidence available, but relatively wide characters would seem to be advantageous, and condensed faces should certainly be avoided.

The optimum stroke width to height ratio will depend on whether the characters are dark on a light ground or vice versa. Ratios of 1:10 (2) and 1:6 (23) have been suggested, and McCormick (18) suggests a ratio of between 1:6 and 1:8 for black lettering on a white ground. For reversed images however, the phenomenon of irradiation, whereby a light image on a dark ground seems larger than the same sized dark image on a light ground, must be taken into account. McCormick recommends a ratio of between 1:8 and 1:10 in this case.

The width to height ratio and the stroke width to height ratio for a number of common type styles are shown in Table 11.4, which is given by Galer (13). Most of the examples fall within the suggested range of 1:1 and 1:1.66 for width to height ratio. For black characters on a white ground, four fall within the recommended range for stroke width to height ratio. These are Futura Medium, Helvetica Light, Univers 55 and Univers 53. For white characters on a black ground, Folio Light and Rotring stencil

Table 11.4 Width to height, and strokewidth to height ratios for a number of type styles. (Figures are approximate, and are based on a measurement of the capital letter E.) (After Galer, 13.)

Style	Width: height ratio	Strokewidth: height ratio
Grotesque 216	1:1.6	1:4.5
Helvetica Medium	1:1.4	1:4.8
Futura Medium	1:1.8	1:6.0
Univers 65	1:1.5	1:4.8
Folio Medium	1:1.5	1:5.7
Folio Medium Extended	1:1.3	1:5.3
Helvetica Light	1:1.4	1:7.1
Folio Light	1:1.6	1:9.1
Univers 55	1:1.6	1:7.4
Univers 53	1:1.3	1:6.8
Rotring stencil system	1:1.4	1:10

system are suitable. From the table it would appear that Helvetica Light, Univers 53 and Universe 55 are among the best styles to use for dark lettering on a light background, and Folio Light or the Rotring stencil system for light on dark.

11.4.7 *Layout of text*

The amount of text on each slide should be kept to a minimum. Recommended maximum numbers of lines for horizontal formats are 6–7 (23) and 10–11 (2).

It has been suggested that the maximum number of characters per line should be approximately 40 (23), which is considerably less than the maximum for printed materials. It is likely that line, word and letter spacing should be somewhat more generous than in printed materials, but there would appear to be no experimental evidence on this point at present. Unjustified setting is recommended for slides. With a line as short as 40 characters, justification could result in very uneven word and letter spacing and excessive hyphenation. Hyphenation should be avoided completely if possible.

11.4.8 *Tables, graphs and diagrams*

Most of the comments made in sections 11.3.4 and 11.3.5 on printed tables and figures also apply to slides, but there are some additional points to be made specifically in relation to slides.

Where photographs are used, some indication of scale should always be given if the size is not immediately obvious from the nature of the subject. With photographs of cell and tissue structures, it is desirable to give a vertical and horizontal scale (in μm for example) rather than a magnification factor. The audience will then be better able to estimate the dimensions of the structure shown.

In preparing graphs, charts and diagrams, the use of colour is a realistic possibility on slides. Colour will certainly increase the attention attracting qualities of a slide, and it can be used very effectively to emphasize certain information (Fig. 11.6) and to code and distinguish between different kinds of information. The colours in a colour coding system need to be markedly different in hue however, and it is inadvisable to use more than about six colours. Colour can be used for lettering, lines or shading, but the effects on legibility of the resulting image and background colour combinations must be borne in mind.

In drawing lines on graphs, charts and diagrams, Simmonds (31) recommends that the difference between thick and thin lines should not be too great. He suggests that the difference should not be more than four units, so that if the thinnest line is 2 mm, the thickest should be about 6 mm.

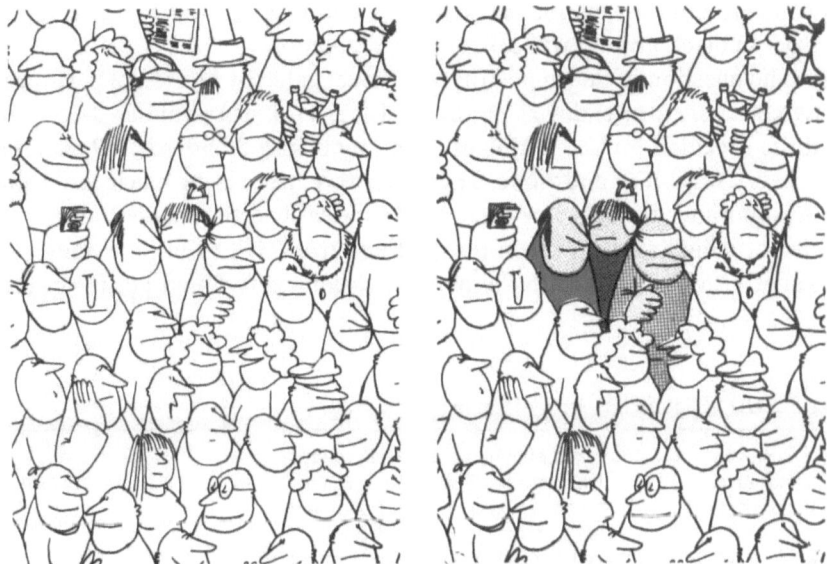

Fig. 11.6 The use of colour coding in drawing attention to certain information in an illustration. (Through the courtesy of OMI, University of Utrecht, The Netherlands).

11.4.9 *Contrast*

The level of contrast on a slide is a very important factor in legibility, and it interacts markedly with other factors such as the size of the lettering.

Metcalf (20) in his experiment found that contrast affected the speed and accuracy with which subjects were able to compare geometric symbols in a projected display. Contrast he defined as:

$$\frac{\text{difference in background and image brightness}}{\text{background brightness.}}$$

He varied contrast from 0.1–0.7 and found that a contrast of 0.1 gave significantly lower visibility than other values. The effects of poor contrast were least when the visual angle subtended by the symbols was high. Background brightness alone did not significantly affect visibility. Wilkinson (44) also found a significant decrease in performance with black characters on a white ground when the contrast ratio was reduced from 1:95 to 1:60 and to 1:30. Like Metcalf he found that there was a strong interaction between contrast and letter height, changes in either having the greatest effect when the other was at its minimum.

Thus contrast must be adequate, but on the other hand, too high a contrast may cause discomfort as a result of glare. Black characters on a bright white ground for example may have this effect. Bright white characters on a dense black ground will suffer from irradiation.

The question of image polarity was investigated by Morton (22). He prepared a series of slides as shown in Table 11.5. The symbols used were those specified for the testing of vision by British Standard 4189, and the artwork was black and white. The slides were made using black and white lith film and diazochrome film. Morton identified the smallest character size at which subjects made recognition errors. His results are shown in Table 11.6. Subjects with the poorest visual acuity found positives most

Table 11.5   Slide forms used by Morton (22). See text.

1. Positive:	black symbols, white background
2. Negative:	white symbols, black background
3. Positive:	black symbols, yellow background
	(as 1, but using yellow filter)
4. Negative:	yellow symbols, black background
	(as 2, but using a yellow filter)
5. Diazochrome:	white on blue

Table 11.6 Ranking for legibility and subject preference, for five slide forms (after Morton, 22).

Legibility	Slide form		Subjective preference
(best)	Positive, black on white	Positive, black on white	(38%)
	Positive, black on yellow	Diazo, white on blue	(27%)
	Diazo, white on blue	Positive, black on yellow	(26%)
	Negative, white on black	Negative, white on black	(5%)
(worst)	Negative, yellow on black	Negative, yellow on black	(4%)
			(100%)

legible, though older subjects tended to be less sensitive to differences between positives and negatives. As a general rule it would therefore seem advisable to avoid negative presentations. In some cases, however, a negative image may be an effective way of drawing attention to small areas containing captions.

Whichever polarity is chosen, slides in a series should be the same. Constant change from dark to light backgrounds will prevent the eyes from becoming visually adapted and will be fatiguing. If colour is used for image or background, a high contrast must still be maintained if legibility is not to be impaired.

11.4.10 *The use of colour*

The use of a coloured background with black lettering has been investigated by Snowberg (33). He used two sets of red, blue, green, yellow, and white backgrounds. One set was matched for luminosity and the other for transmission density. Each slide contained five lines of randomized letters. Five different letter heights were used, giving letter to screen height ratios of 1:20, 1:30, 1:40, 1:50 and 1:60.

Snowberg found that when the slides were matched for luminance there were significant differences between the hues. The affects were particularly marked for the smaller letter sizes of 1:50 and 1:60. White was significantly better than yellow, green, red and blue, yellow was significantly better than green, red and blue, and blue was significantly worse than the other four backgrounds. Researchers in optometry have shown that green, and especially yellow, focus on the fovea of the eye. Red focuses beyond the fovea and blue focuses in front of it. These studies of chromatic aberration explain the fact that even when luminances are equal, there are still differ-

Table 11.7 Slide luminance values arranged hierarchically from highest to lowest overall acuity (after Snowberg, 33). T = equal transmission, L = equal luminance.

Rank	Colour	Brightness level
1	White (L)	15
2	Yellow (L)	15
3	White (T)	10
4	Green (T)	16
5	Blue (T)	19
6	Red (L)	15
7	Yellow (T)	10
8	Green (L)	15
9	Red (T)	10
10	Blue (L)	15

ences in acuity between the colours. No such clear cut hierarchy was detected on the slides matched for transmission density, though there were significant differences between the colours with the smaller letter sizes. Snowberg concluded that the hierarchy depends on the interaction of colour and brightness, or luminance. Table 11.7 shows the rank order for acuity of the 10 backgrounds tested, and their brightness (differences between adjacent ranks 2 to 9 were not significant).

Snowberg therefore recommended white backgrounds for maximum visual acuity, and suggested that blue backgrounds should be avoided where legibility is critical. Yellow and green are currently popular as background colours, though in Snowberg's equal transmission slides they were not superior to red and blue.

11.4.11 *Projection of slides*

The quality of the projected image depends on such factors as the size of the image on the screen, the distance from the audience to the screen, the angle which the screen is viewed, and the illumination in the auditorium during projection.

Image size is determined by the distance from the projector to the screen and the focal distance of the objective in the projector. The relation between the latter two factors and the size of the image is given in Table 11.8. To allow projection of slides in horizontal and vertical formats, the screen should be square. The use of two projectors is often advantageous for making comparisons, for example between normal and pathological situations. In this case the width of the screen must be twice the height.

For optimum legibility the spectators should be sitting at a distance from the screen equal to or larger than twice the height (for vertical images) or

PRESENTATION OF INFORMATION

Table 11.8 Height/width of the image on the screen (m) (after Blaauw, 2).

Distance between projector and screen (m)	Focal distance of the objective (mm)							
	35	50	90	120	150	200	230	300
1	0.95							
2	1.95	1.30						
3	2.95	1.95	1.10					
4	3.95	2.65	1.50	1.10				
5		3.30	1.85	1.40	1.10			
6		4.00	2.25	1.70	1.35			
7			2.65	1.95	1.55			
8			3.05	2.25	1.80	1.35		
9			3.45	2.55	2.05	1.50		
10			3.80	2.85	2.25	1.70	1.35	
11				3.15	2.50	1.85	1.45	
12				3.45	2.75	2.05	1.60	1.35
13				3.70	2.95	2.20	1.75	1.45
14				4.00	3.20	2.40	1.90	1.55
15					3.45	2.55	2.05	1.70
16					3.65	2.75	2.15	1.80
17					3.90	2.90	2.30	1.90
18					4.15	3.10	2.45	2.05
19						3.25	2.60	2.15
20						3.45	2.75	2.25
21						3.60	2.85	2.40
22						3.80	3.00	2.50
25						4.30	3.45	2.85
30							4.15	3.45
35								4.00

the width (for horizontal images) of the image on the screen, and smaller than or equal to five times this height or width. For larger distances special objectives in the projector are required.

Viewing angle has a very important effect on legibility. Wilkinson (44) used nine viewing angles, at 15° intervals on an arc from 60° left and right of the subject. He found a significant detrimental effect on performance as viewing angle deviated from 0°. The effect of changes in viewing angle was greatest when character size and/or contrast were at low values. Blaauw (2) recommends that all spectators should be seated within an arc of 30° to the left and right of the screen.

Partial illumination of the auditorium during projection is desirable so that the audience can make notes. Illumination, however, diminishes the contrast between symbols and background on the slide. To minimize any reduction in legibility, the use of "blue" light (from a xenon light source for example), is often preferred to illumination with white light.

11.4.12 *Overhead projection*

As far as legibility is concerned, the general principles discussed above in relation to 35 mm (5 × 5 cm) slides can be applied to overhead projection transparencies.

One of the main disadvantages of overhead projection is that the method can be used only for small audiences because the size of the image on the screen is smaller than that from a 35 mm slide projector for comparable focal and projection distances. Thirty-five mm slides will therefore normally be preferred for detailed text and figures. The great advantage of the overhead projector, however, is that it can be used to present information sequentially by simply adding more slides or overlays to the display.

The lettering on transparencies may be produced in a number of ways, but the most convenient forms are likely to be transfer lettering, or handwritten or stencilled lettering produced with a felt-tipped pen. A felt-tipped pen with a stencil gives a cheap yet legible result. A further great advantage is that transparencies can be prepared spontaneously during a lecture. It is also possible to produce them from printed materials, and this process is relatively cheap and easy if no enlargement or reduction is required. Text often needs enlarging however, and this will be slightly more expensive.

Lighting during the projection of transparencies is generally less crucial than for 35 mm slides, but the limitations on viewing distance are somewhat greater.

11.5 Posters

11.5.1 *Introduction*

The use of posters as a means of communicating research results and related information is becoming increasingly popular. To be effective, however, posters need to be well designed. Like slides, they have certain characteristics which are very different from those of the printed book, and they must be designed with these special characteristics in mind.

11.5.2 *Content*

In studying a poster, the reader can linger over it as long as he or she wishes, and in this sense a poster presentation is more like a printed page than a slide. In most cases, however, the reader will be confronted with a series of posters, and he will be expected to study them while standing up. Information must therefore be presented clearly and concisely so that the essentials of the message are immediately obvious. Studies in museums (36) have shown that visitors tire of reading long caption panels very quickly, and while they may begin by systematically reading each one, they soon start to wander at random. After this point they will presumably spend most time on the most "accessible" displays. Posters which seek to stimulate interest rather than give precise details are likely to **be** the most succesful. The information should be restricted to a small number of clearly defined logical units whose relation to one another is made immediately clear by the layout and typography of the poster.

11.5.3 *Size*

The overall size of posters must take into account the quantity of information which the reader can be expected to be able to absorb, and the size of lettering and figures which will be necessary for legibility at the average distance from which the poster will be read. The use of International Standard Paper Sizes will have two advantages. Firstly, because the different sizes have similar proportions, posters of different sizes can be successfully displayed together if necessary. Secondly, the proportions are convenient for reproduction purposes (see section 11.4). The most convenient sizes are likely to be A1 (594 × 841 mm) or A0 (841 × 1189 mm).

11.5.4 *Character size*

Posters will generally be read at a distance of, say, between 60 and 180 cm as opposed to the normal reading distance of about 35 cm for books. The character size will therefore need to be correspondingly larger for posters. Typescript, computer printout, or figures taken directly from conventional printed materials will not be adequate.

The exact choice of size can often best be made by experimentation with the chosen style of lettering in the conditions under which it will be read. As a general rule, however, it may be said that the character size for text should be at least 24 point (8.5 mm approximately).

All text and symbols on a poster should be legible from the same distance. The reader should not have to move closer to peer at diagram captions for example. Main titles, of course, may be substantially larger than text in order to attract attention from a distance, and headings may also be differentiated from the text by their size.

11.5.5 *Character style*

The choice of character style will depend to some extent on the method of
lettering chosen. Typesetting and transfer lettering both offer a wide range
of typestyles, while typewriting and stencilled lettering offer a much more
restricted range. A simple sans serif face is generally preferable. It should
not be a condensed face, and the weight should tend towards bold rather
than light.

The text should be in upper and lower case lettering. Headings can be
emphasized by changes in size or weight, or possibly style, but capitals
should be avoided if possible.

11.5.6 *Layout for text*

Many of the comments made in section 11.3.3 also apply here.

Lettering should be spaced tightly enough for the individual letters to hold
together as words, but not so tightly that there is a danger of adjacent letters
fusing when reproduced. Ideally the letters should be proportionally
spaced, though this will not always be possible with typewritten and stencilled materials. Spacing between words should be sufficient to separate
them, but not so great that dazzling vertical rivers of white space are created in the text.

Lines of text should be of moderate length (40–60 characters), and
should be adequately spaced but not excessively far apart.

The use of fairly generous margins will improve the general appearance
of a poster. For aesthetic reasons it is generally desirable that the head
margin should be somewhat less than the foot margin.

11.5.7 *Tables, graphs and diagrams*

The importance of logic and simplicity in the design of tables and figures
has already been stressed. There are, however, some additional points
worth noting in relation to posters.

Wherever possible, captions on graphs and diagrams should be horizontal. In printed materials the vertical axis on a graph will often be labelled
vertically so that the maximum amount of space is available for the graph
itself (Fig. 11.7). This causes no great hardship for the reader as he can
easily turn the book in his hands. Vertical captions are also unlikely to
seriously inconvenience a seated audience looking at slides, but for an audience in a standing position it is desirable that they should not be required

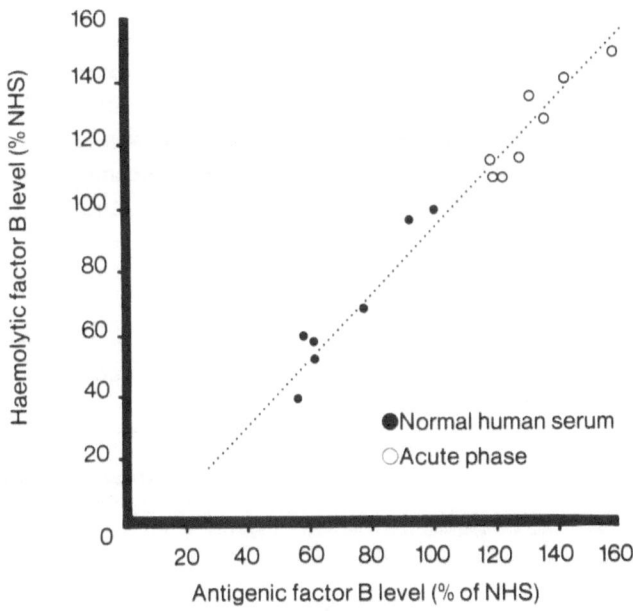

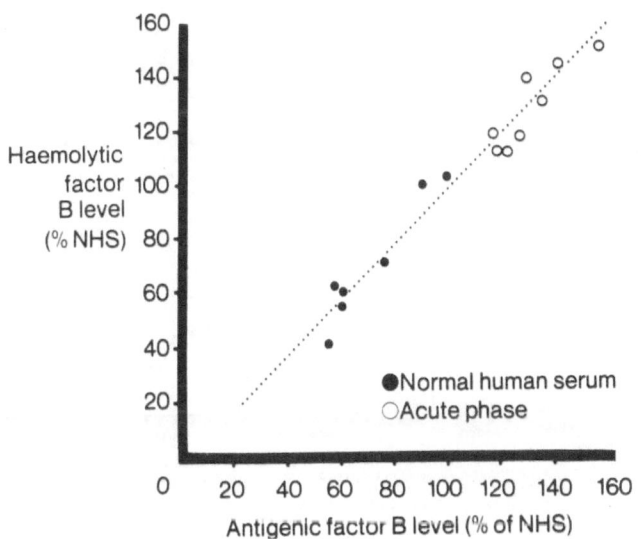

Fig. 11.7 Horizontal labelling of the vertical axis of a graph reduces the area available for the graph itself, but horizontal labelling is nevertheless desirable in an exhibition situation (after Simmonds, 31). If possible, the horizontal label can be placed to the right of the vertical axis (see text).

to frequently twist their necks in order to read diagram captions. To avoid reducing the size of the graph itself however, it may sometimes be possible to place a horizontal label to the right of the vertical axis.

The use of colour on posters can be valuable for attracting attention. Dooley and Harkins (9) found that more attention was paid to coloured posters, regardless of whether the colour was used meaningfully or not. If colour is used, however, it should be used logically, and also sparingly.

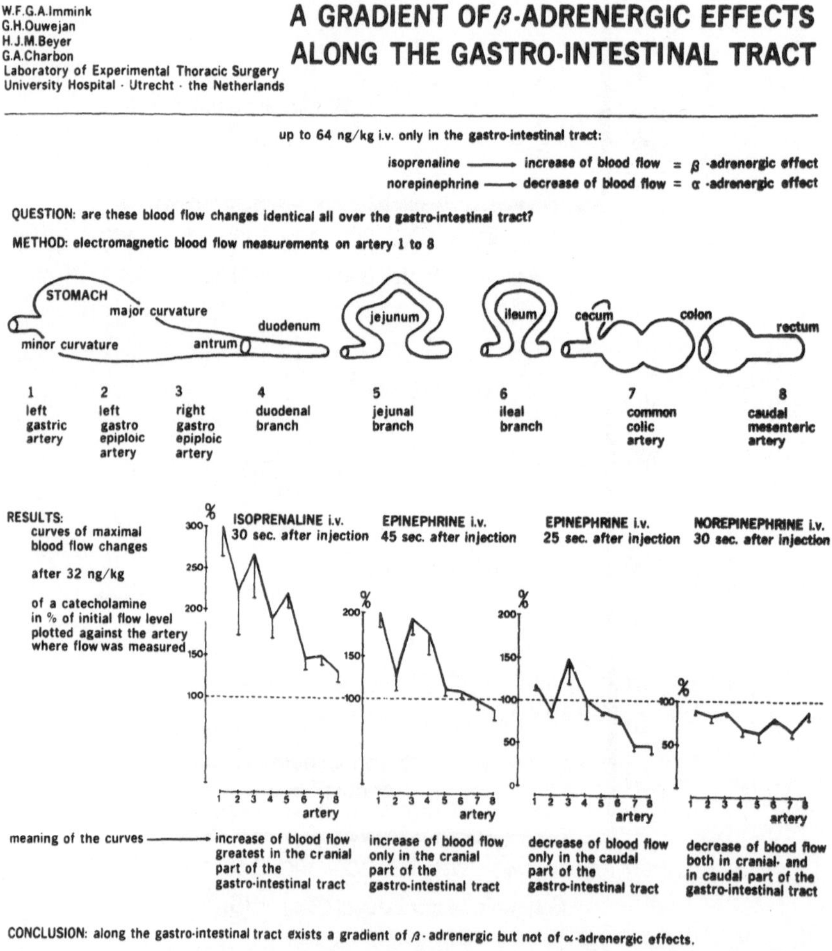

Fig. 11.8 An example of a relatively good poster. In this poster the information is given in a straight-forward and uncomplicated way using a single size of type apart from the heading.

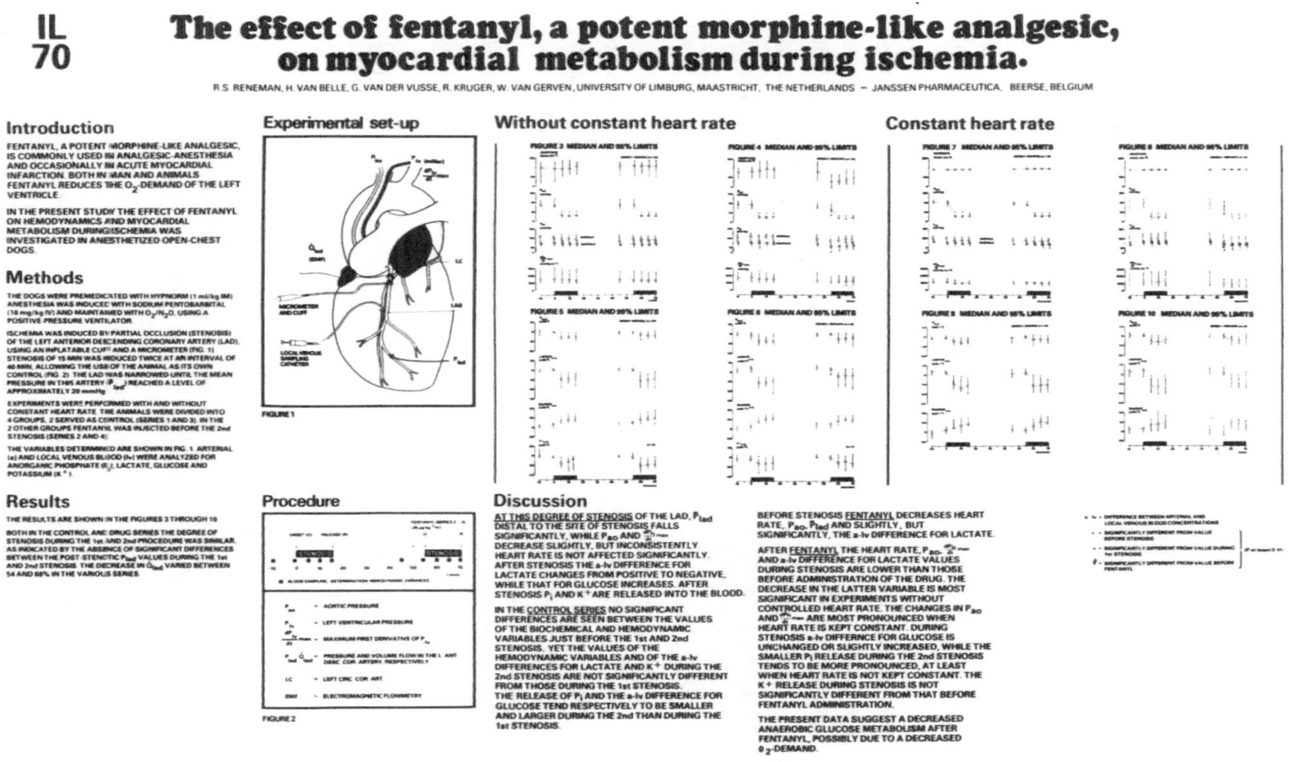

Fig. 11.9 An example of a relatively poor poster. This poster is not as legible as it might be for a number of reasons. The layout does not make clear the relation between the text and the figures, and the differences in the sizes of headings are not immediately obvious or meaningful. The text is made less legible by the use of capitals, and in the larger typesize the lines are rather short. The consistent use of a larger type size in upper and lower case and in a smaller number of wider columns would have been preferable. Finally, the detail on the figures is too small in relation to the text.

11.5.8 *Poster layout*

Text, tables and figures must be brought together in such a way that the intended logical sequence of the poster is immediately obvious. Related text and figures should be linked spatially, and great care should be taken where tables or figures cut across more than one column of text.

11.5.9 *Artwork for posters*

It will rarely be a practical possibility to prepare full size artwork for posters. Typesetting in large sizes tends to be expensive as it is generally done by hand, and if the material is set on an IBM composer, for example, large enough sizes will not be available. The artwork must therefore be prepared at some fraction of the required final size. In some cases it may be possible to use an A4 piece of artwork for a slide and a poster, but it will often be appropriate to include more information on a poster than could reasonably be given on a slide, and A4 will be rather too small an original for an A1 or A0 poster. A3 is likely to be a more suitable size.

Sizes of lettering necessary on originals must be calculated from the sizes required on the final poster and the reduction in size which the artwork represents. The layout should be planned in relation to the full size poster rather than in relation to the original artwork. Typography and layout which appear to be successful on an A4 or A3 original may not necessarily be equally successful when enlarged.

Large areas of light type on a dark ground are not advisable, but if small areas are used to attract attention, the typesize and character spacing should be slightly more generous than usual.

Examples of a relatively good and a relatively poor poster are presented in Figs. 11.8 and 11.9 respectively.

References

1. Beeby, A.W. and Taylor H.P.J. (1973): How well can we use graphs? Communication of Scientific and Technical Information, *17*, 7.
2. Blaauw, H.J.A. (1975): Informatie over het gebruik van audiovisuele media. University of Nijmegen.
3. Burnhill, P. (1970): Typographic education: headings in text. Journal of Typographic Research, *4*(4), 353.
4. Burnhill, P., Hartley, J., and Davies L. (1977): Typographic decision making: the layout of indexes. Applied Ergonomics, *8* (1), 35.

5. Burnhill, P., Hartley, J., and Young, M. (1976): Tables in text. Applied Ergonomics, 7(1), 13.
6. Burt, C.L. (1959): A psychological study of typography. Cambridge: Cambridge University Press.
7. Carter, L.F. (1947): An experiment on the design of tables and graphs used for presenting numerical data. Journal of Applied Psychology, 31, 640.
8. Cattell, J.McK. (1885): Über die Zeit der Erkennung und Benennung von Schriftzeichen, Bildern und Farben. Philosophische Studien, 2, 635.
9. Dooley, R.P. and Harkins, L.E. (1970): Function and attention-getting effects of colour on graphic communications. Perceptual and Motor Skills, 31, 851.
10. Dwyer, F.M. (1976): The effect of IQ level on the instructional effectiveness of black and white and colour illustrations. Audio Visual Communications Review, 24, 49.
11. Erdmann, B. and Dodge, R. (1898): Psychologische Untersuchungen über das Lesen, auf experimenteller Grundlage. Max Niemeyer, Halle.
12. Frase, L.T. (1969): Tabular and diagrammatic presentation of verbal materials. Perceptual and Motor Skills, 29(1), 320.
13. Galer, I.A.R. (1976): Projector slides – preparation, construction and use. Applied Ergonomics 7(4); 190.
14. Goldscheider, A. and Müller, R.F. (1893): Zur Physiologie und Pathologie des Lesens. Zeitschrift für Klinische Medizin, 23, 131.
15. Hailstone, M. (1973): A case for standardisation in the preparation of graphs and diagrams. Medical and Biological Illustration, 23, 8 (1973).
16. Hartley, J., Young, M., and Burnhill, P. (1975): On the typing of tables. Applied Ergonomics, 6, 39.
17. Javal, E. (1905): Physiologie de la lecture et de l'écriture. Felix Alcan, Paris.
18. McCormick, E.J. (1976): Human factors in engineering and design. McGraw Hill Co., New York.
19. Messmer, O. (1903): Zur Psychologie des Lesens bei Kindern und Erwachsenen. Archiv für die gesamte Psychologie, 2, 190.
20. Metcalf, R.M. (1969): The effects of visual angle, brightness and contrast on the visibility of projected material. Dissertation Abstracts, 29(10-A); 3334.
21. Milroy, R., and Poulton, E.C. (1978): Labelling graphs for improved reading speed. Ergonomics, 21(1), 55.
22. Morton, R. (1968): The lantern slide: legibility and production. The Photographic Journal, April, 89.
23. Omi-Information. (1977) University of Utrecht.
24. Operbeck, H. (1970): Effect of paper and ink gloss on legibility. Journal of Typographic Research, 4(2), 187.
25. Poulton, E.C. (1959): Effects of printing types and formats on the comprehension of scientific journals. Nature 184, 1824.
26. Poulton, E.C. (1969): Skimming lists of food ingredients printed in different brightness contrasts. Journal of Applied Psychology, 53 (6), 498.
27. Prince, J.H. (1967): Printing for the visually handicapped. Journal of Typographic Research, 1(1), 31.
28. Rawlinson, G. (1975): How do we recognize words? New Behaviour, (August 28), 336.

29. Schutz, H.G. (1961): An evaluation of methods for presentation of graphic multiple trends. Human Factors, *3*, 108.
30. Schutz, H.G. (1961): An evaluation of formats for graphic trend displays – Experiment 2, Human Factors *3*, 99.
31. Simmonds, D. (1976): Medical chartist's dilemma. Medical and Biological Illustration, *26*, 153.
32. Smith, J.M. and McCombs, M.E. (1971): The graphics of prose. Journalism Quarterly, *48*, 134.
33. Snowberg, R.L. (1973): Bases for the selection of background colours for transparencies, Audio Visual Communication Review, *21*, 191.
34. Spencer, H. (1968): The visible word. Lund Humphries, London.
35. Spencer, H. and Reynolds, L. (1976): Factors affecting the acceptability of microforms as a reading medium. Royal College of Art, Reliability of Print Research Unit, London.
36. Spencer, H. and Reynolds, L. (1977): Directional signing and labelling in libraries and museums: a review of current theory and practice. Royal College of Art, Readability of Print Research Unit, London.
37. Spencer, H., Reynolds, L., and Coe, B. (1973): A comparison of the effectiveness of selected typographic variations. Royal College of Art, Readability of Print Research Unit, London.
38. Spencer, H., Reynolds, L., and Coe, B. (1975): Spatial and typographic coding in printed bibliographical materials. Journal of Documentation, *31*(2), 59.
39. Spencer, H., Reynolds, L., and Coe, B. (1977): The effects of show-through on the legibility of printed text. Royal College of Art, Readability of Print Research Unit, London.
40. Spencer, H., Reynolds, L., and Coe, B. (1977): The effects of image degradation and background noise on the legibility of text and numerals in four different typefaces. Royal College of Art, Readability of Print Research Unit 1975, revised, London.
41. Tinker, M.A. (1963): Legibility of print. Iowa State University Press, Ames.
42. Tinker, M.A. (1965): Bases for effective reading. University of Minnesota Press, Minneapolis.
43. Van Cott, H.P., and Kinkade, R.G. (1972): Human engineering guide to equipment design. Superintendent of Documents, Washington D.C.
44. Wilkinson, G.L. (1976): Projection variables and performance. Audio Visual Communication Review, *24*, 413.
45. Woodward, R.M. (1972): Proximity and direction of arrangement in numeric displays. Human Factors, *14*, 337.
46. Wright, P. (1977): Presenting technical information: a survey of research findings. Instructional Science, *6*, 93.
47. Wright, P. and Fox, K. (1970): Presenting information in tables. Applied Ergonomics, *1*, 234.
48. Zachrisson, B. (1965): Studies in legibility of printed text. Almqvist and Wiksell, Stockholm.

INDEX OF SUBJECTS